# Brain Stimulation and Neuroplasticity

# Brain Stimulation and Neuroplasticity

Editors

**Ulrich Palm**
**Moussa Antoine Chalah**
**Samar S. Ayache**

MDPI • Basel • Beijing • Wuhan • Barcelona • Belgrade • Manchester • Tokyo • Cluj • Tianjin

*Editors*

Ulrich Palm
Department of Psychiatry and Psychotherapy
Klinikum der Universität München
Munich
Germany

Moussa Antoine Chalah
Excitabilité Nerveuse et Thérapeutique
Université Paris-Est-Créteil
Créteil
France

Samar S. Ayache
EA4391 Excitabilité Nerveuse Therapeutique
Université Paris Est Créteil
Creteil
France

*Editorial Office*
MDPI
St. Alban-Anlage 66
4052 Basel, Switzerland

This is a reprint of articles from the Special Issue published online in the open access journal *Brain Sciences* (ISSN 2076-3425) (available at: www.mdpi.com/journal/brainsci/special_issues/Brain_Stimulation_Neuroplasticity).

For citation purposes, cite each article independently as indicated on the article page online and as indicated below:

LastName, A.A.; LastName, B.B.; LastName, C.C. Article Title. *Journal Name* **Year**, *Volume Number*, Page Range.

**ISBN 978-3-0365-5170-8 (Hbk)**
**ISBN 978-3-0365-5169-2 (PDF)**

# Contents

# Preface to "Brain Stimulation and Neuroplasticity"

Non-invasive brain stimulation methods have emerged as diagnostic and therapeutic tools in recent years. This Special Issue aims at gathering pre-clinical and clinical data on brain stimulation techniques (electrical and magnetic stimulation methods).

This Special Issue compiles newest research on clinical and neurophysiological application of brain stimulation methods and the impact of brain stimulation on imaging outcomes, neurobiological markers, and clinical variables (including neurological, affective and cognitive measures).

**Ulrich Palm, Moussa Antoine Chalah, and Samar S. Ayache**
*Editors*

MDPI

*Editorial*

# Brain Stimulation and Neuroplasticity

**Ulrich Palm [1,2,*], Moussa A. Chalah [3,4] and Samar S. Ayache [3,4]**

1 Department of Psychiatry and Psychotherapy, Klinikum der Universität München, 80336 Munich, Germany
2 Medical Park Chiemseeblick, Rasthausstr. 25, 83233 Bernau-Felden, Germany
3 EA4391 Excitabilité Nerveuse & Thérapeutique, Université Paris Est Créteil, 94010 Créteil, France; moussachalah@gmail.com (M.A.C.); samarayache@gmail.com (S.S.A.)
4 Service de Physiologie—Explorations Fonctionnelles, Hôpital Henri Mondor, Assistance Publique—Hôpitaux de Paris, 94010 Créteil, France
* Correspondence: u.palm@medicalpark.de

**Citation:** Palm, U.; Chalah, M.A.; Ayache, S.S. Brain Stimulation and Neuroplasticity. *Brain Sci.* **2021**, *11*, 873. https://doi.org/10.3390/brainsci11070873

Received: 22 June 2021
Accepted: 23 June 2021
Published: 30 June 2021



Electrical or magnetic stimulation methods for brain or nerve modulation have been widely known for centuries, beginning with the Atlantic torpedo fish for the treatment of headaches in ancient Greece, followed by Luigi Galvani's experiments with frog legs in baroque Italy, and leading to the interventional use of brain stimulation methods across Europe in the 19th century. However, actual research focusing on the development of transcranial magnetic stimulation (TMS) is beginning in the 1980s and transcranial electrical brain stimulation methods, such as transcranial direct current stimulation (tDCS), transcranial alternating current stimulation (tACS), and transcranial random noise stimulation (tRNS), are investigated from around the year 2000.

Today, electrical, or magnetic stimulation methods are used for either the diagnosis or exploration of neurophysiology and neuroplasticity functions, or as a therapeutic intervention in neurologic or psychiatric disorders (i.e., structural damage or functional impairment of central or peripheral nerve function).

This Special Issue 'Brain Stimulation and Neuroplasticity' gathers ten research articles and two review articles on various magnetic and electrical brain stimulation methods in healthy populations and in patients with neurologic or psychiatric disorders. Articles were clustered to either belong to the magnetic or electrical stimulation techniques.

Transcranial magnetic stimulation was used by Haeckert et al. [1] to assess neurophysiologic effects in healthy volunteers. They investigated the aftereffects of continuous theta burst stimulation (cTBS) with 300 and 600 pulses and found no relevant changes of motor evoked potentials (MEP) during an observation period of 30 min. This study does not support the findings of some studies reporting that cTBS 300 increases and cTBS 600 decreases MEP in relaxed healthy volunteers. It adds evidence to the broad variability, i.e., the lack of a clear direction of MEP changes after cTBS, which has also been shown in other studies.

Hoonhorst et al. [2] used TMS pulses to detect the central motor conduction time (CMCT) in patients following an ischemic stroke. Therefore, the stimulation was applied over the non-infarcted hemisphere, particularly over the primary motor cortex, to generate MEP. They showed that CMCT was prolonged directly after a stroke in 60% of patients and did not normalize within 11 days. Although the mechanism for this phenomenon remains unclear, the authors not only suggested the contribution of transcallosal, but also reticulospinal, tectospinal, and rubrospinal pathways at its basis.

Repetitive peripheral magnetic stimulation (rPMS) with 25 Hz frequency was used by Malejko et al. [3] and revealed higher pain thresholds in patients with borderline personality disorder (BPD) compared to healthy controls and patients with major depression. Furthermore, patients with BPD did not show a modulation in their emotional reaction to increasing intensity levels of unpleasant somatosensory stimulation. Study results suggest an altered pain processing in BPD and are in line with previous studies.

Finally, in a review article, Klimek and Rogalska [4] elucidate the role of extremely low-frequency magnetic fields (ELF-MF) on human health. These magnetic fields may be caused naturally (e.g., solar activity), or by humans (e.g., electronic devices, transmission lines). Principally, magnetic fields can influence hormones, neurotransmission, inflammation, and cellular signal cascades. The authors reviewed the literature of the last decade dealing with the consequences of magnetic field exposure in daily life and found that ELF-MF may cause both beneficial and detrimental stress to cellular functioning. Due to a mass of confounding factors, a clear distinction of a detrimental threshold is not possible at this stage and standardized measurements are needed for future studies.

Transcranial direct current stimulation (tDCS) was used by Adam et al. [5] in a randomized study to investigate the effects on serum mature Brain Derived Neurotrophic Factor (mBDNF) in patients with schizophrenia and auditory verbal hallucinations. Interestingly, a single session of active left-side prefrontal-temporoparietal stimulation decreased mBDNF levels compared to sham tDCS, suggesting a potential modulation of mBDNF- tropomyosin receptor kinase B pathways in order to promote neuroplasticity in the central nervous system. However, the role of BDNF in tDCS-elicited neuroplasticity remains unclear.

Another study using tDCS investigated the effects of visual cortex stimulation in patients with proliferative diabetic retinopathy [6]. De Venecia and Fresnoza showed that cathodal stimulation decreased reaction time and improved visual acuity, whereas sham stimulation had no effect. The authors suggest that there is an improvement in visual discrimination after reduction of neuronal noise by cathodal stimulation.

The treatment of Parkinson's Disease Related Fatigue (PDRF) with tDCS is proposed by Zaehle in a review article [7]. He showed that PDRF is largely overlooked in the clinical management of Parkinson's Disease and severely impacts the quality of life in these patients. PDRF shows correlation with the symptoms of depression, therefore an anodal stimulation of left prefrontal cortical areas analogously to the treatment of depression is suggested.

A second article dealing with Parkinson's disease evaluated the long-term course of nine patients receiving extradural motor cortex stimulation (EMCS). Piano et al. [8] found that treatment was safe and there was a slight improvement of motor fluctuations and dyskinesias, also reflected by an improvement in the quality of life.

Chen et al. [9] reported the differential effects of 10 Hz and 20 Hz tACS on cerebral activation patterns in patients with chronic stroke. Data acquisition by functional Magnetic Resonance Imaging (fMRI) showed that 20 Hz tACS might facilitate local segregation in motor-related regions and global integration at the whole-brain level. Furthermore, 20 Hz, but not 10 Hz tACS, increased nodal clustering. The authors suggest that 20 Hz tACS might induce higher modulation effects, which could be used in rehabilitation therapies to facilitate neuromodulation.

Home treatment with tACS to improve migraine attacks was proposed by Antal et al. [10]. Patients were trained to perform a visual cortex stimulation when a migraine attack started. If the attack did not resolve within two hours after stimulation, patients were allowed to take their rescue medication. It was calculated that 21% of migraine attacks were terminated by active tACS, compared to 0% in the sham group. The authors suggest that the inhibitory character of 140 Hz tACS could reduce neuronal activity during the occurrence of migraines.

Kim et al. [11] reported the use of electroacupuncture (4 points) in combination with computer-based cognitive rehabilitation (CCR) to improve mild cognitive impairment. Compared to a control group receiving CCR only, electroacupuncture and CCR showed no superiority in terms of cognitive improvement, which was seen as a CCR effect in both groups.

Finally, Ko et al. [12] investigated the effects of noisy galvanic stimulation (GVS) of the mastoid processes in patients with bilateral vestibular hypofunction and in healthy volunteers. They found an improvement of sway in both groups during walking and standing, and an increase in alpha, beta, gamma, and theta band power in the left parietal lobe in both groups. It is postulated that GVS can improve postural stability in patients with vestibular hypofunction.

In summary, the articles in this Special Issue cover a broad range of clinical applications of different (non)invasive stimulation techniques for modulating various disorders or for neurophysiological investigations. Of note, the actual literature presents some limitations related to the methodological differences, the scarcity of studies or the small sample size. This reflects the need for further large-scale studies in the emerging field of novel brain stimulation techniques.

**Author Contributions:** Conceptualization and writing—original draft preparation, U.P.; Writing—review and editing, S.S.A. and M.A.C. All authors have read and agreed to the published version of the manuscript.

**Funding:** This research received no external funding.

**Conflicts of Interest:** Conflict of interest statement M.A.C. declares having received compensation from Janssen Global Services LLC. S.S.A. declares having received travel grants or compensation from Genzyme, Biogen, Novartis, and Roche. U.P. declares no conflict of interest.

## References

1. Haeckert, J.; Rothwell, J.; Hannah, R.; Hasan, A.; Strube, W. Comparative Study of a Continuous Train of Theta-Burst Stimulation for a Duration of 20 s (cTBS 300) versus a Duration of 40 s (cTBS 600) in a Pre-Stimulation Relaxed Condition in Healthy Volunteers. *Brain Sci.* **2021**, *11*, 737. [CrossRef]
2. Hoonhorst, M.H.J.; Nijland, R.H.M.; Emmelot, C.H.; Kollen, B.J.; Kwakkel, G. TMS-Induced Central Motor Conduction Time at the Non-Infarcted Hemisphere Is Associated with Spontaneous Motor Recovery of the Paretic Upper Limb after Severe Stroke. *Brain Sci.* **2021**, *11*, 648. [CrossRef] [PubMed]
3. Malejko, K.; Huss, A.; Schönfeldt-Lecuona, C.; Braun, M.; Graf, H. Emotional Components of Pain Perception in Borderline Personality Disorder and Major Depression—A Repetitive Peripheral Magnetic Stimulation (rPMS) Study. *Brain Sci.* **2020**, *10*, 905. [CrossRef] [PubMed]
4. Klimek, A.; Rogalska, J. Extremely Low-Frequency Magnetic Field as a Stress Factor—Really Detrimental?—Insight into Literature from the Last Decade. *Brain Sci.* **2021**, *11*, 174. [CrossRef] [PubMed]
5. Adam, O.; Psomiades, M.; Rey, R.; Mandairon, N.; Suaud-Chagny, M.-F.; Mondino, M.; Brunelin, J. Frontotemporal Transcranial Direct Current Stimulation Decreases Serum Mature Brain-Derived Neurotrophic Factor in Schizophrenia. *Brain Sci.* **2021**, *11*, 662. [CrossRef] [PubMed]
6. de Venecia, A.B.F., III; Fresnoza, S.M. Visual Cortex Transcranial Direct Current Stimulation for Proliferative Diabetic Retinopathy Patients: A Double-Blinded Randomized Exploratory Trial. *Brain Sci.* **2021**, *11*, 270. [CrossRef] [PubMed]
7. Zaehle, T. Frontal Transcranial Direct Current Stimulation as a Potential Treatment of Parkinson's Disease-Related Fatigue. *Brain Sci.* **2021**, *11*, 467. [CrossRef] [PubMed]
8. Piano, C.; Bove, F.; Mulas, D.; Di Stasio, E.; Fasano, A.; Bentivoglio, A.R.; Daniele, A.; Cioni, B.; Calabresi, P.; Tufo, T. Extradural Motor Cortex Stimulation in Parkinson's Disease: Long-Term Clinical Outcome. *Brain Sci.* **2021**, *11*, 416. [CrossRef] [PubMed]
9. Chen, C.; Yuan, K.; Chu, W.C.W.; Tong, R.K.Y. The Effects of 10 Hz and 20 Hz tACS in Network Integration and Segregation in Chronic Stroke: A Graph Theoretical fMRI Study. *Brain Sci.* **2021**, *11*, 377. [CrossRef] [PubMed]
10. Antal, A.; Bischoff, R.; Stephani, C.; Czesnik, D.; Klinker, F.; Timäus, C.; Chaieb, L.; Paulus, W. Low Intensity, Transcranial, Alternating Current Stimulation Reduces Migraine Attack Burden in a Home Application Set-Up: A Double-Blinded, Randomized Feasibility Study. *Brain Sci.* **2020**, *10*, 888. [CrossRef] [PubMed]
11. Kim, J.-H.; Han, J.-Y.; Park, G.-C.; Lee, J.-S. Cognitive Improvement Effects of Electroacupuncture Combined with Computer-Based Cognitive Rehabilitation in Patients with Mild Cognitive Impairment:A Randomized Controlled Trial. *Brain Sci.* **2020**, *10*, 984. [CrossRef] [PubMed]
12. Ko, L.-W.; Chikara, R.K.; Chen, P.-Y.; Jheng, Y.-C.; Wang, C.-C.; Yang, Y.-C.; Li, L.P.-H.; Liao, K.-K.; Chou, L.-W.; Kao, C.-L. Noisy Galvanic Vestibular Stimulation (Stochastic Resonance) Changes Electroencephalography Activities and Postural Control in Patients with Bilateral Vestibular Hypofunction. *Brain Sci.* **2020**, *10*, 740. [CrossRef] [PubMed]

*Article*

# Comparative Study of a Continuous Train of Theta-Burst Stimulation for a Duration of 20 s (cTBS 300) versus a Duration of 40 s (cTBS 600) in a Pre-Stimulation Relaxed Condition in Healthy Volunteers

**Jan Haeckert [1], John Rothwell [2], Ricci Hannah [3], Alkomiet Hasan [1,4,*] and Wolfgang Strube [1,4]**

[1] Department of Psychiatry, Psychotherapy and Psychosomatics, Medical Faculty, University of Augsburg, BKH Augsburg, Dr.-Mack-Str. 1, 86156 Augsburg, Germany; Jan.Haeckert@bkh-augsburg.de (J.H.); Wolfgang.Strube@bkh-augsburg.de (W.S.)
[2] Sobell Department of Motor Neuroscience and Movement Disorders, UCL Institute of Neurology, Queen Square, London WC1N 3BG, UK; j.rothwell@ion.ucl.ac.uk
[3] Department of Psychology, University of California San Diego, 9500 Gilman Drive, La Jolla, CA 92093, USA; rhannah@ucsd.edu
[4] Department of Psychiatry and Psychotherapy, University Hospital Munich, LMU Munich, Nußbaumstraße 7, 80336 München, Germany
* Correspondence: alkomiet.hasan@med.uni-augsburg.de

**Citation:** Haeckert, J.; Rothwell, J.; Hannah, R.; Hasan, A.; Strube, W. Comparative Study of a Continuous Train of Theta-Burst Stimulation for a Duration of 20 s (cTBS 300) versus a Duration of 40 s (cTBS 600) in a Pre-Stimulation Relaxed Condition in Healthy Volunteers. *Brain Sci.* **2021**, *11*, 737. https://doi.org/10.3390/brainsci11060737

Academic Editors: Moussa Antoine Chalah, Samar S. Ayache and Ulrich Palm

Received: 24 March 2021
Accepted: 28 May 2021
Published: 1 June 2021



**Abstract:** As variable after effects have been observed following phasic muscle contraction prior to continuous theta-burst stimulation (cTBS), we here investigated two cTBS protocols (cTBS300 and cTBS600) in 20 healthy participants employing a pre-relaxed muscle condition including visual feedback on idle peripheral surface EMG activity. Furthermore, we assessed corticospinal excitability measures also from a pre-relaxed state to better understand the potential impact of these proposed contributors to TBS. Motor-evoked potential (MEP) magnitude changes were assessed for 30 min. The linear model computed across both experimental paradigms (cTBS300 and cTBS600) revealed a main effect of TIME COURSE ($p = 0.044$). Separate exploratory analysis for cTBS300 revealed a main effect of TIME COURSE ($p = 0.031$), which did not maintain significance after Greenhouse–Geisser correction ($p = 0.073$). For cTBS600, no main effects were observed. An exploratory analysis revealed a correlation between relative SICF at 2.0 ms ($p = 0.006$) and after effects (relative mean change) of cTBS600, which did not survive correction for multiple testing. Our findings thereby do not support the hypothesis of a specific excitability modulating effect of cTBS applied to the human motor-cortex in setups with pre-relaxed muscle conditions.

**Keywords:** non-invasive brain stimulation; continuous theta burst stimulation (cTBS); cTBS 300 versus cTBS 600; pre-relaxed muscle condition; healthy participants

## 1. Introduction

Noninvasive transcranial brain stimulation (NIBS) is a safe and effective method to investigate neuronal functioning and neuroplasticity changes in the human brain. Different stimulation protocols have been established, which are viewed to induce excitability changes of the motor cortex (M1) that outlast the stimulation interventions themselves. These effects have either been related to so called long-term potentiation (LTP)-like plasticity or long-term depression (LTD)-like plasticity [1,2]. Stimulation protocols such as transcranial magnetic stimulation (TMS), transcranial direct current stimulation (tDCS), and transcranial random noise stimulation (tRNS) typically need a conditioning of at least several minutes to induce after effects [3,4]. However, in 2005, the theta-burst stimulation (TBS) technique was introduced by Huang [5,6], and has gained attention due to its potential to induce long-lasting after effects with relatively short stimulation durations and low stimulation intensities. In TBS, a burst of 3 stimuli at 50 Hz is repeated at intervals of 200 ms. The first TBS protocols applied to humans were intermittent TBS (iTBS) and continuous TBS (cTBS) [6]. In the case of iTBS, the short 50 Hz stimulation trains are interspersed with pauses, while in cTBS, the stimulation is applied continuously. Based on animal studies reporting that short intermittent stimulation trains in the case of iTBS enhanced synaptic efficacy and led to excitability enhancing effects [7], it has been presumed that excitability changes are evoked through increased calcium influx into the postsynaptic neurons. For iTBS, it was described that the applied repeated short trains of stimulation resulted in enhanced cortical excitability in both animal studies and in humans [6,8]. The alternation between short bursts of stimulation and the pauses between them was assumed to result in the predominantly excitatory effect of iTBS. In contrast, Huang et al. described that TBS applied continuously and for a duration of 40 s (resulting in 600 pulses, hence named cTBS 600) induced excitability, diminishing after effects that lasted up to one hour following application [6]. Here, the continuous applied stimulation train was proposed to result in adaptation processes due to the increased influx of calcium, and inhibitory effects were considered to overcome excitatory effects [2,9]. In addition to these findings, continuous TBS stimulation for shorter durations of 20 s (resulting in 300 pulses, hence named cTBS 300) could be viewed as an intermediate of the iTBS and cTBS 600 paradigms, as Gentner et al. described the excitability enhancing effects of the motor-evoked potential (MEP) amplitudes lasting for about 25 min following this shorter variant of continuous TBS application [10].

Although these paradigms provide important insights for the interaction of physiological processes in the development of excitability changes, the observed after-effects following all three introduced TBS paradigms are subject to a considerable amount of variability both within and between individuals [2,11,12]. This hinders its utility as both a research and clinical tool. Various factors such as age, gender, time of day for stimulation application, attention during TBS, genetic and developmental factors as well as network activity are considered relevant contributors to TBS variability [2]. As shown elsewhere, both stimulation intensity and prior voluntary motor activation before TBS stimulation might pose controllable factors, contributing to an observed elevated intra-individual variability [10]. Of note, TBS has usually been delivered by employing a stimulation intensity equivalent to 80% active motor threshold (AMT). To determine the active motor threshold, tonic contraction of the target muscle is necessary prior to applying TBS. It has been reported that TBS applied with 80% AMT following either cTBS 300 or cTBS 600 predominantly induced excitability diminishing effects [10]. However, if participants are completely relaxed more than 10 min prior to applying cTBS 300, a mild facilitatory effect has been described, while cTBS 600 resulted in reduced excitability [10]. Furthermore, bidirectional after effects have been observed following phasic muscle contraction prior to cTBS application [13] and after administration of the L-type $Ca^{2+}$ blocking drug nimodipine [14]. It has been proposed that after a period of rest, cTBS induces an increased $Ca^{2+}$ influx into postsynaptic neurons via NMDA-receptors as well as L-type $Ca^{2+}$ channels, causing excitability enhancing after effects. Furthermore, application of nimodipine prior to TBS stimulation effectively resulting in smaller amounts of $Ca^{2+}$ influx via NMDA channels has been assumed to evoke excitability diminishing effects [2,15]. One could therefore assume that prior muscle contraction causes an activity dependent change in L-type $Ca^{2+}$-entry, resulting in decreased MEP magnitudes. In addition to these neurophysiological mechanisms on the synaptic level, TBS after-effects have been related to contributing factors of inhibitory and facilitatory neuronal circuitry [16–19]. However, it remains unresolved as to what extent different states of the balance between inhibitory and facilitatory cortical networks contribute to TBS after-effects.

Based on this outlined current state of our knowledge, we here directly assessed the after effects of two established cTBS paradigms, cTBS 300 versus cTBS 600, in 20 healthy participants in a so called pre-relaxed muscle condition (including visual feedback of EMG activity). In contrast to previous studies, we used a pre-relaxed muscle condition to control for the potentially impeding contributory effect of muscle activation prior to cTBS. We hypothesized that in a pre-relaxed condition, after effects of both cTBS paradigms would be less variable (compared to findings in previous studies) and that we would thus obtain a clear bidirectional pattern, whereas cTBS 300 would result in significant enhancement of cortical excitability, while significant excitability diminishing after effects would be observed following cTBS 600. Furthermore, we tested a variety of cortical excitability parameters as these pose potential contributors to the expected after-effects of TBS and to better understand the physiological mechanisms underpinning the changes in excitability both following cTBS 300 and cTBS 600, respectively, from a pre-relaxed state. In this regard, we assessed short latency-intracortical inhibition (SICI), intracortical facilitation (ICF), and short-interval intracortical facilitation (SICF) as these parameters allow for assessments of cortical excitability states of the motor-cortex by means of paired-pulse TMS protocols [17]. When a subthreshold conditioning stimulus (S1) precedes a suprathreshold test stimulus (S2) at inter-stimulus intervals (ISI) of 1 to 7 ms, MEP magnitudes are diminished due to intra-cortical inhibition (SICI). Furthermore, increasing the ISI from 8 to 30 ms leads to facilitation (ICF) of MEP magnitudes [17]. In contrast, increasing the S1 intensity toward peri- and suprathreshold levels followed by an S2 stimulus at peri-threshold intensity, leads to synergistic levels of facilitation (SICF) [20]. SICF has been demonstrated to develop over short ISIs (1 to 5 ms) with three distinct peaks at ISI 1–1.5, 2.4–2.9, and >4.5 ms. Given the increasing number of clinical studies employing TBS paradigms as treatment options for psychiatric and neurological conditions such as depression [21–23], further insights about the sources of variability might contribute to relevant improvements in designing individualized treatment paradigms.

## 2. Materials and Methods

### 2.1. Subjects

Twenty healthy subjects (14 female, 16 right-handed, mean age: 25.3 ± 4.3) participated in this study after giving informed consent. None of the subjects had a history of neurological or mental illness or had metallic cerebral implants, nor had a history of alcohol or drug abuse and nobody was taking any neuroactive medication. The study protocol was designed in accordance with the Declaration of Helsinki and was approved by the Ethics Committee of the University College London.

### 2.2. Design

All 20 subjects attended three experimental sessions (all conducted by the same investigator) separated by at least three days to control for carry over effects. Before cTBS was applied, subjects took part in an additional experiment assessing various individual parameters of motor-cortical excitability (see Figure 1). After this first experimental session, participants underwent two cTBS sessions (cTBS 300 and cTBS 600) in a pseudorandomized order.

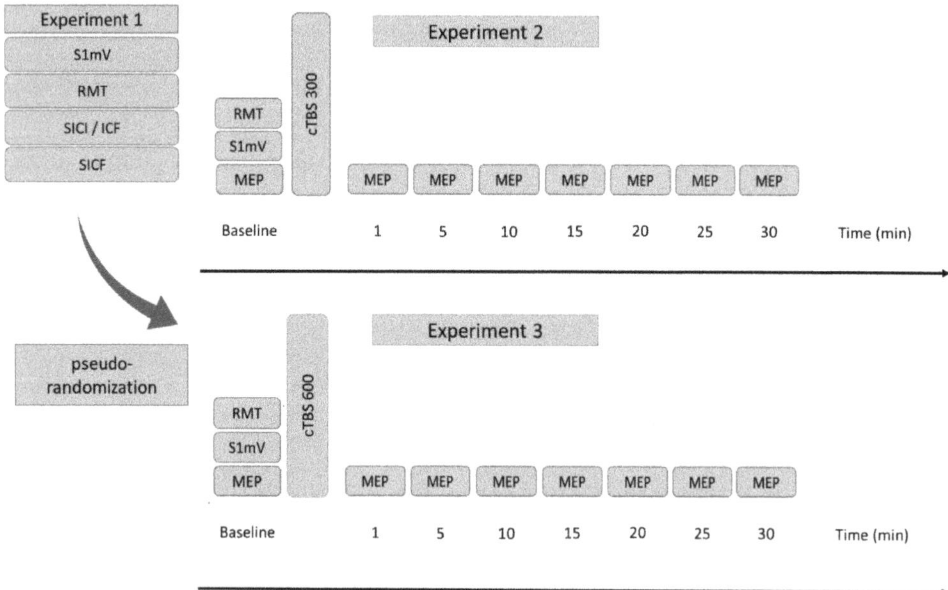


**Figure 1.** Study flowchart. Experiment 1 (**left**): S1 mV: intensity to evoke MEP of approximately 1 mV, RMT: resting motor threshold, SICI/ICF: short-intracortical inhibition/intracortical facilitation, SICF: short-interval intracortical facilitation. Experiment 2 (**upper right**): MEP: motor-evoked potentials. cTBS 300: continuous theta-burst stimulation for a duration of 20 s. Experiment 3 (**lower right**): MEP: motor-evoked potentials. cTBS 600: continuous theta-burst stimulation for a duration of 40 s. Time bins are reported in minutes. In Experiment 1, all measures (MEP, S1 mV, and RMT) were acquired using a monophasic stimulator. In Experiment 2 and Experiment 3, MEP and S1 mV were obtained using a monophasic stimulator for excitability monitoring, while RMT and a second S1 mV were acquired with a biphasic stimulator.

### 2.3. Experimental Procedures

During all experiments, participants were placed in a comfortable chair with their head and arms resting in a convenient position [24]. We recorded electromyographic activity (EMG) via surface electrodes on the right first dorsal interosseous muscle (FDI) of the participant's hand and contralateral to the TMS stimulation site. Raw signals were amplified and bandpass-filtered (3 Hz–2 kHz range) using a Digitimer D-360 amplifier setup (Digitimer Ltd., Welwyn Garden City, UK) and digitized at 5 kHz using a 1401 data acquisition interface (Cambridge Electronic Design Ltd., Cambridge, UK) controlled by Signal Software (Version 5, Cambridge Electronic Design, Cambridge, UK). At the end of the study, all data were analyzed off-line using the Signal Software and NuCursor by an investigator not involved in the experiments. During the experiments, visual feedback of EMG activity was used to help participants maintain complete muscle relaxation. Assessment of motor cortical excitability was performed with a standard figure-of-eight TMS coil (70 mm, The Magstim Company Ltd., Whitland, UK) connected to a monophasic Magstim Bistim$^2$ stimulator (The Magstim Company Ltd., Whitland, UK). Both cTBS protocols were delivered using the same coil design connected to a biphasic Magstim Rapid$^2$ stimulator (The Magstim Company Ltd., Whitland, UK). In all experiments, the coil was held tangentially to the skull above the left primary motor cortex (M1), with the handle pointing in a dorsolateral direction at a 45° angle from the midsagittal line, leading to a posterior-anterior induced current [25]. The stimulation site that produced the largest and most stable motor evoked potentials (MEP) at moderately supra-threshold stimulation intensities was marked for each experimental session separately with a skin marker for consistent coil positioning.

### 2.4. Baseline Excitability and Monitoring of Excitability Changes

In experimental session 1, all measures were performed with a MagStim Bistim$^2$ monophasic transcranial magnetic stimulator. Here, different parameters of corticospinal excitability were assessed (see Figure 1). Single-pulse TMS measurements included resting motor threshold (RMT) and the intensity to evoke MEP of approximately 1 mV peak-to-peak amplitude (S1 mV). Short-latency intracortical inhibition (SICI) and intracortical facilitation (ICF) were recorded with a standardized paired-pulse protocol (Stimulus 1: 90% RMT, Stimulus 2: S1 mV, inter-stimulus intervals (ISI): 2 ms/3 ms/9 ms/12 ms [17]). The test pulse was applied 20 times, and all paired-pulses were applied 10 times in a randomized order at 0.2 Hz. Short-interval intracortical facilitation (SICF) was evaluated with another paired-pulse protocol (Stimulus 1: S1 mV, Stimulus 2: 90% RMT, inter-stimulus intervals (ISI): 1.4 ms/1.6 ms/1.8 ms/2.0 ms/2.2 ms/2.4 ms/2.6 ms/2.8 ms/3.0 ms/3.2 ms [26]). The test pulse was applied 20 times, and all paired-pulses were applied 10 times in a randomized order at 0.2 Hz. There is good evidence that SICI relates to the activation of γ-aminobutyric acid (GABA) inhibitory circuits ($GABA_A$) in the primary motor cortex [27]. The mechanisms for intracortical facilitation (ICF) are less clear, but are considered to involve the activation of excitatory cortico-cortical pyramidal cells and glutamatergic net effects [17–19]. The physiological origin of SICF remains to be clarified, but an intracortical origin was postulated by epidural spinal cord recordings [28]. SICF occurs at specific inter-stimulus intervals of 1.1–1.5 ms, 2.3–2.9 ms, and 4.1–4.4 ms, and if the intensity of both pulses is either around the motor threshold [20] or if a suprathreshold first pulse and a subthreshold second pulse are applied [26]. The intervals of ~1.5 ms between the facilitatory peaks closely matches the latencies between successive I-waves in epidural spinal cord recordings, therefore it was suggested that SICF reflects facilitatory I-wave interaction [20,26,29]. S1 mV for excitability measures (MEP amplitudes) before and after cTBS intervention in experiments 2 and 3 were assessed with the previously described monophasic stimulator. Single pulse MEP measurements using S1 mV stimulator intensity were obtained at baseline (40 stimuli) and after cTBS (20 stimuli each at the following time bins: 1 min/5 min/10 min/15 min/20 min/25 min/30 min) to monitor the induced after-effects (see Figure 1). The conditioned/unconditioned MEP ratio was calculated for each ISI in the case of paired-pulse measures.

### 2.5. Theta-Burst Stimulation

Continuous TBS (cTBS) was applied according to previously published protocols [6,10]. In short, each burst consisted of three stimuli with a repetition rate of 50 Hz, and the bursts were repeated with a frequency of 5 Hz. In our second experimental session, we applied a continuous train of bursts for a duration of 20 s (cTBS 300 [10]) and in the third experimental session, we applied a continuous train of bursts for a duration of 40 s (cTBS 600 [6]). The stimulation intensity for both cTBS protocols was set at 70% of resting motor threshold (RMT) elicited by a biphasic stimulator as detailed above [10]. We decided to use the RMT and not the active motor threshold (AMT) to avoid any influence of prior voluntary motor activation on after effects induced by cTBS [10,30]. Additionally, we assessed S1 mV prior to cTBS with the biphasic stimulator (see Table 1).

**Table 1.** Baseline motor-evoked potentials (MEP, monophasic), resting motor thresholds (RMT, monophasic in Experiment 1 and biphasic in Experiments 2 and 3), and intensities to evoke MEP of approximately 1 mV (S1 mV, monophasic in experiment 1 and biphasic in Experiments 2 and 3). To ensure comparability between both cTBS experiments, we computed paired samples *t*-tests comparing the baseline neurophysiological measures of Experiments 2 and 3. The respective analyses obtained no significant differences. Note: S1 mV data of one participant were excluded due to incomplete data acquisition. Data are shown as mean values ± standard deviation. mV: millivolt; %: percentage of stimulator output.

| | **Experiment 1** | **Experiment 2 cTBS 300** | **Experiment 3 cTBS 600** | **Dependent Samples *t*-Tests (Experiment 2 vs. Experiment 3)** | |
|---|---|---|---|---|---|
| | | | | **Test Statistic** | ***p*-Values** |
| Baseline-MEP [mV] | - | 0.934 ± 0.244 | 0.938 ± 0.262 | $t_{(19)} = 0.067$ | 0.947 |
| RMT [%] | 47.25 ± 8.30 | 50.30 ± 9.69 | 49.05 ± 9.90 | $t_{(19)} = 1.403$ | 0.177 |
| S1 mV [%] | 55.85 ± 10.42 | 58.00 ± 10.58 | 57.68 ± 9.29 | $t_{(18)} = 0.344$ | 0.735 |

### 2.6. Statistical Methods

SPSS 26 for Windows (IBM, Armonk, NY, USA) was used for all analyses and level of significance was set at $\alpha = 0.05$. Descriptive statistics were used to present Experiment 1 data. Two-tailed paired samples *t*-tests were computed to compare the baseline excitability measures between both cTBS experiments. To assess changes in motor-cortical excitability, an omnibus repeated measures ANOVA (RM-ANOVA) with the within-subjects factors "STIMULATION" (cTBS 300 and cTBS 600) and "TIME COURSE" (Baseline/1 min/5 min/10 min/15 min/20 min/25 min/30 min) was performed (2 × 8 design). Due to the potential variability of MEP magnitude changes following both paradigms under the outlined hypotheses, separate explorative RM-ANOVAs for cTBS 300 and cTBS 600 were performed again for the within-subjects factor "TIME COURSE" (Baseline/1 min/5 min/10 min/15 min/20 min/25 min/30 min) as well as for the average of MEP magnitude changes employing the within-subjects factor "TIME" (Baseline/average MEPs after cTBS). Sphericity was tested with the Mauchly's test and, if necessary, Greenhouse–Geisser correction was applied. Comparisons of post-baseline time bins with the baseline were performed using LSD tests (estimated marginal means).

Next, we assessed how many participants showed expected MEP magnitude changes following either cTBS 300 (expected increase) or cTBS 600 (expected decrease), as foregoing investigations had repeatedly demonstrated variable after effects following cTBS [31–34]. For this purpose, we defined 'expected response' as an MEP-magnitude increase >100% relative to the individual baseline following cTBS 300 (expected facilitation [10]) and as an MEP-size decrease <100% relative to the individual baseline following cTBS 600 (expected inhibition [6]). Moreover, we tested 10% and 50% changes from the baseline, respectively, to receive more insight into the potential expected and non-expected MEP changes following cTBS.

As recent findings postulate a possible relationship between cortical excitability parameters (SICI, ICF, SICF) and intra- and inter-subject response variability following cTBS [34], we computed Pearson correlational coefficients between relative mean post MEP magnitude changes of both TBS experiments and neurophysiological factors of cortical excitability obtained in Experiment 1 (SICI, ICF, and SICF). For correlation analyses, 95% CIs were calculated according to Bonett and Wright [35].

## 3. Results

### 3.1. Baseline Characteristics

Twenty healthy subjects (14 female, 16 right-handed, mean age: 25.3 ± 4.3) participated in a total of 60 experimental sessions. To ensure comparability between both cTBS experiments, we computed paired samples *t*-tests comparing the baseline neurophysiological measures of Experiments 2 and 3. The respective analyses obtained no significant differences: biphasic RMT ($p = 0.177$), S1 mV ($p = 0.735$), baseline MEP ($p = 0.947$) (see Table 1). Descriptive analyses of parameters of baseline excitability (monophasic S1 mV, RMT, SICI/ICF, SICF, I/O) are detailed in Table 2.

**Table 2.** Descriptive statistics of baseline excitability measures (Experiment 1: S1 mV, RMT, SICI, ICF, SICF) and for relative mean post stimulation MEP magnitudes of Experiment 2 (cTBS 300) and Experiment 3 (cTBS 600). Relative values (paired-pulse/test pulse) are reported for short-intracortical inhibition (SICI), intracortical facilitation (ICF) and short-interval intracortical facilitation (SICF). All measures were obtained using 90% RMT stimulation intensity for the conditioning stimulus in paired-pulse protocols. ISI: inter-stimulus interval, RMT: resting motor threshold (monophasic), S1 mV: intensity to evoke MEP of approximately 1 mV (monophasic). Data are shown as mean values ± standard deviation.

| | ISI | N | Mean ± SD |
|---|---|---|---|
| S1 mV | | 20 | 55.85 ± 10.42 |
| RMT | | 20 | 47.25 ± 8.30 |
| SICI | 2 ms | 20 | 0.75 ± 0.45 |
| | 3 ms | 20 | 0.89 ± 0.70 |
| ICF | 9 ms | 20 | 2.16 ± 1.28 |
| | 12 ms | 20 | 2.22 ± 1.17 |
| SICF | 1.4 ms | 20 | 3.44 ± 1.47 |
| | 1.6 ms | 20 | 3.43 ± 2.10 |
| | 1.8 ms | 20 | 2.09 ± 1.33 |
| | 2.0 ms | 20 | 1.54 ± 0.69 |
| | 2.2 ms | 20 | 1.36 ± 0.55 |
| | 2.4 ms | 20 | 1.97 ± 0.66 |
| | 2.6 ms | 20 | 2.57 ± 0.76 |
| | 2.8 ms | 20 | 2.13 ± 0.69 |
| | 3.0 ms | 20 | 2.17 ± 0.99 |
| | 3.2 ms | 20 | 1.87 ± 1.01 |
| cTBS 300 | mean post rel. MEP | 20 | 1.12 ± 0.42 |
| cTBS 600 | mean post rel. MEP | 20 | 0.99 ± 0.36 |

### 3.2. MEP Amplitude Changes over Time

The overall 2 × 8 RM-ANOVA across both experimental paradigms (cTBS 300 and cTBS 600) revealed a main effect of TIME COURSE ($F_{(3.7,69.8)} = 2.658$, $p = 0.044$), but no main effect of STIMULATION ($F_{(1,19)} = 0.961$, $p = 0.339$), and no STIMULATION × TIME COURSE interaction ($F_{(4.4,84.3)} = 0.468$, $p = 0.778$). The stimulation-unspecific effect of TIME COURSE was further evaluated with LSD tests showing only an increase in MEP amplitudes at 1 min ($p = 0.043$, all other $p \geq 0.112$).

Despite the lacking effect of STIMULATION in the aforementioned model, we added exploratory RM-ANOVAs separately for cTBS 300 and cTBS 600 based on a relevant body of literature, where both cTBS paradigms are considered to generate divergent after effects on MEP magnitude changes.

In the case of cTBS 300, this exploratory approach revealed a main effect of TIME COURSE ($p = 0.031$, that did not maintain significance after Greenhouse–Geisser correction ($F_{(3.7,70.4)} = 2.293$, $p = 0.073$). In comparison, the explorative RM-ANOVA for cTBS 600 obtained no TIME COURSE effect ($F_{(4.3,82.1)} = 0.911$, $p = 0.467$). To further investigate these observed differences in our exploratory analysis, we next conducted LSD tests in the case of cTBS 300, which showed higher MEP amplitudes compared to baseline only at 1 min after stimulation ($p = 0.033$, all other $p \geq 0.072$) (see Figure 2). However, this exploratory observation did not survive correction for multiple comparisons following Bonferroni correction. As we had not obtained a significant effect of cTBS 600, the same explorative analysis was not extended to our second stimulation paradigm.

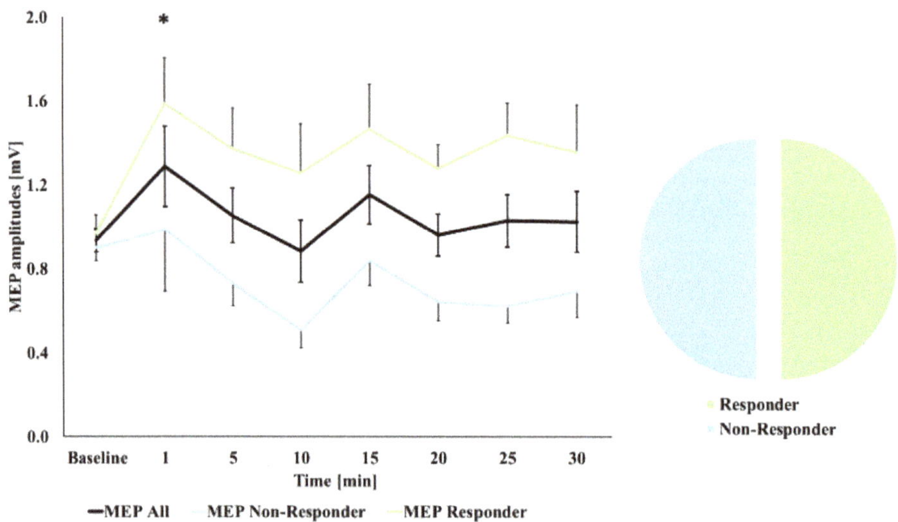


**Figure 2.** Motor-evoked potential (MEP) curves at baseline and following cTBS 300 stimulation. Time bins are reported in minutes. Black curve: MEP curve of all subjects; green curve: MEP curve of participants showing expected after effects (MEP magnitude increase; shown for a >100% threshold here); blue curve: MEP curve of participants showing unexpected after-effects. * indicates the observed significant effect (LSD test without Bonferroni correction) on the level of all participants. Error bars refer to standard error (SEM).

As a next step, we computed the average of MEP magnitude changes across all time bins and compared this measure to MEP magnitudes at baseline. For both stimulation paradigms, the two respective exploratory RM-ANOVAs comparing the baseline to the mean average post-stimulation MEPs obtained no significant main effects of TIME (cTBS 300: $F_{(1,19)} = 1.861$, $p = 0.188$; cTBS 600: $F_{(1,19)} = 0.084$, $p = 0.775$).

Finally, with respect to the above-described response analysis (see statistics section), we observed the expected response in 10 participants (50%) following cTBS 300 (see Figure 2). In the case of cTBS 600, we found an expected decrease in MEP magnitudes in 12 participants (60%, see Figure 3). Different approaches to define response to cTBS by higher/lesser thresholds (namely $\geq$ 110% and $\leq$90%, and $\geq$150% and $\leq$50%, respectively) resulted in gradually decreasing observations of expected response: in the case of a threshold of $\geq$110% and $\leq$90%, respectively, we again observed the expected response in 10 participants (50%) following cTBS 300; however, in the case of cTBS 600, the expected decrease in MEP magnitudes was only observed in seven participants (35%). A more rigorous threshold of $\geq$150% and $\leq$50%, respectively, resulted in only four participants (20%), showing an expected increase in MEP magnitudes following cTBS 300 and only one expected responder (5%) in the case of cTBS 600.

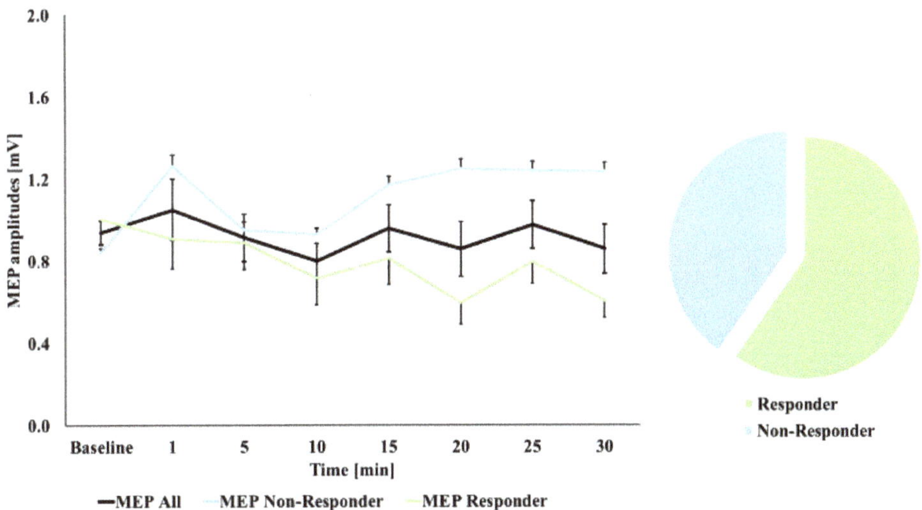


**Figure 3.** Motor-evoked potential (MEP) curves at baseline and following cTBS 600 stimulation. Time bins are reported in minutes. Black curve: MEP curve of all subjects; green curve: MEP curve of participants showing expected after effects (MEP magnitude decrease; shown for a <100% threshold here); blue curve: MEP curve of participants showing unexpected after-effects. Error bars refer to standard error (SEM).

### 3.3. Correlations with Baseline Excitability Measures

For cTBS 300, the Pearson correlational coefficients obtained no significant correlations between relative mean post MEP and SICI (all $|r| \leq 0.269$, all $p \geq 0.251$), ICF (all $|r| \leq 0.072$, all $p \geq 0.762$), or SICF at any ISI (all $|r| \leq 0.202$, all $p \geq 0.392$). In contrast, for cTBS 600, positive correlations were obtained between relative mean post MEPs and SICF in the case of ISI 1.4 ms (r = 0.466, $p = 0.038$, 95% CI [0.030, 0.753]), and ISI 2.0 ms (r = 0.591, $p = 0.006$, 95% CI [0.201, 0.819]) (see Figure 4) (all other ISI: all $|r| \leq 0.398$, all $p \geq 0.082$), while no significant correlations were observed for SICI (all $|r| \leq 0.404$, all $p \geq 0.077$) and ICF (all $|r| \leq 0.132$, all $p \geq 0.580$). However, none of the observed significant correlations survived corrections for multiple comparisons following Bonferroni correction (adjusted $p$-value 0.003, see Table 2 for descriptive statistics of the analyzed variables).

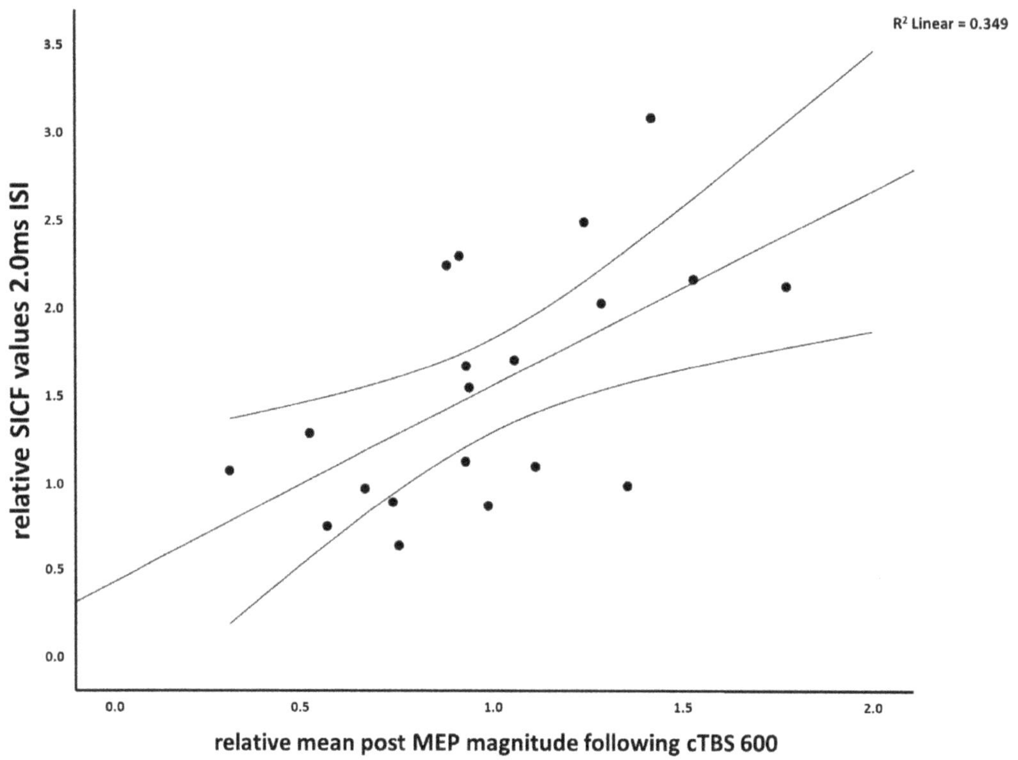


**Figure 4.** Pearson correlational coefficient between relative mean post MEP magnitudes following cTBS 600 and relative SICF values at ISI 2.0 ms (r = 0.591, $p$ = 0.006) including mean 95% confidence intervals.

## 4. Discussion

With this study, we present experiments comparing the after effects induced by cTBS 300 and cTBS 600, respectively, in a cohort of 20 participants using an intensity of 70% RMT and a pre-relaxed muscle condition. Overall, our main analyses did not show a specific stimulation effect of two different cTBS protocols on motor-cortical excitability. Analyses showed a significant effect on the change of MEP amplitudes over time, but subsequent analyses only confirmed a subtle stimulation-independent increase in MEP amplitudes compared to baseline immediately after stimulation. Thus, our overall finding must be considered as a negative finding.

In contrast to our findings, Gentner et al. showed continuously increased MEP magnitudes at 16 min and 24 min after cTBS 300 stimulation using a stimulation intensity of 70% RMT in 16 participants [10]. Similarly, Doeltgen et al. showed a significant increase in MEP magnitude at 30 min following cTBS 300 in their study on 16 subjects using a stimulation intensity of 70% RMT [36]. Interestingly, authors were not able to show an effect when stimulation intensity was adapted to 65% RMT [36]. Contrary to these findings, Stefan et al. described a significant MEP-decrease directly after the stimulation in seven subjects following cTBS 300 with an intensity of 70% RMT [37]. Furthermore, Fang et al. used an intensity of 80% AMT in nine participants and showed a significant MEP-decrease following cTBS 300 [38]. A consecutive meta-analysis summarized that stimulation with cTBS 300 induced mainly inhibitory after effects as it was also first described by Huang [1,6]. However, in this meta-analysis, most of the cTBS 300 protocols were performed with a stimulation intensity set to the AMT and not with RMT as in our experiments. These differences of muscle pre-activation prior to cTBS might have significantly contributed to the inter-study differences [10,13,39].

Furthermore, in the case of cTBS 600, while long-term depression of the MEP amplitude would have been expected [1,2,6,10], we were not able to obtain significant after effects. In contrast, Huang et al. reported reduced excitability after effects that lasted up to one hour following application [6]. Furthermore, Stefan et al. were able to show MEP magnitude decreases at 5 min and at 35 min following cTBS 600 with a stimulation intensity of 70% RMT in a cohort of 18 subjects [37]. Moreover, Di Lazzaro et al. also reported a significant decrease in MEP amplitudes directly after stimulation using an intensity of 80% RMT. However, this effect was abolished 30 min after the stimulation [40]. In addition, Doeltgen et al. tested cTBS 600 with 80% AMT in 14 participants and reported significant MEP magnitude decreases at 20 min and 30 min following stimulation [41]. Finally, Goldsworthy et al. observed significant reductions in MEP amplitudes directly after stimulation (0 min and 10 min) in 16 subjects when using cTBS 600 and a stimulation intensity of 70% RMT. Again, these effects had normalized at the second measurement taken between 30 and 40 min following stimulation [42].

Of note, while some previous experiments reported different after effects, our findings of no excitability-diminishing effect of cTBS 600 were in line with a number of previous experiments. In a study on 56 participants employing cTBS 600 stimulation with an intensity of 80% AMT, Hamada et al. were also not able to show a significant after effect [43]. In this cohort, a high inter-individual variability of the induced after effects was observed and the authors were capable of predicting about 50% of the expected inhibition after cTBS stimulation, when subjects showed a larger MEP latency difference evoked by different current directions, which is thought to reflect the recruitment of different neuronal populations [43]. Additionally, Hordacre et al. did not report a significant after effect of cTBS 600 at an intensity of 70% RMT in a cohort of 34 subjects [44]. At the same time, they observed a correlation of baseline MEP variability and cTBS 600 response, where subjects with higher baseline MEP variability showed a stronger inhibitory response [44]. In a second experiment, the authors further investigated I-wave recruitment and MEP variability and reported a significant correlation between AP-LM latency difference and cTBS response, which did not survive correction for multiple comparison [44].

In our corresponding analyses evaluating the relationship between SICF and the after-effects, we were able to show an exploratory correlation between SICF at 1.4 ms and 2.0 ms and MEP-changes following cTBS 600, which did not survive Bonferroni correction. Further, for the 1.4 ms ISI the lower end of the confidence interval is 0.03, which indicates that this correlation may not be present in the population. The 1.4 ms correlation must thus be interpreted with caution. While exploratory, this analysis indicated for the first time that participants showing less excitation with the SICF protocol might also display the expected excitability diminishing after effects followed by cTBS 600 stimulation. This new finding is in line with previous reports also showing a potential association between I-wave recruitment and cTBS after effects.

Considering the findings that subjects in whom late I-wave circuits were likely activated by TMS were more likely to respond in the expected direction with TBS [43] and the association between cTBS response and late I-wave recruitment [44], our observation of a potential correlation between reduced early SICF responses and excitability diminishing effects of cTBS 600 (see Figure 4) might further contribute to identifying another factor impacting cTBS variability. One could conclude that I-wave recruitment might pose a measurable variable that might help to determine and facilitate expected response to inhibitory motor-cortical cTBS paradigms.

In an overview of our observations, our study bears the limitation that the overall after-effects of both cTBS stimulation paradigms were not consistent with our outlined hypotheses. This could—at least in part—be explained by the so far reported high rates of inter-individual variability following both employed TBS paradigms or our sample size of 20 participants. In context with recent respective considerations, an even larger sample size of ≥30 participants might help to show more reliable subgroup differences and overall group level effects [2] and future research efforts using cTBS should include these considerations. Additionally limiting is the fact that baseline characteristics were measured not at the same day of TBS application. Finally, our main analyses provided a negative result—all subsequent analyses must be considered as exploratory that would not survive corrections for multiple testing. Thus, these findings must be confirmed in independent samples.

However, a specific advantage of our study is that, so far, it composes the largest experimental study comparing the after-effects induced by both cTBS 300 and cTBS 600, respectively, in a cohort of healthy subjects and a pre-relaxed muscle condition, and that we were able to relate our findings in an exploratory analysis to new potential baseline predictors of cortical excitability. In this regard, we obtained a relationship between early SICF responses and expected after effects of cTBS 600. As neurophysiological findings such as these might pose a potential means of identifying people that respond to cTBS in the desired way, further research efforts appear relevant here, especially for clinical applications.

**Author Contributions:** J.H.: Formal analysis, investigation, writing—original draft preparation, visualization, writing—review and editing. J.R.: resources, supervision—review and editing. A.H.: Conceptualization, investigation, methodology, writing—review and editing; R.H.: Investigation, resources, supervision, data analysis—review and editing. W.S.: Formal analysis, investigation, writing—original draft preparation, visualization, writing—review and editing. All authors have read and agreed to the published version of the manuscript.

**Funding:** This research received funding from the German Association for Psychiatry, Psychotherapy, and Psychosomatics (DGPPN) (travel grant to A. Hasan).

**Institutional Review Board Statement:** The study was conducted according to the guidelines of the Declaration of Helsinki, and approved by the Ethics Committee of the University College London Ethical approval number: 03/N108.

**Informed Consent Statement:** Written informed consent was obtained from all individual participants included in the study.

**Data Availability Statement:** Data are available upon reasonable scientific request.

**Conflicts of Interest:** The authors declare that the research was conducted in the absence of any commercial or financial relationships that could be construed as potential conflicts of interest.

## References

1. Chung, S.W.; Hill, A.T.; Rogasch, N.C.; Hoy, K.E.; Fitzgerald, P.B. Use of Theta-Burst Stimulation in Changing Excitability of Motor Cortex: A Systematic Review and Meta-Analysis. *Neurosci. Biobehav. Rev.* **2016**, *63*, 43–64. [CrossRef]
2. Suppa, A.; Huang, Y.-Z.; Funke, K.; Ridding, M.C.; Cheeran, B.; Di Lazzaro, V.; Ziemann, U.; Rothwell, J.C. Ten Years of Theta Burst Stimulation in Humans: Established Knowledge, Unknowns and Prospects. *Brain Stimul.* **2016**, *9*, 323–335. [CrossRef]
3. Cirillo, G.; Di Pino, G.; Capone, F.; Ranieri, F.; Florio, L.; Todisco, V.; Tedeschi, G.; Funke, K.; Di Lazzaro, V. Neurobiological After-Effects of Non-Invasive Brain Stimulation. *Brain Stimul.* **2017**, *10*, 1–18. [CrossRef]
4. Paulus, W. Transcranial Electrical Stimulation (TES—TDCS; TRNS, TACS) Methods. *Neuropsychol. Rehabil.* **2011**, *21*, 602–617. [CrossRef]
5. Huang, Y.-Z.; Rothwell, J.C. The Effect of Short-Duration Bursts of High-Frequency, Low-Intensity Transcranial Magnetic Stimulation on the Human Motor Cortex. *Clin. Neurophysiol.* **2004**, *115*, 1069–1075. [CrossRef]
6. Huang, Y.-Z.; Edwards, M.J.; Rounis, E.; Bhatia, K.P.; Rothwell, J.C. Theta Burst Stimulation of the Human Motor Cortex. *Neuron* **2005**, *45*, 201–206. [CrossRef] [PubMed]
7. Capocchi, G.; Zampolini, M.; Larson, J. Theta Burst Stimulation Is Optimal for Induction of LTP at Both Apical and Basal Dendritic Synapses on Hippocampal CA1 Neurons. *Brain Res.* **1992**, *591*, 332–336. [CrossRef]
8. Papazachariadis, O.; Dante, V.; Verschure, P.F.M.J.; Del Giudice, P.; Ferraina, S. ITBS-Induced LTP-like Plasticity Parallels Oscillatory Activity Changes in the Primary Sensory and Motor Areas of Macaque Monkeys. *PLoS ONE* **2014**, *9*, e112504. [CrossRef] [PubMed]
9. Huang, Y.-Z.; Rothwell, J.C.; Chen, R.-S.; Lu, C.-S.; Chuang, W.-L. The Theoretical Model of Theta Burst Form of Repetitive Transcranial Magnetic Stimulation. *Clin. Neurophysiol.* **2011**, *122*, 1011–1018. [CrossRef]

10. Gentner, R.; Wankerl, K.; Reinsberger, C.; Zeller, D.; Classen, J. Depression of Human Corticospinal Excitability Induced by Magnetic Theta-Burst Stimulation: Evidence of Rapid Polarity-Reversing Metaplasticity. *Cereb. Cortex* **2008**, *18*, 2046–2053. [CrossRef]
11. Jannati, A.; Block, G.; Oberman, L.M.; Rotenberg, A.; Pascual-Leone, A. Interindividual Variability in Response to Continuous Theta-Burst Stimulation in Healthy Adults. *Clin. Neurophysiol.* **2017**, *128*, 2268–2278. [CrossRef]
12. Vallence, A.-M.; Goldsworthy, M.R.; Hodyl, N.A.; Semmler, J.G.; Pitcher, J.B.; Ridding, M.C. Inter- and Intra-Subject Variability of Motor Cortex Plasticity Following Continuous Theta-Burst Stimulation. *Neuroscience* **2015**, *304*, 266–278. [CrossRef] [PubMed]
13. Iezzi, E.; Conte, A.; Suppa, A.; Agostino, R.; Dinapoli, L.; Scontrini, A.; Berardelli, A. Phasic Voluntary Movements Reverse the Aftereffects of Subsequent Theta-Burst Stimulation in Humans. *J. Neurophysiol.* **2008**, *100*, 2070–2076. [CrossRef] [PubMed]
14. Wankerl, K.; Weise, D.; Gentner, R.; Rumpf, J.-J.; Classen, J. L-Type Voltage-Gated $Ca^{2+}$ Channels: A Single Molecular Switch for Long-Term Potentiation/Long-Term Depression-like Plasticity and Activity-Dependent Metaplasticity in Humans. *J. Neurosci.* **2010**, *30*, 6197–6204. [CrossRef]
15. Goldsworthy, M.R.; Müller-Dahlhaus, F.; Ridding, M.C.; Ziemann, U. Inter-Subject Variability of LTD-like Plasticity in Human Motor Cortex: A Matter of Preceding Motor Activation. *Brain Stimul.* **2014**, *7*, 864–870. [CrossRef]
16. Van den Bos, M.A.J.; Menon, P.; Howells, J.; Geevasinga, N.; Kiernan, M.C.; Vucic, S. Physiological Processes Underlying Short Interval Intracortical Facilitation in the Human Motor Cortex. *Front. Neurosci.* **2018**, *12*, 240. [CrossRef]
17. Kujirai, T.; Caramia, M.D.; Rothwell, J.C.; Day, B.L.; Thompson, P.D.; Ferbert, A.; Wroe, S.; Asselman, P.; Marsden, C.D. Corticocortical Inhibition in Human Motor Cortex. *J. Physiol.* **1993**, *471*, 501–519. [CrossRef]
18. Chen, R.; Tam, A.; Bütefisch, C.; Corwell, B.; Ziemann, U.; Rothwell, J.C.; Cohen, L.G. Intracortical Inhibition and Facilitation in Different Representations of the Human Motor Cortex. *J. Neurophysiol.* **1998**, *80*, 2870–2881. [CrossRef]
19. Ni, Z.; Gunraj, C.; Chen, R. Short Interval Intracortical Inhibition and Facilitation during the Silent Period in Human. *J. Physiol.* **2007**, *583*, 971–982. [CrossRef]
20. Tokimura, H.; Ridding, M.C.; Tokimura, Y.; Amassian, V.E.; Rothwell, J.C. Short Latency Facilitation between Pairs of Threshold Magnetic Stimuli Applied to Human Motor Cortex. *Electroencephalogr. Clin. Neurophysiol.* **1996**, *101*, 263–272. [CrossRef]
21. Plewnia, C.; Pasqualetti, P.; Große, S.; Schlipf, S.; Wasserka, B.; Zwissler, B.; Fallgatter, A. Treatment of Major Depression with Bilateral Theta Burst Stimulation: A Randomized Controlled Pilot Trial. *J. Affect. Disord.* **2014**, *156*, 219–223. [CrossRef]
22. Chistyakov, A.V.; Kreinin, B.; Marmor, S.; Kaplan, B.; Khatib, A.; Darawsheh, N.; Koren, D.; Zaaroor, M.; Klein, E. Preliminary Assessment of the Therapeutic Efficacy of Continuous Theta-Burst Magnetic Stimulation (CTBS) in Major Depression: A Double-Blind Sham-Controlled Study. *J. Affect. Disord.* **2015**, *170*, 225–229. [CrossRef] [PubMed]
23. Cheng, C.-M.; Juan, C.-H.; Chen, M.-H.; Chang, C.-F.; Lu, H.J.; Su, T.-P.; Lee, Y.-C.; Li, C.-T. Different Forms of Prefrontal Theta Burst Stimulation for Executive Function of Medication- Resistant Depression: Evidence from a Randomized Sham-Controlled Study. *Prog. Neuropsychopharmacol. Biol. Psychiatry* **2016**, *66*, 35–40. [CrossRef] [PubMed]
24. Hasan, A.; Hamada, M.; Nitsche, M.A.; Ruge, D.; Galea, J.M.; Wobrock, T.; Rothwell, J.C. Direct-Current-Dependent Shift of Theta-Burst-Induced Plasticity in the Human Motor Cortex. *Exp. Brain Res.* **2012**, *217*, 15–23. [CrossRef] [PubMed]
25. Di Lazzaro, V.; Restuccia, D.; Oliviero, A.; Profice, P.; Ferrara, L.; Insola, A.; Mazzone, P.; Tonali, P.; Rothwell, J.C. Effects of Voluntary Contraction on Descending Volleys Evoked by Transcranial Stimulation in Conscious Humans. *J. Physiol.* **1998**, *508 Pt 2*, 625–633. [CrossRef]
26. Ziemann, U.; Tergau, F.; Wassermann, E.M.; Wischer, S.; Hildebrandt, J.; Paulus, W. Demonstration of Facilitatory I Wave Interaction in the Human Motor Cortex by Paired Transcranial Magnetic Stimulation. *J. Physiol.* **1998**, *511 Pt 1*, 181–190. [CrossRef]
27. Ziemann, U.; Lönnecker, S.; Steinhoff, B.J.; Paulus, W. The Effect of Lorazepam on the Motor Cortical Excitability in Man. *Exp. Brain Res.* **1996**, *109*, 127–135. [CrossRef]
28. Di Lazzaro, V.; Rothwell, J.C.; Oliviero, A.; Profice, P.; Insola, A.; Mazzone, P.; Tonali, P. Intracortical Origin of the Short Latency Facilitation Produced by Pairs of Threshold Magnetic Stimuli Applied to Human Motor Cortex. *Exp. Brain Res.* **1999**, *129*, 494–499. [CrossRef]
29. Ziemann, U. I-Waves in Motor Cortex Revisited. *Exp. Brain Res.* **2020**, *238*, 1601–1610. [CrossRef]
30. Goldsworthy, M.R.; Pitcher, J.B.; Ridding, M.C. The Application of Spaced Theta Burst Protocols Induces Long-Lasting Neuroplastic Changes in the Human Motor Cortex. *Eur. J. Neurosci.* **2012**, *35*, 125–134. [CrossRef]
31. Gamboa, O.L.; Antal, A.; Moliadze, V.; Paulus, W. Simply Longer Is Not Better: Reversal of Theta Burst after-Effect with Prolonged Stimulation. *Exp. Brain Res.* **2010**, *204*, 181–187. [CrossRef] [PubMed]
32. López-Alonso, V.; Cheeran, B.; Río-Rodríguez, D.; Fernández-Del-Olmo, M. Inter-Individual Variability in Response to Non-Invasive Brain Stimulation Paradigms. *Brain Stimul.* **2014**, *7*, 372–380. [CrossRef]
33. Tiksnadi, A.; Murakami, T.; Wiratman, W.; Matsumoto, H.; Ugawa, Y. Direct Comparison of Efficacy of the Motor Cortical Plasticity Induction and the Interindividual Variability between TBS and QPS. *Brain Stimul.* **2020**, *13*, 1824–1833. [CrossRef] [PubMed]
34. Guerra, A.; López-Alonso, V.; Cheeran, B.; Suppa, A. Variability in Non-Invasive Brain Stimulation Studies: Reasons and Results. *Neurosci. Lett.* **2020**, *719*, 133330. [CrossRef]
35. Bonett, D.G.; Wright, T.A. Sample Size Requirements for Estimating Pearson, Kendall and Spearman Correlations. *Psychometrika* **2000**, *65*, 23–28. [CrossRef]

36. Doeltgen, S.H.; Ridding, M.C. Low-Intensity, Short-Interval Theta Burst Stimulation Modulates Excitatory but Not Inhibitory Motor Networks. *Clin. Neurophysiol.* **2011**, *122*, 1411–1416. [CrossRef]
37. Stefan, K.; Gentner, R.; Zeller, D.; Dang, S.; Classen, J. Theta-Burst Stimulation: Remote Physiological and Local Behavioral after-Effects. *Neuroimage* **2008**, *40*, 265–274. [CrossRef]
38. Fang, J.-H.; Huang, Y.-Z.; Hwang, I.-S.; Chen, J.-J.J. Selective Modulation of Motor Cortical Plasticity during Voluntary Contraction of the Antagonist Muscle. *Eur. J. Neurosci.* **2014**, *39*, 2083–2088. [CrossRef]
39. Huang, Y.-Z. What Do We Learn from the Influence of Motor Activities on the After-Effect of Non-Invasive Brain Stimulation? *Clin. Neurophysiol.* **2016**, *127*, 1011–1012. [CrossRef] [PubMed]
40. Di Lazzaro, V.; Dileone, M.; Pilato, F.; Capone, F.; Musumeci, G.; Ranieri, F.; Ricci, V.; Bria, P.; Di Iorio, R.; de Waure, C.; et al. Modulation of Motor Cortex Neuronal Networks by RTMS: Comparison of Local and Remote Effects of Six Different Protocols of Stimulation. *J. Neurophysiol.* **2011**, *105*, 2150–2156. [CrossRef]
41. Doeltgen, S.H.; Ridding, M.C. Modulation of Cortical Motor Networks Following Primed θ Burst Transcranial Magnetic Stimulation. *Exp. Brain Res.* **2011**, *215*, 199–206. [CrossRef] [PubMed]
42. Goldsworthy, M.R.; Pitcher, J.B.; Ridding, M.C. Neuroplastic Modulation of Inhibitory Motor Cortical Networks by Spaced Theta Burst Stimulation Protocols. *Brain Stimul.* **2013**, *6*, 340–345. [CrossRef] [PubMed]
43. Hamada, M.; Murase, N.; Hasan, A.; Balaratnam, M.; Rothwell, J.C. The Role of Interneuron Networks in Driving Human Motor Cortical Plasticity. *Cereb. Cortex* **2013**, *23*, 1593–1605. [CrossRef] [PubMed]
44. Hordacre, B.; Goldsworthy, M.R.; Vallence, A.-M.; Darvishi, S.; Moezzi, B.; Hamada, M.; Rothwell, J.C.; Ridding, M.C. Variability in Neural Excitability and Plasticity Induction in the Human Cortex: A Brain Stimulation Study. *Brain Stimul.* **2017**, *10*, 588–595. [CrossRef]

*Article*

# Frontotemporal Transcranial Direct Current Stimulation Decreases Serum Mature Brain-Derived Neurotrophic Factor in Schizophrenia

**Ondine Adam [1,2,†], Marion Psomiades [2,†], Romain Rey [1,2], Nathalie Mandairon [1,3], Marie-Francoise Suaud-Chagny [2], Marine Mondino [1,2] and Jerome Brunelin [1,2,*]**

1 Centre Hospitalier Le Vinatier, F-69500 Bron, France; ondine.adam@ch-le-vinatier.fr (O.A.); romain.rey@ch-le-vinatier.fr (R.R.); nathalie.mandairon@cnrs.fr (N.M.); marine.mondino@ch-le-vinatier.fr (M.M.)
2 INSERM U1028, CNRS UMR5292, PSYR2 Team, Lyon Neuroscience Research Center, Université Claude Bernard Lyon 1, Université Jean Monnet, F-69500 Bron, France; marion.psomiades@ch-le-vinatier.fr (M.P.); marie-francoise.suaud-chagny@ch-le-vinatier.fr (M.-F.S.-C.)
3 INSERM U1028, CNRS UMR5292, NEUROPOP Team, Lyon Neuroscience Research Center, Université Claude Bernard Lyon 1, Université Jean Monnet, F-69500 Bron, France
* Correspondence: jerome.brunelin@ch-le-vinatier.fr
† Share first authorship.

**Abstract:** Although transcranial direct current stimulation (tDCS) shows promise as a treatment for auditory verbal hallucinations in patients with schizophrenia, mechanisms through which tDCS may induce beneficial effects remain unclear. Evidence points to the involvement of neuronal plasticity mechanisms that are underpinned, amongst others, by brain-derived neurotrophic factor (BDNF) in its two main forms: pro and mature peptides. Here, we aimed to investigate whether tDCS modulates neural plasticity by measuring the acute effects of tDCS on peripheral mature BDNF levels in patients with schizophrenia. Blood samples were collected in 24 patients with schizophrenia before and after they received a single session of either active (20 min, 2 mA, $n$ = 13) or sham ($n$ = 11) frontotemporal tDCS with the anode over the left prefrontal cortex and the cathode over the left temporoparietal junction. We compared the tDCS-induced changes in serum mature BDNF (mBDNF) levels adjusted for baseline values between the two groups. The results showed that active tDCS was associated with a significantly larger decrease in mBDNF levels (mean −20% ± standard deviation 14) than sham tDCS (−8% ± 21) (F = 5.387; $p$ = 0.030; $\eta^2$ = 0.205). Thus, mature BDNF may be involved in the beneficial effects of frontotemporal tDCS observed in patients with schizophrenia.

**Keywords:** tDCS; schizophrenia; plasticity; mature BDNF

**Citation:** Adam, O.; Psomiades, M.; Rey, R.; Mandairon, N.; Suaud-Chagny, M.-F.; Mondino, M.; Brunelin, J. Frontotemporal Transcranial Direct Current Stimulation Decreases Serum Mature Brain-Derived Neurotrophic Factor in Schizophrenia. *Brain Sci.* **2021**, *11*, 662. https://doi.org/10.3390/brainsci11050662

Academic Editor: Juha Veijola

Received: 8 April 2021
Accepted: 15 May 2021
Published: 19 May 2021

**Publisher's Note:** MDPI stays neutral with regard to jurisdictional claims in published maps and institutional affiliations.



## 1. Introduction

Auditory verbal hallucinations (AVH) are disabling and frequent symptoms in patients with schizophrenia. Among the available therapeutic strategies, some evidence suggests that transcranial direct current stimulation (tDCS) with the anode placed over the left prefrontal cortex and the cathode placed over the left temporoparietal junction may alleviate treatment-resistant AVH [1] up to three months after tDCS treatment [2]. However, although tDCS has shown encouraging clinical results [3], the brain mechanisms through which tDCS may have sustainable beneficial effects on AVH remain unclear. During tDCS, a weak direct current is circulating between two electrodes placed over the scalp of the subject. Physiological effects of tDCS are thought to be mediated by a modulation of the cortical excitability of brain regions situated under the location of electrodes [4]. The effects do not seem limited to the targeted cortical area and the modulatory effect may also spread to a large network of inter-connected brain regions [5,6]. These local and regional effects can outlast the stimulation period, suggesting long-term potentiation (LTP)/long-term depression (LTD)-like mechanisms [7], which require activity-dependent brain-derived neurotrophic factor (BDNF) secretion [8,9].

BDNF is a neurotrophic factor involved in neuronal plasticity including neuronal growth, cortical excitability, and neuronal regeneration mechanisms [10]. Several lines of evidence from animal research have suggested that the effects of tDCS depend on endogenous BDNF levels [11], and that tDCS promotes LTP mechanisms by increasing the expression of BDNF [12]. However, clinical studies investigating the effects of tDCS on BDNF levels in patients with neuropsychiatric conditions have produced controversial results. Although some studies showed that tDCS can induce an increase in serum BDNF levels in opioid-addicted patients [13] or in patients with Parkinson's disease [14], others did not detect any effects of tDCS on BDNF levels in patients with major depressive disorder [15,16] and bipolar disorder [17]. To the best of our knowledge, no studies have investigated the effect of tDCS on peripheral BDNF levels in patients with schizophrenia. Although some methodological issues regarding the tDCS electrode montage or the diagnosis of included patients should be taken into account to explain the observed discrepancies between studies, the methods used to measure BDNF levels also varied between studies. The classical peripheral measure of BDNF (whether in plasma or serum) includes a combination of the three BDNF isoforms that co-exist in the blood: the BDNF precursor protein (proBDNF) and the results of its proteolytic cleavage: the mature BDNF (mBDNF) and the BDNF prodomain [18]. Although the role of BDNF prodomain remains unclear [19], mBDNF and proBDNF exhibit opposite effects on neural plasticity through their binding to specific receptors. ProBDNF preferentially binds to p75 receptors and is involved in LTD mechanisms, pruning of dendritic arborization, cone retractation, and negatively regulates cell survival [20–23]. Conversely, mBDNF preferentially binds to tropomyosin receptor kinase B (TrkB) receptors and mBDNF-TrkB signaling is involved in LTP mechanisms, survival of neuronal networks, development of dendritic arborization, and neuronal cone growth [24–27]. The proportion of each isoform may vary depending on internal (such as age, sex, or drugs [28–30]) and external factors [24], and on neuronal activity-dependent mechanisms. For instance, low frequency stimulation of cultured hippocampal neurons preferentially induces proBDNF secretion, whereas high-frequency stimulation increases extracellular mBDNF leading to LTP [31]. Despite the opposing implications of the different isoforms on neural plasticity, a large majority of studies in humans measured only total BDNF, whereas one may hypothesize that the lasting beneficial effects observed after tDCS would be supported by a modulation of mBDNF-TrkB signaling involved in LTP mechanisms rather than pathways of other BDNF isoforms.

The aim of this study was therefore to investigate the effects of tDCS on mBDNF levels in patients with schizophrenia. To achieve our goal, we compared mean changes in serum mBDNF levels before and after a single session of either active or sham frontotemporal tDCS in patients with schizophrenia and AVH.

## 2. Materials and Methods

### 2.1. Study Design

This study occurred from 2011 to 2015 at the Centre Hospitalier le Vinatier, psychiatric hospital, Bron, France. In a randomized sham-controlled double-blind trial, participants were randomized to receive either a single session of active or sham tDCS. The study consisted of collecting blood samples once before and once after a single session of tDCS to measure the acute changes in serum mBDNF levels evoked by tDCS. The study was approved by a local ethics committee (Comité de Protection des Personnes Sud-Est VI, France, AU872, 02/02/2011) and complied with international standards for testing with human participants (Declaration of Helsinki). The study was pre-registered in a public database (http://clinicaltrials.gov, registration number NCT02652832, on 12 January 2016). The current study was an ancillary study to a clinical trial previously published elsewhere (NCT00870909; on 27 March 2009) [2]. All participants provided written informed consent before participation.

## 2.2. Participants

Twenty-seven patients with a DSM-IV-TR diagnosis of schizophrenia agreed to participate in the study. Patients presented with refractory auditory hallucinations, defined as the persistence of daily AVH without remission despite antipsychotic medication at an adequate dosage for at least 3 months and after the failure of at least one other previous treatment with an antipsychotic of different class at adequate dose and duration for the current episode. Exclusion criteria included significant neurological illness; head trauma, history of a seizure not induced by drug withdrawal, current alcohol or drug abuse, or inability to provide informed consent. Since one post-tDCS blood sample was missing, the final sample consisted of data from 26 participants and the analyzed sample from 24 (see Figure 1). Current antipsychotic medication was calculated as chlorpromazine clinically equivalent dose in mg/day following the instructions of Gardner et al. [32].

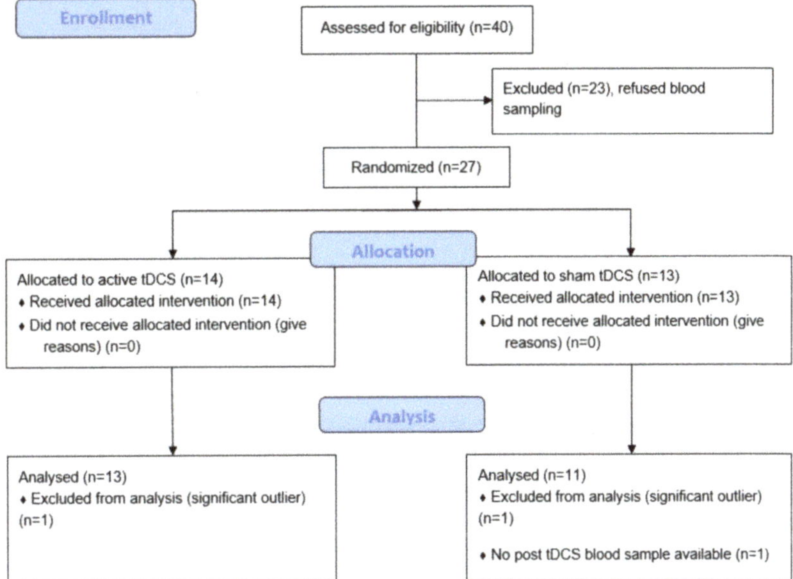


**Figure 1.** CONSORT flow diagram of the study.

## 2.3. tDCS Treatment

Stimulation was administered using a DC Plus Stimulator (NeuroConn, GmBH) with an intensity set at 2 mA. The stimulation was delivered during 20 min (ramp-up/ramp-down of 30 s). Two 7 × 5 cm surface electrodes (35 cm$^2$) placed in saline-soaked sponges were positioned over the scalp according to the 10/20 placement system for electroencephalogram. The anode was placed over the left prefrontal cortex (midway between F3 and FP1) and the cathode was placed over the left temporo-parietal junction (midway between T3 and P3). Rationale for this montage is justified in a previously published article [2].

The sham stimulation consisted of delivering active stimulation with the same electrode montage but only during the first 40 s of the 20 min period (2 mA, 30 s ramp-up/down). For double-blinding, the study mode of the DC-stimulator was used. It consisted of entering a predefined individual 5-digit code into the tDCS stimulator corresponding to active or sham stimulation, so that both the patient and the tDCS operator were blind to the tDCS condition.

To assess the blinding of participants, four participants were randomly chosen to complete a questionnaire that required them to guess whether they received either active or sham stimulation.

### 2.4. Measures of Serum BDNF

Two blood samples were collected, one before the tDCS session (between 8:00 a.m. and 9:00 a.m.) and one other after the end of the tDCS session (at 11:30 a.m.) on a Monday morning. Since sport practice, tobacco smoking, and alcohol consumption may modulate BDNF levels, participants were asked to avoid physical exercise and alcohol consumption during the 24 h prior to the experiment (Table 1). Tobacco smoking, which is also known to interfere with tDCS-induced aftereffects in patients with schizophrenia [33], was not allowed on the morning of the experiment. Blood samples of 5 mL were collected in fasting participants in a Vacutainer SST™ II Advance tube. After 20 min of clotting time, the whole blood sample was centrifuged at 3500× $g$ for 20 min at 4 °C to isolate the serum. The serum was then collected, aliquoted (200 mL), and stored at −24 °C until analysis.

**Table 1.** Baseline sociodemographic and clinical data of participants in active and sham groups.

| | Active Group (Mean ± SD) | Sham Group (Mean ± SD) | $p$-Value |
|---|---|---|---|
| $n$ total | 13 | 11 | |
| Age (years) | 33.08 ± 8.96 | 37.18 ± 9.38 | 0.285 |
| Illness duration (years) | 10.38 ± 9.51 | 14.73 ± 7.79 | 0.132 |
| Sex ($n$) | 6F/7M | 5F/6M | 0.973 |
| (%) | 46%/54% | 45%/55% | |
| Handedness ($n$) | 11R/1L/1 both | 9R/2L | 0.484 |
| Smokers (%) | 58% | 40% | 0.392 |
| Alcohol intake [1] (%) | 0% | 9% | 0.267 |
| Physical exercise [1] (%) | 8% | 9% | 0.902 |
| PANSS Total | 66.00 ± 14.89 | 69.78 ± 14.89 | 0.565 |
| PANSS Positive | 18.00 ± 4.81 | 19.44 ± 3.47 | 0.450 |
| PANSS Negative | 19.00 ± 5.63 | 17.33 ± 5.31 | 0.493 |
| PANSS General | 29.00 ± 5.31 | 33.00 ± 7.62 | 0.262 |
| mBDNF (pg·mL$^{-1}$) | 16,510.80 ± 4346.98 | 13,257.50 ± 3274.58 | 0.054 |
| Antipsychotic dose (CPZeq) | 930.41 ± 415.13 | 1192.91 ± 449.45 | 0.151 |
| Molecule | | | |
| Typical antipsychotics | 4 | 2 | 0.649 |
| Atypical antipsychotics | 12 | 11 | 1.000 |
| Clozapine | 4 | 4 | 1.000 |
| Antidepressants | 3 | 4 | 0.659 |
| Benzodiazepines | 5 | 2 | 0.386 |
| Anxiolytics | 3 | 6 | 0.206 |

CPZeq, chlorpromazine clinically equivalent dose in mg/day [32]; F, female; M, male; mBDNF, mature brain-derived neurotrophic factor; L, left-hander; PANSS, positive and negative syndrome scale; R, right-hander; SD, standard deviation. [1] Alcohol consumption and physical exercise were controlled for the 24 h prior to the experiment. $p$-values were obtained using independent samples $t$-tests for age, PANSS scores, CPZeq and mBDNF levels, and Fischer's exact test tests for other variables.

mBDNF levels were quantified by enzyme-linked immunosorbent assay (ELISA), according to the manufacturers' instructions (mature BDNF Immunoassay, Aviscera Bioscience, Santa-Clara, CA, USA). Serum samples were applied on precoated 96-well plates and allowed to incubate for two hours at room temperature. Plates were successively incubated with anti-human BDNF antibodies, streptavidin-HRP conjugate, and substrate. The reaction was shut down by stop solutions provided by the manufacturer. The absorbance was read at 450 nm with a micro-plate reader (Perkin Elmer Wallac 1420 Victor2, Winpact Scientific Inc, Saratoga, CA, USA). According to a reference curve, mBDNF levels are expressed in pg·mL$^{-1}$. Intertrial reproducibility was controlled with an external standard.

### 2.5. Statistical Analysis

All statistical analyses were performed in Jasp version 0.14. Statistical significance was defined as $p < 0.05$. Baseline demographic and clinical data were compared between groups (active vs. sham) using independent sample $t$-tests for continuous variables and Fischer's exact tests for categorical variables. The primary outcome was the change in mBDNF levels induced by tDCS (ΔBDNF), calculated as mBDNF level after the tDCS session minus mBDNF level before the tDCS session. Potential outliers were identified with Grubb's test based on the value of the changes in mBDNF levels (https://www.graphpad.com, accessed on 19 May 2021). To compare the effect of active versus sham tDCS on mBDNF level changes, a one-way ANCOVA was performed with tDCS condition (active vs. sham) as a between-subjects factor and baseline mBDNF levels as a covariate. The choice of introducing baseline mBDNF levels as a covariate was made to reduce possible effects of baseline levels on tDCS-induced changes since the two groups showed a trend toward a significant difference in baseline mBDNF levels (see Section 3). Effect size was estimated using eta squared ($\eta^2$). To investigate the potential influence of other clinical and demographic variables known to have an impact on baseline serum BDNF levels such as age [28], sex [29], illness duration, or medication [30], exploratory Pearson's correlations were undertaken.

## 3. Results

Two participants were identified as significant outliers regarding our primary outcome, one in each group, and were thus excluded from the analysis. The final analyzed sample consisted of 24 patients: 13 in the active group and 11 in the sham group.

There were no significant differences between the two groups regarding sociodemographic and clinical characteristics (Table 1). However, since baseline mBDNF levels tended to differ between the two groups ($p = 0.054$), we added baseline mBDNF level as a covariate in the analysis to control for this factor. The distribution of the different treatments in terms of dose and molecules between the two groups was relatively balanced and no significant differences were observed (Table 1).

### 3.1. Effect of tDCS on mBDNF Levels

A mean 20% (±14) decrease in mBDNF levels was observed after active tDCS (ΔBDNF $= -3181.221 \pm 1862$ pg·mL$^{-1}$), whereas a mean 8% (±21) decrease was observed after sham tDCS (ΔBDNF $= -912.988 \pm 2643$ pg·mL$^{-1}$) (Figure 2). The ANCOVA revealed a significant main effect of tDCS condition on changes in serum mBDNF levels (F = 5.387; $p = 0.030$; $\eta^2 = 0.205$). In other words, the decrease in serum mBDNF levels was significantly greater after active than after sham tDCS when adjusted for baseline mBDNF levels. There was no significant effect of the covariate baseline mBDNF level on mBDNF changes (F = 0.080, $p = 0.780$, $\eta^2 = 0.003$).

Exploratory Pearson's correlation analyses undertaken to investigate the influence of clinical and demographic variables on baseline serum mBDNF levels revealed no significant effect of age ($r = 0.238$; $p = 0.262$), dose of antipsychotic medication measured as chlorpromazine equivalent in mg/day ($r = 0.093$; $p =$ 0. 666), or illness duration ($r = 0.178$; $p = 0.404$). No significant differences were observed for sex proportion or type of treatments between the two groups at baseline. No difference was observed at baseline between men and women regarding mBDNF levels (men 15,400.287 ± 5126.064 versus women 14569.925 ± 2786.04; $p = 0.636$).

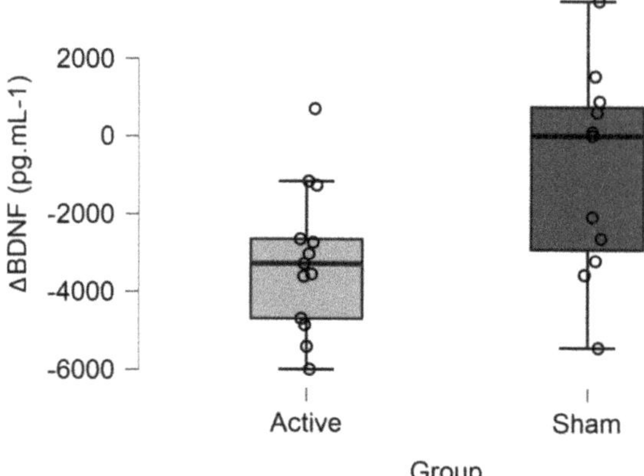


**Figure 2.** Changes in mBDNF levels (ΔBDNF) after active or sham tDCS. The decrease in serum mBDNF levels was significantly greater after active than after sham tDCS when adjusted for baseline mBDNF levels ($F = 5.387$; $p = 0.030$; $\eta^2 = 0.205$).

### 3.2. *Tolerability and Blinding*

All participants tolerated the tDCS session well and no serious adverse events were observed. Only two participants from the sham group reported medium adverse effects after the tDCS session (neck suffering, headache, and concentration difficulties).

Regarding the blinding, all four of the participants who were evaluated belonged to the sham group. Among them, two thought they had received active stimulation, one thought he had received the sham stimulation, and one was not able to guess the condition he received. These results suggest that participants were unable to identify which stimulation they were receiving and that the blinding of participants was well respected.

## 4. Discussion

The aim of the current study was to investigate the effects of a single session of frontotemporal tDCS on serum mBDNF levels in patients with schizophrenia and AVH. Active stimulation led to a significantly greater reduction in mBDNF levels compared with sham stimulation, with a small to medium effect size.

Our findings are consistent with previous studies using other forms of noninvasive brain stimulation (NIBS) techniques that reported a significant decrease in peripheral BDNF levels after stimulation of the prefrontal cortex [34–36]. Along this line, it was reported that a single session of repetitive transcranial magnetic stimulation (rTMS) applied over the left prefrontal cortex can decrease BDNF levels in healthy volunteers [34], and that multiple sessions of either low- or high-frequency rTMS applied over the prefrontal cortex can also decrease peripheral BDNF levels in both healthy volunteers [35] and in patients with amyotrophic lateral sclerosis [36]. However, our current findings do not support results from other studies reporting that NIBS did not modulate BDNF levels in patients with depression receiving high frequency rTMS [37], and in patients with uni- or bipolar depression receiving repeated sessions of frontal tDCS with the anode placed over the left prefrontal cortex [15–17]. Moreover, the present results are opposite to those from studies reporting that NIBS may induce an increase in BDNF levels in patients with depression receiving high-frequency rTMS over the left prefrontal cortex [38] or in patients with opioid addiction receiving tDCS over the dorsolateral prefrontal cortex [13]. Although all these studies targeted the PFC, the discrepancies between observed results may be related to variations in other NIBS parameters such as the total number of sessions delivered (which varies from 1 to 22 between studies), the type of the NIBS itself (rTMS or tDCS), and the number of sessions delivered by day, all of which are important to take into account [39–42].

The effects of NIBS on BDNF levels may also be influenced by participant diagnosis, further explaining the heterogeneous results observed in the literature. In line with this hypothesis, low frequency rTMS applied to the prefrontal cortex may induce a decrease in BDNF levels in healthy volunteers but not in patients with amyotrophic lateral sclerosis, who showed a decrease in BDNF levels only after high-frequency rTMS [36]. Thus, NIBS may either increase, decrease, or have no effect on peripheral BDNF levels depending on NIBS parameters and on the diagnosis of participants. To the best of our knowledge, no previous study has investigated the effects of tDCS on BDNF levels in patients with schizophrenia, although some evidence suggests abnormal BDNF regulation in this illness [43,44]. The present findings suggest that a single session of frontotemporal tDCS may decrease mBDNF in patients with schizophrenia and treatment-resistant AVH. Our results cannot be extended to patients with schizophrenia with other types of prominent symptoms (such as those with prominent negative symptoms or those with non-treatment-resistant schizophrenia) or to patients with another psychiatric condition that may have another baseline BDNF status.

The methods used to measure BDNF levels may also contribute to the heterogeneous results observed in the NIBS literature. Firstly, BDNF can be collected from either serum or plasma, while there is no consensus on their relative interpretations. Secondly, to date, the majority of studies have measured the effects of NIBS on total BDNF levels. To the best of our knowledge, the present study is one of the first to specifically explore tDCS effects on mBDNF rather than on total BDNF. Such methodological differences prevent any direct comparison between the current results and the literature. However, one may hypothesize that measuring the effects of one NIBS session on each of the BDNF isoforms is a particularly relevant approach to better understand NIBS-induced neuroplasticity, since proBDNF and mBDNF exhibit opposite effects on plasticity [27]. Due to its beneficial effects on LTP via the BDNF-TrkB signaling pathway, the mBDNF isoform may better reflect the beneficial effects of tDCS on neural plasticity. Nevertheless, BDNF is first synthesized as proBDNF, stored in dense core vesicles in neurons, and thereafter can be cleaved in mBDNF, either in the intracellular or extracellular compartment [45]. Thus, the observed increase, decrease, or null effect of tDCS on total BDNF after tDCS in the literature may be partly explained by compensation phenomena between the isoform concentrations. In this regard, one potential explanation for our results is that reduction in serum mBDNF may reflect an increased use of mBDNF in the central nervous system (CNS) after one tDCS session, resulting in the enhancement in neuronal plasticity induced by the stimulation [11,46].

Additionally, it was reported that BDNF-Val66Met-polymorphism interacts with tDCS dose to predict neurocognitive outcomes in patients with depression [47]. Similarly, BDNF-Val66Met-polymorphism may interact with tDCS to predict cortical plasticity in patients with schizophrenia [48]. Such gene–environment interactions support the hypothesis of a close relationship between tDCS-induced neural plasticity and BDNF, even in patients with schizophrenia.

Finally, we observed a trend toward a statistically significant difference in baseline mBDNF levels between the active and sham groups that may have influenced the present results. To rule out this potentially confounding effect, we introduced baseline mBDNF levels as a covariate in the primary analysis and we undertook exploratory analyses to control for the role of different moderators that may have influenced baseline mBDNF levels in patients with schizophrenia. It was reported that: (i) age negatively correlates with total BDNF levels [28], (ii) men exhibit higher levels of BDNF compared with women in a sample including patients with depression and healthy controls [29], and (iii) antipsychotics may increase BDNF levels in patients with schizophrenia [30] in a dose-dependent manner with clozapine but not with first generation antipsychotics [49]. No differences have been identified between active and sham groups regarding these parameters, and there was no correlation between these parameters and mBDNF levels at baseline. Further analyses did not reveal significant associations between baseline mBDNF and the variables identified above, and no differences were observed between the active and sham groups, suggesting that baseline mBDNF did not influence current results. Another important interaction that may have influenced the results observed in patients with schizophrenia under antipsychotic medication is the close relationship between BDNF synthesis and central dopamine release. It was observed in heterozygous BDNF (BDNF+/−) mice that endogenous BDNF may influence central dopamine neurotransmission by regulating the release and uptake dynamics of pre-synaptic dopamine transmission [50]. The interaction between antipsychotics binding on dopamine receptors, tDCS-induced dopamine release [6], and endogenous levels of BDNF needs further investigation in patients with schizophrenia.

The current study has some limitations that should be acknowledged. Firstly, the present results were obtained from an ancillary study to an RCT investigating the clinical effects of 10 sessions of tDCS on hallucinations in patients with schizophrenia [2]. The current study therefore has a relatively low statistical power, and no a priori sample size calculation was performed regarding the current outcome. Secondly, we investigated here the acute effect of a single session of tDCS on mBDNF, but it would have been interesting to examine the effect of repeated sessions of tDCS on BDNF levels since repeated sessions are usually needed to obtain a sustainable clinical effect. Thirdly, we only measured serum mBDNF levels, which prevented us from investigating the effects of tDCS on other isoforms of BDNF or on total BDNF. Fourthly, we only investigated acute changes after a single session of tDCS, thus no delayed effects of tDCS on mBDNF were investigated. Notably, in animal models, tDCS could either increase or decrease both serum and central BDNF levels depending on the delay between the tDCS sessions and the BDNF measure [51], suggesting that this parameter can be of major interest. Fifthly, while serum BDNF may be a good indicator of BDNF regulation in the central nervous system [52], it may not directly reflect brain fluctuations. However, it was shown that BDNF can cross the blood–brain barrier [53] and that BDNF blood levels correlate with BDNF levels in brain tissues [54]. Finally, we did not investigate the effect of BDNF single nucleotide polymorphism on the response to tDCS, although it was shown that BDNF-Val66Met-polymorphism interacts with tDCS aftereffects [48]. This specific polymorphism possibly interferes with the secretory pathway, resulting in a decreased secretion of BDNF [55,56]. More studies are needed to investigate the effect of this polymorphism on tDCS-induced mBDNF level variation.

## 5. Conclusions

In patients with schizophrenia and AVH, the current study highlighted that one session of active frontotemporal tDCS induces a significantly larger decrease in serum mBDNF levels compared with the sham. Beneficial effects of tDCS observed in patients with schizophrenia may be underpinned by a regulation of the mBDNF-TrkB signaling-related pathway that modulates neural plasticity. This study is a first step toward a better understanding of the mechanisms through which tDCS modulates brain plasticity in patients with schizophrenia. Further studies are needed to explore the effects of tDCS on each BDNF isoform and to investigate the respective roles of BDNF polymorphisms in the mechanisms underlying the physiological effects of tDCS in the CNS.

**Author Contributions:** J.B., M.M., and M.-F.S.-C. conceived and designed the study. N.M. and M.P. designed and performed the ELISA analysis. O.A., J.B., M.-F.S.-C., and M.M. contributed to the analysis and interpretation of the results. O.A. wrote the first draft of the manuscript. M.-F.S.-C., N.M., R.R., and M.P. critically revised the manuscript. O.A. and M.P. share first authorship. All authors have read and agreed to the published version of the manuscript.

**Funding:** This research was funded by the scientific council of Le Vinatier Hospital Center, grant number CSR B04; and the Fondation de l'Avenir #RMA 2015 (MFSC).

**Institutional Review Board Statement:** The study was conducted according to the guidelines of the Declaration of Helsinki, and approved by the Ethics Committee of Comité de Protection des Personnes Sud-Est VI, France AU872 (2 February 2011), AFSSAPS protocol code ID RCB 2010-A01249-30 (24 October 2008).

**Informed Consent Statement:** Informed consent was obtained from all subjects involved in the study.

**Data Availability Statement:** Data will be available upon reasonable request by contacting the corresponding author.

**Acknowledgments:** The authors are very grateful to Caroline Damasceno who carried out the stimulation sessions and collected the blood samples.

**Conflicts of Interest:** The authors declare no conflict of interest. The funders had no role in the design of the study; in the collection, analyses, or interpretation of data; in the writing of the manuscript, or in the decision to publish the results.

## References

1. Mondino, M.; Sauvanaud, F.; Brunelin, J. A Review of the Effects of Transcranial Direct Current Stimulation for the Treatment of Hallucinations in Patients with Schizophrenia. *J. ECT* **2018**, *34*, 164–171. [CrossRef] [PubMed]
2. Brunelin, J.; Mondino, M.; Gassab, L.; Haesebaert, F.; Gaha, L.; Suaud-Chagny, M.-F.; Saoud, M.; Mechri, A.; Poulet, E. Examining Transcranial Direct-Current Stimulation (TDCS) as a Treatment for Hallucinations in Schizophrenia. *Am. J. Psychiatry* **2012**, *169*, 719–724. [CrossRef] [PubMed]
3. Fregni, F.; El-Hagrassy, M.M.; Pacheco-Barrios, K.; Carvalho, S.; Leite, J.; Simis, M.; Brunelin, J.; Nakamura-Palacios, E.M.; Marangolo, P.; Venkatasubramanian, G.; et al. Evidence-Based Guidelines and Secondary Meta-Analysis for the Use of Transcranial Direct Current Stimulation (TDCS) in Neurological and Psychiatric Disorders. *Int. J. Neuropsychopharmacol.* **2021**, *24*, 256–313. [CrossRef] [PubMed]
4. Nitsche, M.A.; Paulus, W. Sustained Excitability Elevations Induced by Transcranial DC Motor Cortex Stimulation in Humans. *Neurology* **2001**, *57*, 1899–1901. [CrossRef]
5. Keeser, D.; Meindl, T.; Bor, J.; Palm, U.; Pogarell, O.; Mulert, C.; Brunelin, J.; Möller, H.-J.; Reiser, M.; Padberg, F. Prefrontal Transcranial Direct Current Stimulation Changes Connectivity of Resting-State Networks during FMRI. *J. Neurosci.* **2011**, *31*, 15284–15293. [CrossRef]
6. Fonteneau, C.; Redoute, J.; Haesebaert, F.; Le Bars, D.; Costes, N.; Suaud-Chagny, M.-F.; Brunelin, J. Frontal Transcranial Direct Current Stimulation Induces Dopamine Release in the Ventral Striatum in Human. *Cereb. Cortex* **2018**, *28*, 2636–2646. [CrossRef]
7. Stagg, C.J.; Nitsche, M.A. Physiological Basis of Transcranial Direct Current Stimulation. *Neuroscientist* **2011**, *17*, 37–53. [CrossRef]
8. Lu, B. BDNF and Activity-Dependent Synaptic Modulation. *Learn. Mem.* **2003**, *10*, 86–98. [CrossRef]
9. Mowla, S.J.; Pareek, S.; Farhadi, H.F.; Petrecca, K.; Fawcett, J.P.; Seidah, N.G.; Morris, S.J.; Sossin, W.S.; Murphy, R.A. Differential Sorting of Nerve Growth Factor and Brain-Derived Neurotrophic Factor in Hippocampal Neurons. *J. Neurosci.* **1999**, *19*, 2069–2080. [CrossRef]
10. Sasi, M.; Vignoli, B.; Canossa, M.; Blum, R. Neurobiology of Local and Intercellular BDNF Signaling. *Pflug. Arch. Eur. J. Physiol.* **2017**, *469*, 593–610. [CrossRef]
11. Fritsch, B.; Reis, J.; Martinowich, K.; Schambra, H.M.; Ji, Y.; Cohen, L.G.; Lu, B. Direct Current Stimulation Promotes BDNF-Dependent Synaptic Plasticity: Potential Implications for Motor Learning. *Neuron* **2010**, *66*, 198–204. [CrossRef]
12. Podda, M.V.; Cocco, S.; Mastrodonato, A.; Fusco, S.; Leone, L.; Barbati, S.A.; Colussi, C.; Ripoli, C.; Grassi, C. Anodal Transcranial Direct Current Stimulation Boosts Synaptic Plasticity and Memory in Mice via Epigenetic Regulation of Bdnf Expression. *Sci. Rep.* **2016**, *6*, 1–19. [CrossRef]
13. Eskandari, Z.; Dadashi, M.; Mostafavi, H.; Armani Kia, A.; Pirzeh, R. Comparing the Efficacy of Anodal, Cathodal, and Sham Transcranial Direct Current Stimulation on Brain-Derived Neurotrophic Factor and Psychological Symptoms in Opioid-Addicted Patients. *Basic Clin. Neurosci.* **2019**, *10*, 641–650. [CrossRef]
14. Hadoush, H.; Banihani, S.A.; Khalil, H.; Al-Qaisi, Y.; Al-Sharman, A.; Al-Jarrah, M. Dopamine, BDNF and Motor Function Postbilateral Anodal Transcranial Direct Current Stimulation in Parkinson's Disease. *Neurodegener. Dis. Manag.* **2018**, *8*, 171–179. [CrossRef]
15. Palm, U.; Fintescu, Z.; Obermeier, M.; Schiller, C.; Reisinger, E.; Keeser, D.; Pogarell, O.; Bondy, B.; Zill, P.; Padberg, F. Serum Levels of Brain-Derived Neurotrophic Factor Are Unchanged after Transcranial Direct Current Stimulation in Treatment-Resistant Depression. *J. Affect. Disord.* **2013**, *150*, 659–663. [CrossRef]
16. Brunoni, A.R.; Padberg, F.; Vieira, E.L.M.; Teixeira, A.L.; Carvalho, A.F.; Lotufo, P.A.; Gattaz, W.F.; Benseñor, I.M. Plasma Biomarkers in a Placebo-Controlled Trial Comparing TDCS and Escitalopram Efficacy in Major Depression. *Prog. Neuro-Psychopharmacol. Biol. Psychiatry* **2018**, *86*, 211–217. [CrossRef]
17. Goerigk, S.; Cretaz, E.; Sampaio-Junior, B.; Vieira, É.L.M.; Gattaz, W.; Klein, I.; Lafer, B.; Teixeira, A.L.; Carvalho, A.F.; Lotufo, P.A.; et al. Effects of TDCS on Neuroplasticity and Inflammatory Biomarkers in Bipolar Depression: Results from a Sham-Controlled Study. *Prog. Neuro-Psychopharmacol. Biol. Psychiatry* **2021**, *105*, 110119. [CrossRef]
18. Wang, M.; Xie, Y.; Qin, D. Proteolytic Cleavage of ProBDNF to MBDNF in Neuropsychiatric and Neurodegenerative Diseases. *Brain Res. Bull.* **2020**, 172–184. [CrossRef]
19. Ambigapathy, G.; Zheng, Z.; Li, W.; Keifer, J. Identification of a Functionally Distinct Truncated BDNF MRNA Splice Variant and Protein in *Trachemys scripta elegans*. *PLoS ONE* **2013**, *8*, e67141. [CrossRef]
20. Woo, N.H.; Teng, H.K.; Siao, C.-J.; Chiaruttini, C.; Pang, P.T.; Milner, T.A.; Hempstead, B.L.; Lu, B. Activation of P75 NTR by ProBDNF Facilitates Hippocampal Long-Term Depression. *Nat. Neurosci.* **2005**, *8*, 1069–1077. [CrossRef]

21. Koshimizu, H.; Kiyosue, K.; Hara, T.; Hazama, S.; Suzuki, S.; Uegaki, K.; Nagappan, G.; Zaitsev, E.; Hirokawa, T.; Tatsu, Y.; et al. Multiple Functions of Precursor BDNF to CNS Neurons: Negative Regulation of Neurite Growth, Spine Formation and Cell Survival. *Mol. Brain* **2009**, *2*, 27. [CrossRef] [PubMed]
22. Teng, H.K.; Teng, K.K.; Lee, R.; Wright, S.; Tevar, S.; Almeida, R.D.; Kermani, P.; Torkin, R.; Chen, Z.-Y.; Lee, F.S.; et al. ProBDNF Induces Neuronal Apoptosis via Activation of a Receptor Complex of P75NTR and Sortilin. *J. Neurosci.* **2005**, *25*, 5455–5463. [CrossRef] [PubMed]
23. Rösch, H.; Schweigreiter, R.; Bonhoeffer, T.; Barde, Y.-A.; Korte, M. The Neurotrophin Receptor P75NTR Modulates Long-Term Depression and Regulates the Expression of AMPA Receptor Subunits in the Hippocampus. *Proc. Natl. Acad. Sci. USA* **2005**, *102*, 7362–7367. [CrossRef] [PubMed]
24. Kowiański, P.; Lietzau, G.; Czuba, E.; Waśkow, M.; Steliga, A.; Moryś, J. BDNF: A Key Factor with Multipotent Impact on Brain Signaling and Synaptic Plasticity. *Cell Mol. Neurobiol.* **2018**, *38*, 579–593. [CrossRef] [PubMed]
25. Pang, P.T.; Teng, H.K.; Zaitsev, E.; Woo, N.T.; Sakata, K.; Zhen, S.; Teng, K.K.; Yung, W.-H.; Hempstead, B.L.; Lu, B. Cleavage of ProBDNF by TPA/Plasmin Is Essential for Long-Term Hippocampal Plasticity. *Science* **2004**, *306*, 487–491. [CrossRef] [PubMed]
26. Patterson, S.L.; Abel, T.; Deuel, T.A.S.; Martin, K.C.; Rose, J.C.; Kandel, E.R. Recombinant BDNF Rescues Deficits in Basal Synaptic Transmission and Hippocampal LTP in BDNF Knockout Mice. *Neuron* **1996**, *16*, 1137–1145. [CrossRef]
27. Deinhardt, K.; Chao, M.V. Shaping Neurons: Long and Short Range Effects of Mature and ProBDNF Signalling upon Neuronal Structure. *Neuropharmacology* **2014**, *76*, 603–609. [CrossRef]
28. Lommatzsch, M.; Zingler, D.; Schuhbaeck, K.; Schloetcke, K.; Zingler, C.; Schuff-Werner, P.; Virchow, J.C. The Impact of Age, Weight and Gender on BDNF Levels in Human Platelets and Plasma. *Neurobiol. Aging* **2005**, *26*, 115–123. [CrossRef]
29. Ozan, E.; Okur, H.; Eker, C.; Eker, Ö.; Gönül, A.S.; Akarsu, N. The Effect of Depression, BDNF Gene Val66met Polymorphism and Gender on Serum BDNF Levels. *Brain Res. Bull.* **2010**, *81*, 61–65. [CrossRef]
30. Fernandes, B.S.; Steiner, J.; Berk, M.; Molendijk, M.L.; Gonzalez-Pinto, A.; Turck, C.W.; Nardin, P.; Gonçalves, C.-A. Peripheral Brain-Derived Neurotrophic Factor in Schizophrenia and the Role of Antipsychotics: Meta-Analysis and Implications. *Mol. Psychiatry* **2015**, *20*, 1108–1119. [CrossRef]
31. Nagappan, G.; Zaitsev, E.; Senatorov, V.V.; Yang, J.; Hempstead, B.L.; Lu, B. Control of Extracellular Cleavage of ProBDNF by High Frequency Neuronal Activity. *Proc. Natl. Acad. Sci. USA* **2009**, *106*, 1267–1272. [CrossRef]
32. Gardner, D.M.; Murphy, A.L.; O'Donnell, H.; Centorrino, F.; Baldessarini, R.J. International Consensus Study of Antipsychotic Dosing. *Am. J. Psychiatry* **2010**, *167*, 686–693. [CrossRef] [PubMed]
33. Brunelin, J.; Hasan, A.; Haesebaert, F.; Nitsche, M.A.; Poulet, E. Nicotine Smoking Prevents the Effects of Frontotemporal Transcranial Direct Current Stimulation (TDCS) in Hallucinating Patients with Schizophrenia. *Brain Stimul.* **2015**, *8*, 1225–1227. [CrossRef] [PubMed]
34. Gaede, G.; Hellweg, R.; Zimmermann, H.; Brandt, A.U.; Dörr, J.; Bellmann-Strobl, J.; Zangen, A.; Paul, F.; Pfueller, C.F. Effects of Deep Repetitive Transcranial Magnetic Stimulation on Brain-Derived Neurotrophic Factor Serum Concentration in Healthy Volunteers. *Neuropsychobiology* **2014**, *69*, 112–119. [CrossRef]
35. Schaller, G.; Sperling, W.; Richter-Schmidinger, T.; Mühle, C.; Heberlein, A.; Maihöfner, C.; Kornhuber, J.; Lenz, B. Serial Repetitive Transcranial Magnetic Stimulation (RTMS) Decreases BDNF Serum Levels in Healthy Male Volunteers. *J. Neural. Transm.* **2014**, *121*, 307–313. [CrossRef]
36. Angelucci, F.; Oliviero, A.; Pilato, F.; Saturno, E.; Dileone, M.; Versace, V.; Musumeci, G.; Batocchi, A.P.; Tonali, P.A.; Lazzaro, V.D. Transcranial Magnetic Stimulation and BDNF Plasma Levels in Amyotrophic Lateral Sclerosis. *NeuroReport* **2004**, *15*, 717–720. [CrossRef]
37. Lang, U.E.; Bajbouj, M.; Gallinat, J.; Hellweg, R. Brain-Derived Neurotrophic Factor Serum Concentrations in Depressive Patients during Vagus Nerve Stimulation and Repetitive Transcranial Magnetic Stimulation. *Psychopharmacology* **2006**, *187*, 56–59. [CrossRef]
38. Zhao, X.; Li, Y.; Tian, Q.; Zhu, B.; Zhao, Z. Repetitive Transcranial Magnetic Stimulation Increases Serum Brain-Derived Neurotrophic Factor and Decreases Interleukin-1β and Tumor Necrosis Factor-α in Elderly Patients with Refractory Depression. *J. Int. Med. Res.* **2019**, *47*, 1848–1855. [CrossRef]
39. Monte-Silva, K.; Kuo, M.-F.; Liebetanz, D.; Paulus, W.; Nitsche, M.A. Shaping the Optimal Repetition Interval for Cathodal Transcranial Direct Current Stimulation (TDCS). *J. Neurophysiol.* **2010**, *103*, 1735–1740. [CrossRef]
40. Alonzo, A.; Brassil, J.; Taylor, J.L.; Martin, D.; Loo, C.K. Daily Transcranial Direct Current Stimulation (TDCS) Leads to Greater Increases in Cortical Excitability than Second Daily Transcranial Direct Current Stimulation. *Brain Stimul.* **2012**, *5*, 208–213. [CrossRef]
41. Paulus, W. Transcranial Electric and Magnetic Stimulation. *Brain Stimul.* **2013**, *116*, 329–342.
42. Stagg, C.J.; Antal, A.; Nitsche, M.A. Physiology of Transcranial Direct Current Stimulation. *J. ECT* **2018**, *34*, 144–152. [CrossRef] [PubMed]
43. Green, M.J.; Matheson, S.L.; Shepherd, A.; Weickert, C.S.; Carr, V.J. Brain-Derived Neurotrophic Factor Levels in Schizophrenia: A Systematic Review with Meta-Analysis. *Mol. Psychiatry* **2011**, *16*, 960–972. [CrossRef] [PubMed]
44. Ahmed, A.O.; Kramer, S.; Hofman, N.; Flynn, J.; Hansen, M.; Martin, V.; Pillai, A.; Buckley, P.F. A Meta-Analysis of Brain-Derived Neurotrophic Factor Effects on Brain Volume in Schizophrenia: Genotype and Serum Levels. *Neuropsychobiology* **2021**, *11*, 1–14. [CrossRef] [PubMed]

45. Leßmann, V.; Brigadski, T. Mechanisms, Locations, and Kinetics of Synaptic BDNF Secretion: An Update. *Neurosci. Res.* **2009**, *65*, 11–22. [CrossRef]
46. Vallence, A.M.; Ridding, M.C. Non-Invasive Induction of Plasticity in the Human Cortex: Uses and Limitations. *Cortex* **2014**, *58*, 261–271. [CrossRef]
47. McClintock, S.M.; Martin, D.M.; Lisanby, S.H.; Alonzo, A.; McDonald, W.M.; Aaronson, S.T.; Husain, M.M.; O'Reardon, J.P.; Weickert, C.S.; Mohan, A.; et al. Neurocognitive Effects of Transcranial Direct Current Stimulation (TDCS) in Unipolar and Bipolar Depression: Findings from an International Randomized Controlled Trial. *Depress. Anxiety* **2020**, *37*, 261–272. [CrossRef]
48. Strube, W.; Nitsche, M.A.; Wobrock, T.; Bunse, T.; Rein, B.; Herrmann, M.; Schmitt, A.; Nieratschker, V.; Witt, S.H.; Rietschel, M.; et al. BDNF-Val66Met-Polymorphism Impact on Cortical Plasticity in Schizophrenia Patients: A Proof-of-Concept Study. *Int. J. Neuropsychopharmacol.* **2015**, *18*, pyu040. [CrossRef]
49. Grillo, R.W.; Ottoni, G.L.; Leke, R.; Souza, D.O.; Portela, L.V.; Lara, D.R. Reduced Serum BDNF Levels in Schizophrenic Patients on Clozapine or Typical Antipsychotics. *J. Psychiatr. Res.* **2007**, *41*, 31–35. [CrossRef]
50. Bosse, K.E.; Maina, F.K.; Birbeck, J.A.; France, M.M.; Roberts, J.J.P.; Colombo, M.L.; Mathews, T.A. Aberrant Striatal Dopamine Transmitter Dynamics in Brain-Derived Neurotrophic Factor-Deficient Mice. *J. Neurochem.* **2012**, *120*, 385–395. [CrossRef]
51. Filho, P.R.M.; Vercelino, R.; Cioato, S.G.; Medeiros, L.F.; de Oliveira, C.; Scarabelot, V.L.; Souza, A.; Rozisky, J.R.; Quevedo, A.d.S.; Adachi, L.N.S.; et al. Transcranial Direct Current Stimulation (TDCS) Reverts Behavioral Alterations and Brainstem BDNF Level Increase Induced by Neuropathic Pain Model: Long-Lasting Effect. *Prog. Neuro-Psychopharmacol. Biol. Psychiatry* **2016**, *64*, 44–51. [CrossRef]
52. Klein, A.B.; Williamson, R.; Santini, M.A.; Clemmensen, C.; Ettrup, A.; Rios, M.; Knudsen, G.M.; Aznar, S. Blood BDNF Concentrations Reflect Brain-Tissue BDNF Levels across Species. *Int. J. Neuropsychopharmacol.* **2011**, *14*, 347–353. [CrossRef]
53. Pan, W.; Banks, W.A.; Fasold, M.B.; Bluth, J.; Kastin, A.J. Transport of Brain-Derived Neurotrophic Factor across the Blood–Brain Barrier. *Neuropharmacology* **1998**, *37*, 1553–1561. [CrossRef]
54. Sartorius, A.; Hellweg, R.; Litzke, J.; Vogt, M.; Dormannn, C.; Vollmayr, B.; Danker-Hopfe, H.; Gass, P. Correlations and Discrepancies between Serum and Brain Tissue Levels of Neurotrophins after Electroconvulsive Treatment in Rats. *Pharmacopsychiatry* **2009**, 270–276. [CrossRef]
55. Egan, M.F.; Kojima, M.; Callicott, J.H.; Goldberg, T.E.; Kolachana, B.S.; Bertolino, A.; Zaitsev, E.; Gold, B.; Goldman, D.; Dean, M.; et al. The BDNF Val66met Polymorphism Affects Activity-Dependent Secretion of BDNF and Human Memory and Hippocampal Function. *Cell* **2003**, *112*, 257–269. [CrossRef]
56. Chen, Z.-Y.; Patel, P.D.; Sant, G.; Meng, C.-X.; Teng, K.K.; Hempstead, B.L.; Lee, F.S. Variant Brain-Derived Neurotrophic Factor (BDNF) (Met66) Alters the Intracellular Trafficking and Activity-Dependent Secretion of Wild-Type BDNF in Neuro-secretory Cells and Cortical Neurons. *J. Neurosci.* **2004**, *24*, 4401–4411. [CrossRef]

*Article*

# TMS-Induced Central Motor Conduction Time at the Non-Infarcted Hemisphere Is Associated with Spontaneous Motor Recovery of the Paretic Upper Limb after Severe Stroke

**Maurits H. J. Hoonhorst [1], Rinske H. M. Nijland [2], Cornelis H. Emmelot [3], Boudewijn J. Kollen [4] and Gert Kwakkel [2,5,6,7,*]**

1 Rehabilitation Center Vogellanden, 8013 XZ Zwolle, The Netherlands; m.hoonhorst@vogellanden.nl
2 Amsterdam Rehabilitation Research Center | Reade, 1054 HW Amsterdam, The Netherlands; r.nijland@reade.nl
3 Department of Rehabilitation Medicine, Isala, 8025 AB Zwolle, The Netherlands; c.h.emmelot@home.nl
4 Department of General Practice and Elderly Care Medicine, University of Groningen, University Medical Center Groningen, 9712 CP Groningen, The Netherlands; b.kollen@home.nl
5 Amsterdam University Medical Center, Department of Rehabilitation Medicine, Amsterdam Movement Sciences, 1081 BT Amsterdam, The Netherlands
6 Amsterdam Neurosciences, Amsterdam University Medical Centre, 1081 HV Amsterdam, The Netherlands
7 Department of Physical Therapy and Human Movement Sciences, Feinberg School of Medicine, Northwestern University of Chicago, Evanston, IL 60208, USA
* Correspondence: g.kwakkel@amsterdamumc.nl; Tel.: +31-204-441-940

**Citation:** Hoonhorst, M.H.J.; Nijland, R.H.M.; Emmelot, C.H.; Kollen, B.J.; Kwakkel, G. TMS-Induced Central Motor Conduction Time at the Non-Infarcted Hemisphere Is Associated with Spontaneous Motor Recovery of the Paretic Upper Limb after Severe Stroke. *Brain Sci.* **2021**, *11*, 648. https://doi.org/10.3390/brainsci11050648

Academic Editors: Ulrich Palm, Moussa Antoine Chalah and Samar S. Ayache

Received: 5 April 2021
Accepted: 11 May 2021
Published: 15 May 2021



**Abstract:** Background: Stroke affects the neuronal networks of the non-infarcted hemisphere. The central motor conduction time (CMCT) induced by transcranial magnetic stimulation (TMS) could be used to determine the conduction time of the corticospinal tract of the non-infarcted hemisphere after a stroke. Objectives: Our primary aim was to demonstrate the existence of prolonged CMCT in the non-infarcted hemisphere, measured within the first 48 h when compared to normative data, and secondly, if the severity of motor impairment of the affected upper limb was significantly associated with prolonged CMCTs in the non-infarcted hemisphere when measured within the first 2 weeks post stroke. Methods: CMCT in the non-infarcted hemisphere was measured in 50 patients within 48 h and at 11 days after a first-ever ischemic stroke. Patients lacking significant spontaneous motor recovery, so-called non-recoverers, were defined as those who started below 18 points on the FM-UE and showed less than 6 points (10%) improvement within 6 months. Results: CMCT in the non-infarcted hemisphere was prolonged in 30/50 (60%) patients within 48 h and still in 24/49 (49%) patients at 11 days. Sustained prolonged CMCT in the non-infarcted hemisphere was significantly more frequent in non-recoverers following FM-UE. Conclusions: The current study suggests that CMCT in the non-infarcted hemisphere is significantly prolonged in 60% of severely affected, ischemic stroke patients when measured within the first 48 h post stroke. The likelihood of CMCT is significantly higher in non-recoverers when compared to those that show spontaneous motor recovery early post stroke.

**Keywords:** stroke; recovery; upper limb; prognosis; transcranial magnetic stimulation

## 1. Introduction

There is growing evidence that stroke in one of the hemispheres also affects the so-called 'non-lesioned hemisphere' [1–3]. Recent studies indicate that these affected neuronal networks depend on the severity of post-stroke motor impairment [2,3], which may differ between primary and secondary motor networks in the non-infarcted hemisphere [2]. Support for these assumptions has been found in several serial kinematic studies investigating the quality of movement (QoM) of the less-affected limb [2,4,5]. For example, the reaching performance of the less-affected limb in patients with a stroke has been shown to be significantly slower and involving less smooth movements than that in healthy subjects, and stroke severity had a significant impact on this discoordination [4,5]. Several explanations have been proposed in the literature for the reduced QoM of the less-affected limb. Anatomically, it is known that about 10 to 15% of all corticospinal descending pathways are uncrossed or double-crossed and innervate the ipsilesional less-affected limb [6–10]. An alternative explanation may be that the reduced QoM is caused by the affected reticulo-, tecto- and possibly rubrospinal pathways [11,12], since these multisynaptic pathways mainly project bilaterally to the trunk and fore-arm and less to the hand muscles [13–15]. In support of this latter assumption, the study by Benecke and colleagues found that patients with hemispherectomy showed ipsilateral responses to transcranial magnetic stimulation (TMS) of the non-infarcted cortex suggesting activation of cortico-reticulospinal pathways [16]. In addition, the reduced QoM of the less-affected limb found in these kinematic studies may also be caused by other concomitant neurological impairments of perception and planning [2]. The reduced QoM of the less-affected limb can also possibly be explained more indirectly by a transcallosal suppression of anatomically related networks in the non-infarcted hemisphere early after a stroke [17–20]. However, the time course of this transhemispheric diaschisis in the non-infarcted hemisphere, as well as its association with the severity of stroke in the infarcted hemisphere, remain unknown so far.

Interestingly, a recent serial kinematic study showed that in most patients, the synergy-dependent coordination of the less-affected limb gradually normalizes within the first 3 months post stroke [5]. It is unknown, however, if this normalization of the QoM of the less-affected limb parallels the concomitant underlying processes of spontaneous neurobiological recovery in the infarcted hemisphere. Moreover, serial kinematic studies of the affected and less-affected limbs during a reaching or pointing task are limited to those patients who can understand and are able to perform such a task early post stroke, while it is actually the most severely affected patients with low FM-UE baseline scores early after stroke who are likely to show no spontaneous motor recovery after stroke onset [21]. In the present study, we investigated if these so-called non-recoverers are the same patients who also suffer most from a reduced central motor conduction time (CMCT) in the non-infarcted hemisphere after stroke. Unfortunately, the impact of a first-ever ischemic stroke on conductivity properties such as the CMCT of the non-infarcted hemisphere is unclear, as is the association between the absence of spontaneous motor recovery of the affected limb and the CMCT in the non-infarcted hemisphere in patients with an acute hemispheric stroke. Previous prognostic studies suggest that TMS-induced motor evoked potentials (MEPs) of the non-infarcted hemisphere are unaffected after a first-ever hemispheric stroke, regardless of stroke severity [22–25]. However, it remains unclear if the CMCT in the non-infarcted hemisphere is significantly prolonged and associated with a lack of spontaneous motor recovery of the affected limb early post stroke.

Therefore, our primary aim was to demonstrate the existence of prolonged CMCT in the non-infarcted hemisphere, measured within the first 48 h, when compared to normative data and secondly, if the severity of motor impairment of the affected upper limb is significantly associated with CMCT at the non-infarcted hemisphere within 48 h and 11 days post stroke.

## 2. Materials and Methods

### 2.1. Subjects

Patients with a first-ever ischemic hemispheric stroke as revealed by MRI or CT were prospectively screened for eligibility, and if eligible, recruited by the Isala neurology department (the Netherlands), between August 2004 and July 2007. An experienced neurologist (PvdB) classified their stroke severity and etiology within 24 h after stroke onset, using the Trial of Org 10172 in Acute Stroke Treatment (TOAST) criteria [26]. To participate in this study, patients had to show symptoms of unilateral paralysis or significant paresis of the affected upper limb (Medical Research Council score, 0–3). Patients were excluded if they (1) had a loss of consciousness, (2) had peripheral nerve pathology, including diabetes or neuromuscular disease, or (3) were unable to undergo rehabilitation because of severe co-morbidity. They were also excluded if the following contraindications to TMS were present: cochlear implants, pregnancy, metal in the brain or skull, implanted neurostimulator, cardiac pacemaker or intracardiac lines and medication infusion devices [27,28].

All patients gave their verbal and/or written informed consent (themselves or by proxy) and all patients were treated according to the Dutch physical therapy guidelines for rehabilitation as soon as possible post stroke, including early mobilization and daily physical therapy interventions involving upper limb training, gait and mobility-related functions and activities [29]. The study protocol was approved by the local medical ethics board of the Isala (the Netherlands) (No 04.0318P).

### 2.2. Stimulation Procedure

The TMS technique and EMG recordings (Nihon Kohden Neuropack 8, Nihon Kohden, Tokyo, Japan) were performed according to the published guidelines [30,31]. The primary motor cortex (M1) was stimulated with a calibrated Magstim Dantec Maglite® (Dantec Dynamics, Bristol, UK). Cortical TMS was applied through a figure-of-eight-shaped coil. Stimuli were administered over the non-infarcted hemisphere, see Figure S1. The hotspot that produced the highest MEP amplitude of the abductor digiti minimi (ADM) muscle of the less-affected upper limb was determined by moving the coil over the scalp in the hand area of M1 of the non-infarcted hemisphere, with the stimulator at the submaximal output. Intra- and inter-observer reliability of assessing the TMS-induced MEP of the non-infarcted hemisphere and total motor conduction time for the less-affected upper limb had been found to be good to excellent ($0.45 < \kappa < 0.87$) in 18 chronic stroke patients and 8 healthy volunteers [32].

A positive MEP was identified independently by two experienced laboratory examiners as the presence of at least three responses to three stimuli per site, producing an MEP amplitude of >50 μV [32]. When no MEP could be elicited at a given site, the coil was slightly moved to find a hotspot at adjoining sites. Cervical stimulation was performed with a 90 mm circular coil in order to activate motor roots at the exit foramina, centered over the C7/C8 cervical spine [31].

The ADM is one of the preferred intrinsic muscles that can be reliably examined as a target muscle for TMS analysis. The ADM was also chosen because of the availability of the largest collection of CMCT values in healthy subjects, which could be used as a normative reference [31]. All recordings were performed with the upper limb relaxed in order to obtain homogeneous data and to avoid variations in the muscle response due to different levels of pre-innervation [28,32]. CMCT (in milliseconds) was estimated by subtracting the peripheral motor conduction time (PMCT) from the shortest total latency time of MEPs [31,33]. PMCT was measured with the direct spinal root stimulation technique [27,31]. We defined prolonged CMCT relative to normative data obtained using the direct spinal root stimulation technique, by taking the mean (i.e., 7.1 ms) plus 2 times the standard deviation (SD) of the error, i.e., 1.1 ms [31]. Consequently, prolonged CMCT was defined as a prolongation exceeding 8.2 ms.

### 2.3. Measurements

Demographic and clinical characteristics of the participants and the TMS-induced MEPs and CMCT values were obtained within the first 48 h after a stroke (i.e., hyperacute phase) and at an average of 11 days post stroke (early subacute phase). Recently, we had found that the value of the presence of TMS-induced MEPs and the CMCT of the infarcted hemisphere in predicting upper limb motor function early after a stroke is mainly determined by the time of assessment [34]. The timing of the measurement at 11 days was chosen for practical reasons, as many patients were subsequently discharged from the department of neurology. Baseline characteristics included age, gender, left/right hemispheric stroke, etiology and stroke severity (TOAST classification) [26], motor impairment of the affected paretic upper limb (motricity index, MI) [35], presence of voluntary finger extension using the FM-UE score [36], activities of daily living using the Barthel Index (BI) [37], presence of dysphagia according to the water-swallowing test [38], and visuospatial neglect (VSN) defined as having two or more omissions in the letter cancellation task [39]. We defined patients with a lack of spontaneous motor recovery, so-called 'non-recoverers', in the present study as those who started at 18 points or lower on the FM-UE within the first 2 days after stroke [21,40,41] and failed to show spontaneous upper limb motor improvement beyond the smallest detectable difference (SDD) of 6 points (10%) on the FM-UE [42,43] during the first 6 months post stroke. Otherwise, subjects were classified as 'recoverers' in the context of spontaneous neurological recovery.

### 2.4. Data Analysis

All data were summarized using descriptive statistics. Categorical variables were described using frequencies and percentages. Continuous variables were described using means, SDs, and ranges, except for skewed variables, which are presented as medians (interquartile ranges). CMCT scores were compared between 48 h and 11 days using the Paired Student *t*-test. For processing the analysis regarding the first aim of the study, differences between recoverers and non-recoverers in CMCT scores at 48 h, categorized into normal or prolonged based on the available normative data [31], were compared using the Fischer exact test. The Mann–Whitney U test was used to describe differences between recoverers and non-recoverers in FMA motor scores at 48 h and again at 11 days. All statistical tests were conducted with IBM SPSS Statistics for Windows, version 25.0 (IBM Corp., Armonk, NY, USA). In addition, differences in FM-UE scores between recoverers and non-recoverers were assessed by plotting individual time series. For the second aim of the study, multilevel analysis was used to assess the difference between recoverers and non-recoverers regarding the longitudinal CMCT response data at 48 h and 11 days after a stroke. The iterative generalized least squares algorithm was used to estimate the regression coefficients. Patient ID was defined as level $j$ and time of measurement as level $i$. A fixed slope and random intercept were selected. The Wald test was used to obtain a $p$-value for each regression coefficient. Outcome scores in the linear multilevel analysis were plotted to check for compliance with model assumptions. In multilevel logistic analysis, the outcome variable presents the logit of the probability (i.e., natural log of the odds) of prolonged longitudinal CMCT scores. Regression coefficients were subsequently transformed into odds ratios by taking the EXP [regression coefficient]. Tests were conducted using MLwiN version 3.04 (University of Bristol., Bristol, UK). A two-tailed significance level of $p < 0.05$ was used for all tests.

## 3. Results

Figure 1 shows the flowchart of screened and recruited stroke patients admitted to our hospital. After informed consent had been obtained, 51 patients with first-ever ischemic hemispheric stroke enrolled in the present prospective observational study. Fifty of them were eligible for analysis, as one patient was censored due to progressive stroke. One patient was lost to follow-up after the first assessment, but the available data of their first assessment was included in the analysis.

The main demographic and clinical characteristics are presented in Table 1. As shown in this Table, we included 29 (58%) men and 21 (42%) women, with a mean age of 70.3 years. The average age in the subgroup of recoverers ($N$ = 33) was 69.7 years compared to 71.4 years for those who failed to show significant recovery beyond the 6 points improvement on the FM-UE ($N$ = 17). As shown in Table 2 and illustrated in Figure 2, the recoverers improved their median FM-UE total score from 35 points within 48 h to 61 points at 6 months. The median FM-UE score of non-recoverers was 3 points within 48 h and showed a median improvement of 2 points at 6 months after stroke. Eighteen out of the 50 (36%) patients showed some voluntary finger extension within 48 h, whereas 32 out of 50 (64%) patients did not. All non-recoverers failed to show voluntary finger extension within the first 48 h after stroke. In addition, 14 out of the 17 (87.5%) non-recoverers failed to show active finger extension at 11 days, compared with 9 out of 33 (27%) recoverers. Finally, recoverers were significantly less likely to have visuospatial neglect ($p < 0.001$), had shorter hospital stays ($p < 0.001$) and significantly higher BI scores ($p < 0.001$) compared to the non-recoverers.

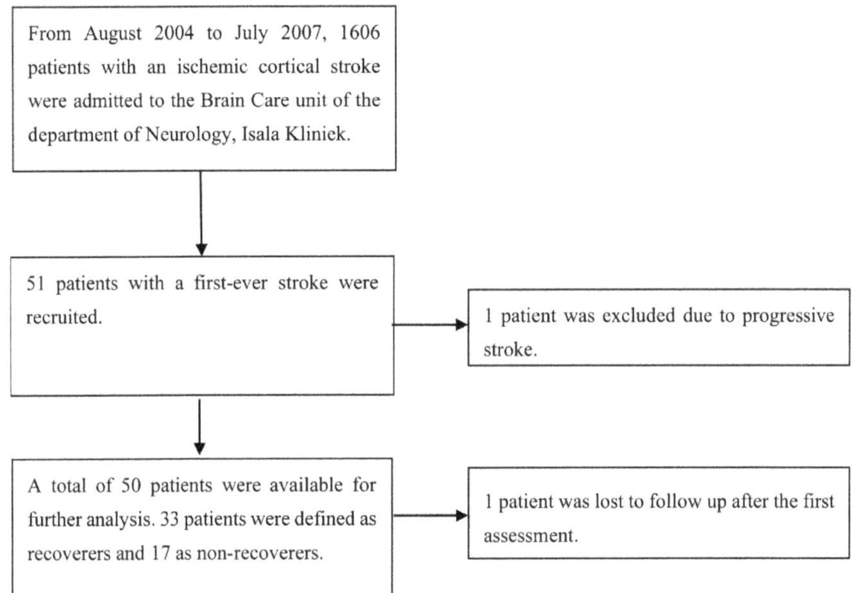


**Figure 1.** Patient exclusion flowchart.

**Table 1.** Patient characteristics of the total group ($N = 50$, second column), and subgroups (right-hand columns) in relation to spontaneous neurological upper limb motor recovery early after stroke.

| | Total Group | Recoverers ($N = 33$) | Non-Recoverers ($N = 17$) | $p$ |
|---|---|---|---|---|
| Gender, F/M | 29/21 | 17/16 | 12/5 | 0.571 |
| Age, mean (SD), yr | 70.3 (12.3) | 69.7 (12.8) | 71.4 (11.4) | <0.0001 * |
| Hemisphere of stroke, L/R | 25/25 | 16/17 | 9/8 | 0.170 |
| Length of hospital stay, mean (range), d | 4.9 (6–38) | 13.1 (6–25) | 15.7 (6–38) | <0.0001 * |
| Type of Stroke (TOAST): LVD/SVD/undetermined | 34/14/2 | 22/10/1 | 12/4/1 | 0.804 |
| Dysphagia at 48 h, yes/no | 33/17 | 20/13 | 13/4 | 1.00 |
| VSN at 48 h, yes/no | 12/38 | 7/26 | 5/12 | <0.0001 * |
| Barthel index score (0–20) at 48 h, median (IQR) | 5.0 (5.5) | 7 (5) | 2 (4) | <0.0001 * |

The subgroups are stroke patients with ('recoverers') or without ('non-recoverers') upper limb motor recovery at 6 months post stroke. Non-recoverers were defined as those patients who started at 18 points or lower on the Fugl-Meyer upper extremity scores for the affected limb within the first 48 h, and showed improvements of 6 points or less within the first 6 months post stroke. Abbreviations: F, female; M, male; L/R, left or right affected hemisphere; TOAST, Trial of Org 10172 in Acute Stroke Treatment; LVD, large vessel disease; SVD, small vessel disease; VSN, visuospatial neglect; IQR, interquartile range. * $p < 0.05$.

**Table 2.** Transcranial Magnetic Stimulation and Fugl-Meyer upper extremity score characteristics for the total group ($N = 50$, second column), and subgroups (third and fourth columns) in relation to spontaneous neurological upper limb motor recovery after stroke.

| | Total Group | Recoverers ($N = 33$) | Non-Recoverers ($N = 17$) | $p$ |
|---|---|---|---|---|
| CMCT at 48 h, mean (SD), ms | 10.27 (4.35) | 9.54 (2.95) | 11.7 (6.10) | <0.0001 * |
| CMCT at 11 days, mean (SD), ms | 9.44 (3.99) | 8.81 (3.17) | 10.7 (5.18) | <0.0001 * |
| FM-UE finger extension at 48 h, yes/no | 18/32 | 18/15 | 0/17 | <0.0001 * |
| FM-UE finger extension at 11 days, yes/no | 26/23 § | 24/9 | 2/14 | 0.065 |
| FM-UE total score (0–66), median (IQR) at 48 h | 7.5 (43) | 35 (53.5) | 3 (2.5) | <0.0001 * |
| FM-UE total score (0–66), median (IQR) at 6 months | 54 (58) § | 61 (12) | 5 (3.8) | <0.0001 * |

The third and fourth columns show subgroup comparisons between stroke patients with ('recoverers') or without ('non-recoverers') upper limb motor recovery at 6 months post stroke. Non-recoverers were defined as those patients who started at 18 points or lower on the FM-UE within the first 48 h and showed improvements of 6 points or less within the first 6 months post stroke. Abbreviations: CMCT, central motor conduction time of the non-infarcted hemisphere; Prolonged CMCT was defined as a latency > 8.2 ms. FM-UE, Fugl-Meyer upper limb motor score. IQR, interquartile range.* $p < 0.05$. § 1 lost to follow-up.

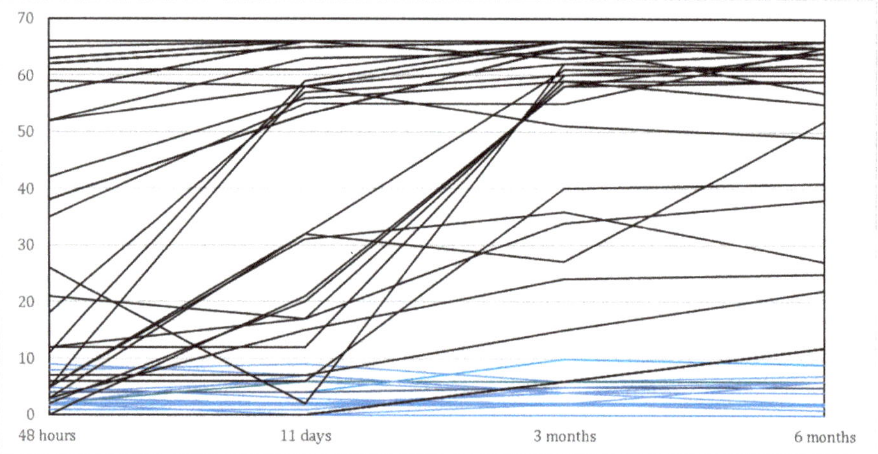


**Figure 2.** Time course of individual FM-UE scores for the affected upper limb of all patients ($N = 50$), between baseline (i.e., within 48 h after stroke) and 6 months post stroke. Blue lines represent the non-recoverers ($N = 17$) and black lines those who showed spontaneous neurological recovery ($N = 33$).

Within 48 h post stroke, 30 out of 50 (60%) patients showed prolonged CMCT in the non-infarcted hemisphere. After 11 days, the number of patients with prolonged CMCT was reduced to 24 out of 49 (49%). Mean CMCT (±SD) of all 50 patients within 48 h post stroke was 10.3 ms (±4.35), which was statistically significantly longer than the CMCT at 11 days after a stroke: 9.44 ms (±3.99); $p = 0.044$. At 48 h after stroke, significantly more patients with spontaneous motor recovery after 6 months showed a normal CMCT in the non-infarcted hemisphere when compared to the non-recoverers ($p = 0.032$).

Multilevel analysis of CMCT (as a continuous variable) between recoverers and non-recoverers, comparing two measurements at 48 h and 11 days after stroke, revealed no statistically significant differences between these subgroups (lower CI = −4.146; upper CI = 0.228; $p = 0.079$). Multilevel logistic analysis yielded statistically significant differences between the recoverers and non-recoverers as regards the longitudinal association with a normal and prolonged CMCT response, respectively, from 48 h to 11 days post stroke (lower CI = 0.086; upper CI = 0.763; $p = 0.015$), with an odds ratio of 0.256 (95%CI: 0.086–0.763).

## 4. Discussion

In the present study, we showed for the first time that in the majority (60%) of patients with a severe upper limb impairment, CMCTs in the non-infarcted hemisphere are significantly prolonged when measured within the first 48 h post stroke. In the hyperacute phase after stroke, prolonged CMCT is associated with the severity of motor impairment of the affected upper limb. More importantly, about half of the patients with prolonged CMCTs in the non-infarcted hemisphere within the first 48 h after stroke will have persistent, prolonged conduction times within the first 11 days. In patients with a severe, first-ever, ischemic hemispheric stroke, the likelihood that CMCTs remain prolonged during this 11 day time window and do not normalize to conduction times seen in healthy age-matched subjects is significantly associated with the absence of significant spontaneous motor recovery (i.e., non-recovery) of the affected upper limb according to FM-UE scores. The odds of prolonged CMCT in the first 11 days are 0.256 for recoverers, compared with the odds of a prolonged CMCT for non-recoverers. This finding suggests that the odds of prolonged CMCT in the subgroup of recoverers are, on average, 74.4% lower than in the subgroup of non-recoverers. Finally, our findings confirm the results of previous work by Byblow and colleagues [44], showing that the integrity of the CST for the less-affected limb remains functionally intact irrespective of stroke severity. Our finding of a significantly prolonged CMCT is also in agreement with recent kinematic studies, in which significant changes in, for example, speed and intralimb coordination were found for the less-affected limb in the first weeks after stroke [4,5].

Question of how to explain the above findings regarding prolonged CMCTs in the non-infarcted hemisphere and their association with the proportional level of spontaneous motor recovery remains unanswered in the present study. As a first hypothesis, one may assume that anatomically related networks in the non-infarcted hemisphere of severely affected stroke patients are temporarily suppressed in the acute phase, probably by transcallosal diaschisis [17,19,45,46]. Although poorly understood, this disturbed function in anatomically associated areas is believed to be caused by inflammation and oxidative stress due to an upregulation of cytokines and increased activity of macrophages and glial cells in areas anatomically associated with the infarcted area [47–49]. In line with this hypothesis, recent longitudinal diffusion tension imaging [50] and fMRI-resting state studies investigating functional connectivity [51] also found evidence for transhemispheric cortical and white matter changes in anatomically related areas very early post stroke. For example, Visser and colleagues showed significant changes in white matter integrity in the ipsi- as well as contralesional brain areas, such as the primary motor, pre-motor and visual cortices, during the first months after a stroke [50]. Theoretically, an alternative explanation may be that the decline of CMCT in the non-infarcted hemisphere between 48 h and 11 days reflects the underlying recovery of the anatomically uncrossed or double-crossed CST and the reticulo-, tecto- and possibly rubrospinal descending pathways. However, the latter explanation is unlikely, as the ipsilateral CST mainly projects to the trunk and upper arm [11,12,52], whereas the recovery of multisynaptic reticulospinal pathways is too slow for CMCT speeds below 8.2 ms [16,53].

Our observational study had some limitations. First, the sample of stroke subjects ($N = 50$) is too small to find robust findings, even though the current study is one of the largest prospective cohorts in this field to start within 48 h after a stroke. Second, we did not combine the present findings with neuroimaging techniques such as CT or MRI angiography or CT/MRI perfusion imaging, which would have allowed us to identify the severity of irreversible brain damage in the hyperacute phase post stroke [54], or to detect the existence of mass effects by central or cingulate shifts affecting the hemodynamics in the non-infarcted hemisphere early post stroke. However, this latter mechanism is unlikely, in our opinion, since the evolution of mass effects by vasogenic oedema requires several days after hemispheric stroke [55] and is accompanied by reduced consciousness of the patient. Third, the cut-off score for prolonged CMCT was obtained from normative laboratory values [31]. Although this is a rather conservative CMCT criterion, it can still be criticized as being susceptible to variation and bias. The used normative data of healthy subjects were specifically obtained with the ADM as target muscle [31]. Age could potentially be a confounder in the association between CMCT and observed improvement. However, the significant association between CMCT of the non-infarcted hemisphere and improvements in FM-UE scores was not significantly influenced in our data after (partial) correcting for age, acknowledging that age was statistically different between recoverers and non-recoverers at baseline in our sample.

Finally, we did not investigate the impact of the type and intensity of rehabilitation as possible confounders of motor recovery in the first 6 months post stroke. All patients received usual care according to the current Dutch guidelines for stroke rehabilitation [29]. However, there is currently no evidence that exercise therapy can significantly influence the time course of spontaneous motor recovery early post stroke [44,56]. In the present study, we provide additional evidence that the CMCT in the non-infarcted hemisphere is prolonged in a large proportion of patients when measured in the hyperacute phase after a severe stroke. In addition, we found that a prolonged CMCT in the non-infarcted hemisphere persisted during the first 11 days in those who showed no spontaneous motor recovery in the most affected upper limb post stroke.

## 5. Conclusions

The current study suggests that CMCT in the non-infarcted hemisphere is significantly prolonged in 60% of severely affected, ischemic stroke patients when measured within the first 48 h post stroke. The likelihood of prolonged CMCT is significantly higher in those patients that show no spontaneous motor recovery (i.e., 'non-recoverers') when compared to those that show spontaneous motor recovery (i.e., ' recoverers') early after a first-ever ischemic hemispheric stroke.

**Supplementary Materials:** The following are available online at https://www.mdpi.com/article/10.3390/brainsci11050648/s1, Figure S1: TMS at the brain.

**Author Contributions:** Conceptualization, M.H.J.H., R.H.M.N. and G.K.; methodology, M.H.J.H., R.H.M.N., G.K., B.J.K.; formal analysis, B.J.K.; investigation, M.H.J.H.; data curation, M.H.J.H.; writing—original draft preparation, M.H.J.H., R.H.M.N., G.K.; writing—review and editing, M.H.J.H., R.H.M.N., C.H.E., B.J.K., G.K.; supervision, G.K. All authors have read and agreed to the published version of the manuscript.

**Funding:** This study was supported by the European Research Council (ERC) under the European Union's Seventh Framework Program (FP/2007–2013)/ERC Advanced Grant no. 291339-4D-EEG awarded to GK.

**Institutional Review Board Statement:** The study was conducted according to the guidelines of the Declaration of Helsinki, and approved by the local medical ethics board of the Isala Kliniek, the Netherlands (No 04.0318P).

**Informed Consent Statement:** All patients gave their verbal and/or written informed consent (themselves or by proxy) and all patients were treated according to the Dutch physical therapy guidelines for rehabilitation.

**Data Availability Statement:** The data presented in this study are available on request from the corresponding author. The data are not publicly available due to practical, ethical and privacy reasons.

**Acknowledgments:** The authors would like to thank Johan Bisschop and Jan Middendorp from the Department of Clinical Neurophysiology of the Isala Kliniek for performing all TMS assessments. This article is dedicated to the memory of Peter van den Berg, a neurologist at Isala Kliniek and the person who inspired this study.

**Conflicts of Interest:** The authors declare no conflict of interest.

## References


1. Morris, J.H.; Van Wijck, F. Responses of the Less Affected Arm to Bilateral Upper Limb Task Training in Early Rehabilitation After Stroke: A Randomized Controlled Trial. *Arch. Phys. Med. Rehabil.* **2012**, *93*, 1129–1137. [CrossRef] [PubMed]
2. Harrington, R.M.; Chan, E.; Rounds, A.K.; Wutzke, C.J.; Dromerick, A.W.; Turkeltaub, P.E.; Harris-Love, M.L. Roles of Lesioned and Nonlesioned Hemispheres in Reaching Performance Poststroke. *Neurorehabilit. Neural Repair* **2020**, *34*, 61–71. [CrossRef] [PubMed]
3. Maenza, C.; Good, D.C.; Winstein, C.J.; Wagstaff, D.A.; Sainburg, R.L. Functional Deficits in the Less-Impaired Arm of Stroke Survivors Depend on Hemisphere of Damage and Extent of Paretic Arm Impairment. *Neurorehabilit. Neural Repair* **2020**, *34*, 39–50. [CrossRef] [PubMed]
4. Van Dokkum, L.E.H.; Le Bars, E.; Mottet, D.; Bonafé, A.; De Champfleur, N.M.; Laffont, I. Modified Brain Activations of the Nondamaged Hemisphere During Ipsilesional Upper-Limb Movement in Persons with Initial Severe Motor Deficits Poststroke. *Neurorehabilit. Neural Repair* **2018**, *32*, 34–45. [CrossRef] [PubMed]
5. Bustrén, E.-L.; Sunnerhagen, K.S.; Murphy, M.A. Movement Kinematics of the Ipsilesional Upper Extremity in Persons with Moderate or Mild Stroke. *Neurorehabilit. Neural Repair* **2017**, *31*, 376–386. [CrossRef] [PubMed]
6. Lawrence, D.G.; Kuypers, H.G.J.M. Pyramidal and Non-Pyramidal Pathways in Monkeys: Anatomical and Functional Correlation. *Science* **1965**, *148*, 973–975. [CrossRef]
7. Kuypers, H.; Brinkman, J. Precentral projections to different parts of the spinal intermediate zone in the rhesus monkey. *Brain Res.* **1970**, *24*, 29–48. [CrossRef]
8. Lawrence, E.S.; Coshall, C.; Dundas, R.; Stewart, J.; Rudd, A.G.; Howard, R.; Wolfe, C.D.A. Estimates of the Prevalence of Acute Stroke Impairments and Disability in a Multiethnic Population. *Stroke* **2001**, *32*, 1279–1284. [CrossRef] [PubMed]
9. Grefkes, C.; Fink, G.R. Connectivity-based approaches in stroke and recovery of function. *Lancet Neurol.* **2014**, *13*, 206–216. [CrossRef]
10. Nathan, P.W.; Smith, M.C.; Deacon, P. The corticospinal tracts in man. *Brain* **1990**, *113*, 303–324. [CrossRef]
11. Lawrence, D.G.; Kuypers, H.G.J.M. The functional organization of the motor system in the monkey: II. *Brain* **1968**, *91*, 15–36. [CrossRef]
12. Lawrence, D.G.; Kuypers, H.G.J.M. The functional organization of the motor system in the monkey: I. *Brain* **1968**, *91*, 1–14. [CrossRef] [PubMed]
13. Baker, S.N. The primate reticulospinal tract, hand function and functional recovery. *J. Physiol.* **2011**, *589*, 5603–5612. [CrossRef] [PubMed]
14. Baker, S.N.; Zaaimi, B.; Fisher, K.M.; Edgley, S.A.; Soteropoulos, D.S. Pathways mediating functional recovery. *Prog. Brain Res.* **2015**, *218*, 389–412. [CrossRef] [PubMed]
15. Zaaimi, B.; Soteropoulos, D.S.; Fisher, K.M.; Riddle, C.N.; Baker, S.N. Classification of Neurons in the Primate Reticular Formation and Changes after Recovery from Pyramidal Tract Lesion. *J. Neurosci.* **2018**, *38*, 6190–6206. [CrossRef]
16. Benecke, R.; Meyer, B.-U.; Freund, H.-J. Reorganisation of descending motor pathways in patients after hemispherectomy and severe hemispheric lesions demonstrated by magnetic brain stimulation. *Exp. Brain Res.* **1991**, *83*, 419–426. [CrossRef]
17. Feeney, D.M.; Baron, J.C. Diaschisis. *Stroke* **1986**, *17*, 817–830. [CrossRef]
18. Bütefisch, C.M.; Weβling, M.; Netz, J.; Seitz, R.J.; Hömberg, V. Relationship between Interhemispheric Inhibition and Motor Cortex Excitability in Subacute Stroke Patients. *Neurorehabilit. Neural Repair* **2008**, *22*, 4–21. [CrossRef]
19. Carrera, E.; Tononi, G. Diaschisis: Past, present, future. *Brain* **2014**, *137*, 2408–2422. [CrossRef]
20. Grefkes, C.; Fink, G.R. Reorganization of cerebral networks after stroke: New insights from neuroimaging with connectivity approaches. *Brain* **2011**, *134*, 1264–1276. [CrossRef]
21. Van Der Vliet, R.; Selles, R.W.; Andrinopoulou, E.; Nijland, R.; Ribbers, G.M.; Frens, M.A.; Meskers, C.; Kwakkel, G. Predicting Upper Limb Motor Impairment Recovery after Stroke: A Mixture Model. *Ann. Neurol.* **2020**, *87*, 383–393. [CrossRef]
22. Traversa, R.; Cicinelli, P.; Pasqualetti, P.; Filippi, M.; Rossini, P.M. Follow-up of interhemispheric differences of motor evoked potentials from the 'affected' and 'unaffected' hemispheres in human stroke. *Brain Res.* **1998**, *803*, 1–8. [CrossRef]
23. Barker, R.N.; Brauer, S.G.; Barry, B.K.; Gill, T.J.; Carson, R.G. Training-induced modifications of corticospinal reactivity in severely affected stroke survivors. *Exp. Brain Res.* **2012**, *221*, 211–221. [CrossRef]
24. McDonnell, M.N.; Stinear, C.M. TMS measures of motor cortex function after stroke: A meta-analysis. *Brain Stimul.* **2017**, *10*, 721–734. [CrossRef]
25. Hammerbeck, U.; Hoad, D.; Greenwood, R.; Rothwell, J.C. The unsolved role of heightened connectivity from the unaffected hemisphere to paretic arm muscles in chronic stroke. *Clin. Neurophysiol.* **2019**, *130*, 781–788. [CrossRef] [PubMed]
26. Adams, H.P.; Bendixen, B.H.; Kappelle, L.J.; Biller, J.; Love, B.B.; Gordon, D.L.; Marsh, E. Classification of subtype of acute ischemic stroke. Definitions for use in a multicenter clinical trial. TOAST. Trial of Org 10172 in Acute Stroke Treatment. *Stroke* **1993**, *24*, 35–41. [CrossRef]
27. Rossini, P.; Burke, D.; Chen, R.; Cohen, L.; Daskalakis, Z.; Di Iorio, R.; Di Lazzaro, V.; Ferreri, F.; Fitzgerald, P.; George, M.; et al. Non-invasive electrical and magnetic stimulation of the brain, spinal cord, roots and peripheral nerves: Basic principles and procedures for routine clinical and research application. An updated report from an I.F.C.N. Committee. *Clin. Neurophysiol.* **2015**, *126*, 1071–1107. [CrossRef]

28. Di Lazzaro, V.; Oliviero, A.; Pilato, F.; Saturno, E.; Dileone, M.; Mazzone, P.; Insola, A.; Tonali, P.; Rothwell, J. The physiological basis of transcranial motor cortex stimulation in conscious humans. *Clin. Neurophysiol.* **2004**, *115*, 255–266. [CrossRef]
29. Van Peppen, R.P.S.; Kwakkel, G.; Wood-Dauphinee, S.; Hendriks, H.J.M.; Van Der Wees, P.J.; Dekker, J. The impact of physical therapy on functional outcomes after stroke: What's the evidence? *Clin. Rehabil.* **2004**, *18*, 833–862. [CrossRef] [PubMed]
30. Rothwell, J. Transcranial Electrical and Magnetic Stimulation of the Brain: Basic Physiological Mechanisms. In *Magnetic Stimulation in Clinical Neurophysiology*, 2nd ed.; Hallet, M., Chokroverty, S., Eds.; Elsevier: Philadelphia, PA, USA, 2005; pp. 43–60.
31. Groppa, S.; Oliviero, A.; Eisen, A.; Quartarone, A.; Cohen, L.; Mall, V.; Kaelin-Lang, A.; Mima, T.; Rossi, S.; Thickbroom, G.; et al. A practical guide to diagnostic transcranial magnetic stimulation: Report of an IFCN committee. *Clin. Neurophysiol.* **2012**, *123*, 858–882. [CrossRef] [PubMed]
32. Hoonhorst, M.H.W.J.; Kollen, B.J.; Berg, P.S.P.V.D.; Emmelot, C.H.; Kwakkel, G. How Reproducible Are Transcranial Magnetic Stimulation–Induced MEPs in Subacute Stroke? *J. Clin. Neurophysiol.* **2014**, *31*, 556–562. [CrossRef]
33. Samii, A.; Luciano, C.; Dambrosia, J.; Hallet, M. Central motor conduction time: Reproducibility and discomfort of different methods. *Muscle Nerve* **1998**, *21*, 1445–1450. [CrossRef]
34. Hoonhorst, M.H.J.; Nijland, R.H.M.; Berg, P.J.S.V.D.; Emmelot, C.H.; Kollen, B.J.; Kwakkel, G. Does Transcranial Magnetic Stimulation Have an Added Value to Clinical Assessment in Predicting Upper-Limb Function Very Early After Severe Stroke? *Neurorehabilit. Neural Repair* **2018**, *32*, 682–690. [CrossRef]
35. Collin, C.; Wade, D. Assessing motor impairment after stroke: A pilot reliability study. *J. Neurol. Neurosurg. Psychiatry* **1990**, *53*, 576–579. [CrossRef] [PubMed]
36. Nijland, R.; Van Wegen, E.; Der Wel, B.H.-V.; Kwakkel, G. Presence of Finger Extension and Shoulder Abduction Within 72 Hours After Stroke Predicts Functional Recovery. *Stroke* **2010**, *41*, 745–750. [CrossRef] [PubMed]
37. Collin, C.; Wade, D.T.; Davies, S.; Horne, V. The Barthel ADL Index: A reliability study. *Int. Disabil. Stud.* **1988**, *10*, 61–63. [CrossRef] [PubMed]
38. Osawa, A.; Maeshima, S.; Tanahashi, N. Water-Swallowing Test: Screening for Aspiration in Stroke Patients. *Cerebrovasc. Dis.* **2013**, *35*, 276–281. [CrossRef]
39. Nijboer, T.C.; Kollen, B.J.; Kwakkel, G. Time course of visuospatial neglect early after stroke: A longitudinal cohort study. *Cortex* **2013**, *49*, 2021–2027. [CrossRef]
40. Prabhakaran, S.; Zarahn, E.; Riley, C.; Speizer, A.; Chong, J.Y.; Lazar, R.M.; Marshall, R.S.; Krakauer, J.W. Inter-individual Variability in the Capacity for Motor Recovery After Ischemic Stroke. *Neurorehabilit. Neural Repair* **2007**, *22*, 64–71. [CrossRef] [PubMed]
41. Winters, C.; Van Wegen, E.E.H.; Daffertshofer, A.; Kwakkel, G. Generalizability of the Proportional Recovery Model for the Upper Extremity After an Ischemic Stroke. *Neurorehabilit. Neural Repair* **2014**, *29*, 614–622. [CrossRef]
42. Sanford, J.; Moreland, J.; Swanson, L.R.; Stratford, P.W.; Gowland, C. Reliability of the Fugl-Meyer Assessment for Testing Motor Performance in Patients Following Stroke. *Phys. Ther.* **1993**, *73*, 447–454. [CrossRef]
43. Gladstone, D.J.; Danells, C.J.; Black, S.E. The Fugl-Meyer Assessment of Motor Recovery after Stroke: A Critical Review of Its Measurement Properties. *Neurorehabilit. Neural Repair* **2002**, *16*, 232–240. [CrossRef] [PubMed]
44. Byblow, W.D.; Stinear, C.M.; Barber, P.A.; Petoe, M.A.; Ackerley, S.J. Proportional recovery after stroke depends on corticomotor integrity. *Ann. Neurol.* **2015**, *78*, 848–859. [CrossRef] [PubMed]
45. Nudo, R.J.; Wise, B.M.; Sifuentes, F.; Milliken, G.W. Neural Substrates for the Effects of Rehabilitative Training on Motor Recovery after Ischemic Infarct. *Science* **1996**, *272*, 1791–1794. [CrossRef]
46. von Monakow, C. *Die Lokalisation im Grosshirn und der Abbau der Funktion Durch Kortikale Herde*; Bergmann: Wiesbaden, Germany, 1914.
47. Block, F.; Dihné, M.; Loos, M. Inflammation in areas of remote changes following focal brain lesion. *Prog. Neurobiol.* **2005**, *75*, 342–365. [CrossRef] [PubMed]
48. Jones, K.A.; Zouikr, I.; Patience, M.; Clarkson, A.N.; Isgaard, J.; Johnson, S.J.; Spratt, N.; Nilsson, M.; Walker, F.R. Chronic stress exacerbates neuronal loss associated with secondary neurodegeneration and suppresses microglial-like cells following focal motor cortex ischemia in the mouse. *Brain Behav. Immun.* **2015**, *48*, 57–67. [CrossRef]
49. Weishaupt, N.; Zhang, A.; DeZiel, R.A.; Tasker, R.A.; Whitehead, S.N. Prefrontal Ischemia in the Rat Leads to Secondary Damage and Inflammation in Remote Gray and White Matter Regions. *Front. Neurosci.* **2016**, *10*, 81. [CrossRef]
50. Visser, M.M.; Yassi, N.; Campbell, B.C.; Desmond, P.M.; Davis, S.M.; Spratt, N.; Parsons, M.; Bivard, A. White Matter Degeneration after Ischemic Stroke: A Longitudinal Diffusion Tensor Imaging Study. *J. Neuroimaging* **2018**, *29*, 111–118. [CrossRef] [PubMed]
51. Volz, L.J.; Rehme, A.K.; Michely, J.; Nettekoven, C.; Eickhoff, S.B.; Fink, G.R.; Grefkes, C. Shaping Early Reorganization of Neural Networks Promotes Motor Function after Stroke. *Cereb. Cortex* **2016**, *26*, 2882–2894. [CrossRef]
52. Soteropoulos, D.S.; Williams, E.R.; Baker, S.N. Cells in the monkey ponto-medullary reticular formation modulate their activity with slow finger movements. *J. Physiol.* **2012**, *590*, 4011–4027. [CrossRef]
53. Ziemann, U.; Ishii, K.; Borgheresi, A.; Yaseen, Z.; Battaglia, F.; Hallett, M.; Cincotta, M.; Wassermann, E.M. Dissociation of the pathways mediating ipsilateral and contralateral motor-evoked potentials in human hand and arm muscles. *J. Physiol.* **1999**, *518*, 895–906. [CrossRef] [PubMed]
54. Lui, Y.; Tang, E.; Allmendinger, A.; Spektor, V. Evaluation of CT Perfusion in the Setting of Cerebral Ischemia: Patterns and Pitfalls. *Am. J. Neuroradiol.* **2010**, *31*, 1552–1563. [CrossRef]

55. Nakano, S.; Iseda, T.; Kawano, H.; Yoneyama, T.; Ikeda, T.; Wakisaka, S. Correlation of early CT signs in the deep middle cerebral artery territories with angiographically confirmed site of arterial occlusion. *Am. J. Neuroradiol.* **2001**, *22*, 654–659. [PubMed]
56. Kwakkel, G.; Winters, C.; Van Wegen, E.E.H.; Nijland, R.H.M.; Van Kuijk, A.A.A.; Visser-Meily, A.; De Groot, J.; De Vlugt, E.; Arendzen, J.H.; Geurts, A.C.H.; et al. Effects of Unilateral Upper Limb Training in Two Distinct Prognostic Groups Early After Stroke. *Neurorehabilit. Neural Repair* **2016**, *30*, 804–816. [CrossRef] [PubMed]

Review

# Frontal Transcranial Direct Current Stimulation as a Potential Treatment of Parkinson's Disease-Related Fatigue

Tino Zaehle [1,2]

1 Department of Neurology, Otto-von-Guericke-University Magdeburg, 39120 Magdeburg, Germany; tino.zaehle@ovgu.de
2 Center for Behavioral Brain Sciences (CBBS), 39106 Magdeburg, Germany

**Abstract:** In contrast to motor symptoms, non-motor symptoms in Parkinson's disease (PD) are often poorly recognized and inadequately treated. Fatigue is one of the most common non-motor symptoms in PD and affects a broad range of everyday activities, causes disability, and substantially reduces the quality of life. It occurs at every stage of PD, and once present, it often persists and worsens over time. PD patients attending the 2013 World Parkinson Congress voted fatigue as the leading symptom in need of further research. However, despite its clinical significance, little progress has been made in understanding the causes of Parkinson's disease-related fatigue (PDRF) and developing effective treatment options, which argues strongly for a greater effort. Transcranial direct current stimulation (tDCS) is a technique to non-invasively modulate cortical excitability by delivering low electrical currents to the cerebral cortex. In the past, it has been consistently evidenced that tDCS has the ability to induce neuromodulatory changes in the motor, sensory, and cognitive domains. Importantly, recent data present tDCS over the frontal cortex as an effective therapeutic option to treat fatigue in patients suffering from multiple sclerosis (MS). The current opinion paper reviews recent data on PDRF and the application of tDCS for the treatment of fatigue in neuropsychiatric disorders to further develop an idea of using frontal anodal tDCS as a potential therapeutic strategy to alleviate one of the most common and severe non-motor symptoms of PD.

**Keywords:** fatigue; Parkinson's disease (PD); tDCS

**Citation:** Zaehle, T. Frontal Transcranial Direct Current Stimulation as a Potential Treatment of Parkinson's Disease-Related Fatigue. *Brain Sci.* **2021**, *11*, 467. https://doi.org/10.3390/brainsci11040467

Academic Editors: Ulrich Palm, Moussa Antoine Chalah and Samar S. Ayache

Received: 26 February 2021
Accepted: 30 March 2021
Published: 8 April 2021



## 1. Introduction

Fatigue is a complex symptom and a multifaceted construct that leads to a general feeling of exhaustion, loss of motivation, and behavioral performance problems [1]. It is a major cause of traffic accidents [2] or accidents in other work-related settings [3]. Importantly, fatigue is often comorbid to a variety of neuropsychiatric disorders, such as depression, cancer, multiple sclerosis (MS), and Parkinson's disease (PD).

In patients with PD, fatigue is one of the most common non-motor symptoms affecting a wide range of daily activities, leading to disability, and significantly reducing the quality of life [4]. Despite its clinical importance, progress in understanding and treating fatigue is still remarkably limited. Some therapeutic approaches for fatigue in PD have been tested, but none are effective against fatigue. Conventional therapies for the motor symptoms of PD do not significantly improve fatigue [5].

While transcranial direct current stimulation (tDCS) has recently been shown to alleviate fatigue in multiple sclerosis (MS) effectively [6–8], data on fatigue in PD are sparse.

In the current opinion paper, I will propose that NIBS approaches can contribute to a better understanding of the fatigue syndrome and stimulate the development of efficient treatments based on rational hypotheses about the underlying pathophysiology, and, finally, argue for frontal anodal tDCS as a potential therapeutic option in Parkinson's disease-related fatigue (PDRF).

## 2. Parkinson's Disease-Related Fatigue (PDRF)

Parkinson's disease (PD) is the second most common neurodegenerative disorder, affecting approximately 1% of the population over 50 years of age [9]. PD is traditionally defined as a basic motor disorder. However, many non-motor symptoms (NMS) also commonly occur in PD. These NMS include pain, cognitive decline, delusions, and notable fatigue. Among the NMS deficits, Parkinson's disease-related fatigue (PDRF) in particular is one of the most common symptoms in PD. It affects up to 58% of patients [10,11], and 30% of PD patients report that PDRF is the symptom with the greatest negative impact on their daily lives [4]. Accordingly, PDRF is an important stressor with a tremendous negative impact on the patients' quality of life and an essential contributor to disease burden [12–14]. Moreover, PDRF already impacts patients at an early untreated stage of the disease and is an important consideration in patient management [15].

In general, from the Latin *fatigare*, fatigue describes an overwhelming feeling of tiredness, weakness, lack of energy, and exhaustion unrelated to physical activity [16]. In various neurological diseases, fatigue is an important but often underappreciated complaint [17,18].

Nowadays, patients suffering from PD are usually appropriately treated for their motor symptoms, whereas a significant proportion of NMS still remains unrecognized or unreported [19]. However, despite the current diagnostic underrepresentation, NMS were described at the very beginning of the clinical description of the syndrome [20]. The first description of PDRF likely came from J. M. Charcot, who described fatigue as early as in the 1870s, in addition to several other typical non-motor aspects of PD [21]. Thus, although recognizing the importance of PDRF seems to be a relatively recent development, it was already recognized in the nineteenth century by the most important clinical neurologists of their time.

Although fatigue is a common and debilitating symptom in PD, the exact etiology and underlying pathophysiology of fatigue in PD remain unclear [22], and—accordingly—there is a significant lack of available effective treatments for PDRF [5]. This considerable lack of progress in understanding fatigue's pathophysiology and its treatment is, in part, due to the fact that fatigue still lacks a universally accepted definition and classification [23,24].

This lack of a consistent fatigue taxonomy complicates its understanding, measurement, and consequently its treatment [25]. To date, fatigue is mostly assessed subjectively using self-report questionnaires. However, because patients assess their perceived fatigue symptoms retrospectively, self-assessments of fatigue are subject to regression to the mean and recall errors that may reduce their accuracy. For example, available fatigue questionnaires for disease-related fatigue in multiple sclerosis (MS) showed low correlations with each other and heterogeneous associations with patients' functional impairments, disease duration, or cognitive deficits [26–28]. In contrast to these subjective fatigue measures, a fatigue-related decline in performance—also known as fatigability—could be quantified using objective indices [29]. Thus, to overcome the subjective nature of fatigue measures and the associated limitations for diagnosis and intervention of MS-related fatigue, we and others [25,30] proposed a generalized fatigue taxonomy that is disease nonspecific and universally applicable. Here, fatigue was broadly classified into physical, psychosocial, and cognitive fatigue. While psychosocial fatigue can only be assessed subjectively, physical and cognitive fatigue concepts imply that fatigue can be assessed both qualitatively as a subjective phenomenon and quantitatively as an objective phenomenon [25]. Specifically, subjective cognitive fatigue refers to a persistently perceived feeling of exhaustion. In contrast, objective cognitive fatigue—also referred to as fatigability—refers to a decline in performance on cognitive tasks, quantifiable as a change in cognitive performance relative to a baseline [23]. Finally, subjective and objective cognitive fatigue can be further subdivided. Subjective fatigue is divided into a trait component and a state component. Trait fatigue refers to a global status that changes slowly over time, whereas state fatigue refers to the change in subjectively perceived fatigue level over time [31]. Accordingly, subjective trait fatigue can be assessed by self-questionnaires and subjective state fatigue by visual analog scales (VAS) or numerical rating scales. In contrast, objective fatigue (fatigability) is, by definition, state-dependent and allows an objective assessment by behavioral or electrophysiological parameters.

Analogous to the assessment in patients with MS, an objective fatigue diagnosis appears to be a prerequisite for information and education in the early disease management of patients with PD [15,32] and ultimately for effective treatment of PDRF.

### 2.1. Ethology of PDRF

The inconsistencies in fatigue definitions also negatively affected the understanding of the pathophysiology of PDRF [23]. Despite the enormous negative impact of fatigue in PD, it remains challenging to delineate the pathophysiology of PDRF from other NMS in PD. In general, proposed physiologic mechanisms include increased circulating proinflammatory cytokines, dysfunction in nigrostriatally and extrastriatally dopaminergic pathways, involvement of non-dopaminergic (especially serotonergic) pathways, autonomic nervous system involvement, and, importantly, underlying prefrontal pathology [33–35].

Previously, PDRF was often assumed to be a reactive phenomenon [36]. In fact, PDRF is highly related to the severity of depressive symptoms [15,37]. Therefore, the understanding of PDRF is significantly biased by its co-occurrence with affective disorders [38]. However, a recent comprehensive review of PDRF [39] summarized clinical and experimental findings that support the view that fatigue is a primary manifestation of PD and not a secondary phenomenon. Accordingly, although PDRF is consistently associated with depression in PD, depression and fatigue often exist independently, and fatigue may persist after a successful depression treatment [40]. In fact, PDRF is present in over 50% of non-depressed PD patients [10]. Moreover, PDRF may precede motor symptoms [41] and does not necessarily correlate with PD duration or motor disability [36]. Thus, PDRF does not appear to be systematically associated with disease duration, stage, or motor symptoms; does not correlate with objective motor fatigability; and is distinguishable from other affective symptoms such as depression, apathy, and somnolence. In addition, PDRF does not respond reliably to dopaminergic or surgical therapies [42–44]. This evidence suggests that PDRF is a primary symptom in PD and is related to pathological nonmotor networks [36].

Recent hypotheses on pathophysiological mechanisms suggested that specific dysfunctions in the frontal cortex may play a significant role in fatigue. Evidence for the involvement of frontal lobe dysfunctions came from observations of fatigue-related executive impairment in patients with PD [22,45]. Accordingly, PDRF was associated with decreased frontal lobe blood flow [22] and prefrontal hypoperfusion [46]. Additionally, impaired connectivity within the frontal lobe was associated with PDRF [47]. This observed hypoactivation of the frontal lobe fitted well with a general model of pathological fatigue [32] that assumed central fatigue as a consequence of dysfunction in a circuit involving the basal ganglia and the frontal cortex. Analogously, Clayton and colleagues [48] introduced an oscillatory model of sustained attention, in which frontomedial theta power supported cognitive control processes while alpha power over task-relevant cortical areas suppressed task-irrelevant processes. They postulated that when a person becomes fatigued, both frontomedial theta and alpha power over task-relevant areas increase. The increase in frontomedial theta power may reflect the reactive engagement of theta-driven cognitive control processes via low-frequency phase synchronization. In contrast, the increase in alpha power over task-relevant cortical areas (e.g., occipital in a visual attention task) suppressed information processing and caused attentional deficits. According to Clayton et al. (2015), the increase of frontomedial theta power reflected the detection of a mismatch between current and desired levels of attention and, in turn, acted as a compensatory control mechanism to enlarge top-down control processes in a fatiguing brain.

### 2.2. Treatment of PDRF

According to the 2018 review of the International Parkinson and Movement Disorder Society (MDS) Evidence-Based Medicine (EBM) committee, which regularly publishes recommendations on treating Parkinson's disease nonmotor symptoms, only the monoamine oxidase (MAO)-B inhibitor rasagiline was considered possibly useful for the management of PDRF when other secondary causes of fatigue were excluded [49]. The efficacy of methylphenidate and modafinil remained investigational. However, a recent comprehensive review on current pharmacologic and non-pharmacologic treatment options for PDRF came to a less positive evaluation. The authors concluded that there was insufficient evidence for all treatment strategies. However, among the available options, the best evidence appeared to be for doxepin, rasagiline, and levodopa infusion therapy [50]. Finally, some studies indicated supportive effects of deep brain stimulation of the subthalamic nucleus (STN-DBS) on PDRF. In an open multicenter study including 60 patients [51], as well as in a subsequent international multicenter, observational study on 173 PD [52], STN-DBS could significantly improve NMS, including fatigue. However, there were also contradicting reports showing that fatigue could also be commonly caused by DBS surgery in PD [43] or at least could not be excluded on an individual level [53].

## 3. Transcranial Direct Current Stimulation (tDCS)

As PDRF drastically affects the patients' quality of life, the development of efficient therapeutic methods for fatigue treatment is of high clinical relevance. Furthermore, for a systematic treatment evaluation and optimization, a reliable and valid assessment of the individual fatigue level by objective parameters is essential.

Transcranial direct current stimulation (tDCS) may offer a unique opportunity to manipulate the maladaptive neuronal activity underlying PD-associated fatigue. The neuromodulatory potential of tDCS was widely demonstrated for cognitive, perceptual, and motor processes [54]. In a clinical context, tDCS could be used to restore pathological brain functions and improve associated symptoms [55,56].

TDCS can generally be considered safe and well-tolerated. The safety of this technique was studied and tested by several researchers who concluded that tDCS, when used and monitored in accordance with international safety guidelines, was a safe and well-tolerated intervention [57]. Due to its relatively low costs and risks, it could be made available to a broad group of patients. Thus, tDCS has the potential to improve and enhance the quality of life by granting less limited access to a wider group of patients. Especially since costs are generally a key element limiting access to medicines, tDCS can substantially improve fairness in medical care.

In general, tDCS delivers small electrical currents to the cerebral cortex. The current flows between an active electrode and a reference electrode. While the scalp shunts some of this current, the majority enters the brain tissue (e.g., [58]), modulating cortical excitability [59]. Based on animal data [60] and seminal work on the human motor domain [61], a rather heuristic model for the mechanism of action was established. According to this somatic doctrine, the direction of the tDCS-induced effect depended on the current polarity. Anodal tDCS had an excitatory effect, while cathodal tDCS decreased cortical excitability in the region under the electrode [62]. These effects were mediated by depolarization of the resting membrane potential. Thus, anodal tDCS increased the neuronal firing rate due to a hyperpolarization of the resting membrane potential, while cathodal tDCS decreased the firing rate due to hypopolarization of the resting membrane potential. However, in contrast to studies examining tDCS effects on the primary motor cortex, the majority of tDCS studies challenged the somatic doctrine with conflicting [63,64] or opposing [65–67] anodal/cathodal effects. While these effects could be partially attributed to the nonlinear nature of the stimulation effects [59], neuroanatomy and, more specifically, the orientation of the somatodendritic axis within the stimulated cortical areas also seemed to be crucial [68]. Indeed, the somatic doctrine was based only on radially directed electric currents [69], but tDCS always generated significant tangential current flow due to cortical folding [70]. Thus, results from several tDCS studies underscored that findings of the underlying neural mechanisms obtained at the primary motor cortex could not simply be generalized to the broader cortical area (e.g., [71]). Interestingly, using a human neuronal in vitro model with a dopaminergic phenotype, a recent study showed that DCS exerts on-line and of-line effects on the expression, aggregation, and autophagic degradation of alpha-synuclein, indicating a potential neuroprotective role of tDCS [72].

## 4. Transcranial Direct Current Stimulation as a Therapeutic Option for Fatigue

The majority of the stimulation studies, designed to counteract the development of fatigue, applied anodal tDCS over the dorsolateral prefrontal cortex (DLPFC), as this area had proven to be most affected by fatigue [73–79].

In healthy participants, positive effects of anodal tDCS over the left DLPFC were consistently demonstrated. A single dose of anodal tDCS was able to reduce fatigue-related vigilance performance decrements over time [73,80], even more effectively than caffeine consumption was able to do [75,76]. Moreover, anodal tDCS over the left DLPFC could successfully counteract fatigability development and reduce the fatigability-related increase in occipital alpha power as well as the decline in sensory gating [77]. In this recent study, we demonstrated that a single session of prefrontal tDCS attenuated the fatigue-induced increase in occipital alpha power. We hypothesized that this effect might be related to a tDCS-induced increase in prefrontal theta power, as previously shown [79,80], supporting the proposed accentuated role of frontomedial theta power in compensatory control mechanisms to augment top-down control processes in a fatigued brain [48].

For MS-related fatigue, positive stimulation effects on subjective fatigue assessed with self-report scales were also reported after five consecutive days of anodal tDCS over the bilateral motor cortex or somatosensory cortex [81,82], over the left DLPFC [83], and bifrontal over the left and right DLPFC [8]. The observed tDCS-related improvement was greater in patients with a higher lesion load in the left frontal cortex [8]). Accordingly, long-term studies in which left frontal tDCS was applied consecutively for 4–6 weeks showed improvement in subjective fatigue that persisted up to 3 weeks thereafter [74,83]. Finally, also a single dose of tDCS over the left prefrontal cortex was an effective therapeutic option for treating fatigue-related deterioration in MS patients' cognitive performance [84]. In this study, we investigated the effects of tDCS on fatigue development in patients with MS and demonstrated a positive effect of frontal tDCS. Anodal tDCS counteracted fatigue-associated performance decrements and improved patients' ability to cope with sustained cognitive demands. The results suggested that tDCS-induced modulations of frontal activity may be an effective therapeutic option for treating fatigue-related deterioration of cognitive performance in patients with MS (see [74,85] for recent reviews).

In PD, applications of tDCS showed to be able to produce transient beneficial effects, both in the motor [86] as well in the non-motor domains, particularly on cognition [87].

Nowadays, however, reports of positive effects of tDCS on PDRF are only very sparse. In a first experiment, Forogh and colleagues [88] investigated the effect of multisession anodal tDCS over the left DLPFC on fatigue and daytime sleepiness in patients with PD. The authors applied a bilateral stimulation scheme with an anode over the left and a cathode over the right DLPFC and performed eight sessions of 20 min stimulation at a current of 0.06 mA/cm$^2$ in 12 patients in an active treatment group and 11 patients in a placebo group. The data showed that anodal tDCS reduced fatigue immediately after treatment and also after a 3-month follow-up. As a further development of this approach, Dobbs and colleagues [89] proposed applying a remotely supervised tDCS protocol (RS-tDCS) to treat PDRF. The authors showed that a repeated application of anodal tDCS over the left DLPFC in a home-treatment context was well tolerated and positively affected subjective fatigue in patients with PD. Interestingly, the administration of repetitive transcranial magnetic stimulation (TMS) was found to improve motor and non-motor symptoms in patients with PD as well [90]. However, the majority of these studies assessed the effects of TMS on the excitability and plasticity of the motor cortex in patients with PD. Only sparse data also indicated supportive effects of TMS over the DLPFC on cognition [91] and depression [92].

In summary, PDRF is one of the most common non-motor symptoms occurring in the majority of patients and affecting a wide range of daily activities. PDRF results in a significant disability and markedly reduces the quality of life. The underlying pathophysiological mechanism in fatigue includes specific dysfunctions of the frontal cortex. However, despite its clinical importance, progress in developing an effective treatment for PDRF is still remarkably limited. Frontal anodal tDCS has proven to be effective for treating fatigue in both healthy participants and patients with neurological disorders such as multiple sclerosis. Moreover, anodal tDCS has been shown to raise hypofunctionality within stimulated cortical areas, including the DLPFC. Accordingly, the use of frontal anodal tDCS holds the promise of a potential therapeutic option for the treatment of PDRF. Further research is needed to determine the parameters of an optimal stimulation as well as to complement the purely subjective measures of fatigue with ones that provide an objective and valid assessment of fatigue and its potential reduction during treatment to make it useful in clinical settings. The concurrent use of neuroimaging methods such as EEG/MEG and fMRI in combination with tDCS is warranted and may be helpful in both target identification and outcome assessment for future tDCS trials for the treatment of PDRF. To conclude that frontal anodal tDCS can be an effective approach for the treatment of PDRF in a clinical setting, further data are needed that convincingly demonstrate (I) that a single session of anodal tDCS over the left DLPFC positively affects PDRF (transient effects), (II) that multisession tDCS can stabilize and/or enhance this effect, (III) that theses stimulation regimens lead to long-term effects of adequate duration, (IV) the specific conditions for a pronounced effect on the patient's subjective as well as objective fatigue, and, finally, (IV) the specific parameters of a successful home-application.

**Funding:** This research received no external funding.

**Conflicts of Interest:** The author declares no conflict of interest.

## References

1. Boksem, M.A.; Tops, M. Mental fatigue: Costs and benefits. *Brain Res. Rev.* **2008**, *59*, 125–139. [CrossRef]
2. Philip, P.; Sagaspe, P.; Taillard, J.; Valtat, C.; Moore, N.; Akerstedt, T.; Charles, A.; Bioulac, B. Fatigue, sleepiness, and performance in simulated versus real driving conditions. *Sleep* **2005**, *28*, 1511–1516. [CrossRef]
3. Caldwell, J.A.; Caldwell, J.L.; Thompson, L.A.; Lieberman, H.R. Fatigue and its management in the workplace. *Neurosci. Biobehav. Rev.* **2019**, *96*, 272–289. [CrossRef]
4. Herlofson, K.; Larsen, J.P. Measuring fatigue in patients with Parkinson's disease—The Fatigue Severity Scale. *Eur. J. Neurol.* **2002**, *9*, 595–600. [CrossRef] [PubMed]
5. Mendonça, D.A.; Menezes, K.; Jog, M.S. Methylphenidate improves fatigue scores in Parkinson disease: A randomized controlled trial. *Mov. Disord.* **2007**, *22*, 2070–2076. [CrossRef] [PubMed]
6. Ayache, S.S.; Chalah, M.A. Fatigue and Affective Manifestations in Multiple Sclerosis—A Cluster Approach. *Brain Sci.* **2019**, *10*, 10. [CrossRef] [PubMed]
7. Ayache, S.S.; Chalah, M.A. Transcranial direct current stimulation: A glimmer of hope for multiple sclerosis fatigue? *J. Clin. Neurosci.* **2018**, *55*, 10–12. [CrossRef]
8. Chalah, M.A.; Grigorescu, C.; Padberg, F.; Kümpfel, T.; Palm, U.; Ayache, S.S. Bifrontal transcranial direct current stimulation modulates fatigue in multiple sclerosis: A randomized sham-controlled study. *J. Neural Transm.* **2020**, *127*, 953–961. [CrossRef] [PubMed]
9. Dorsey, E.R.; Constantinescu, R.; Thompson, J.P.; Biglan, K.M.; Holloway, R.G.; Kieburtz, K.; Marshall, F.J.; Ravina, B.M.; Schifitto, G.; Siderowf, A.; et al. Projected number of people with Parkinson disease in the most populous nations, 2005 through 2030. *Neurology* **2007**, *68*, 384–386. [CrossRef]
10. Friedman, J.H.; Friedman, H. Fatigue in Parkinson's disease: A nine-year follow-up. *Mov. Disord.* **2001**, *16*, 1120–1122. [CrossRef]
11. Barone, P.; Antonini, A.; Colosimo, C.; Marconi, R.; Morgante, L.; Avarello, T.P.; Bottacchi, E.; Cannas, A.; Ceravolo, G.; Ceravolo, R.; et al. The PRIAMO study: A multicenter assessment of nonmotor symptoms and their impact on quality of life in Parkinson's disease. *Mov. Disord.* **2009**, *24*, 1641–1649. [CrossRef] [PubMed]
12. Herlofson, K.; Larsen, J.P. The influence of fatigue on health-related quality of life in patients with Parkinson's disease. *Acta Neurol. Scand.* **2003**, *107*, 1–6. [CrossRef] [PubMed]
13. Karlsen, K.H.; Larsen, J.P.; Tandberg, E.; Maeland, J.G. Influence of clinical and demographic variables on quality of life in patients with Parkinson's disease. *J. Neurol. Neurosurg. Psychiatry* **1999**, *66*, 431–435. [CrossRef]
14. Witjas, T.; Kaphan, E.; Azulay, J.P.; Blin, O.; Ceccaldi, M.; Pouget, J.; Poncet, M.; Chérif, A.A. Nonmotor fluctuations in Parkinson's disease: Frequent and disabling. *Neurology* **2002**, *59*, 408–413. [CrossRef]
15. Herlofson, K.; Ongre, S.O.; Enger, L.K.; Tysnes, O.B.; Larsen, J.P. Fatigue in early Parkinson's disease. Minor inconvenience or major distress? *Eur. J. Neurol.* **2012**, *19*, 963–968. [CrossRef]

16. Bruno, A.E.; Sethares, K.A. Fatigue in Parkinson disease: An integrative review. *J. Neurosci. Nurs.* **2015**, *47*, 146–153. [CrossRef] [PubMed]
17. Krupp, L.B.; Coyle, P.K.; Doscher, C.; Miller, A.; Cross, A.H.; Jandorf, L.; Halper, J.; Johnson, B.; Morgante, L.; Grimson, R. Fatigue therapy in multiple sclerosis: Results of a double-blind, randomized, parallel trial of amantadine, pemoline, and placebo. *Neurology* **1995**, *45*, 1956–1961. [CrossRef] [PubMed]
18. Shulman, L.M.; Taback, R.L.; Rabinstein, A.A.; Weiner, W.J. Non-recognition of depression and other non-motor symptoms in Parkinson's disease. *Parkinsonism Relat. Disord.* **2002**, *8*, 193–197. [CrossRef]
19. Chaudhuri, K.R.; Prieto-Jurcynska, C.; Naidu, Y.; Mitra, T.; Frades-Payo, B.; Tluk, S.; Ruessmann, A.; Odin, P.; Macphee, G.; Stocchi, F.; et al. The nondeclaration of nonmotor symptoms of Parkinson's disease to health care professionals: An international study using the nonmotor symptoms questionnaire. *Mov. Disord.* **2010**, *25*, 704–709. [CrossRef]
20. Garcia-Ruiz, P.J.; Chaudhuri, K.R.; Martinez-Martin, P. Non-motor symptoms of Parkinson's disease A review . . . from the past. *J. Neurol. Sci.* **2014**, *338*, 30–33. [CrossRef]
21. Charcot, J.M. *Lectures on the Diseases of the Nervous System: Delivered at La Salpetriere*; New Sydenham Society: London, UK, 1877; Volume 1.
22. Abe, K.; Takanashi, M.; Yanagihara, T. Fatigue in patients with Parkinson's disease. *Behav. Neurol.* **2000**, *12*, 103–106. [CrossRef] [PubMed]
23. Kluger, B.M.; Krupp, L.B.; Enoka, R.M. Fatigue and fatigability in neurologic illnesses: Proposal for a unified taxonomy. *Neurology* **2013**, *80*, 409–416. [CrossRef] [PubMed]
24. Finsterer, J.; Mahjoub, S.Z. Fatigue in healthy and diseased individuals. *Am. J. Hosp. Palliat. Care* **2014**, *31*, 562–575. [CrossRef] [PubMed]
25. Linnhoff, S.; Fiene, M.; Heinze, H.J.; Zaehle, T. Cognitive Fatigue in Multiple Sclerosis: An Objective Approach to Diagnosis and Treatment by Transcranial Electrical Stimulation. *Brain Sci.* **2019**, *9*, 100. [CrossRef]
26. Flachenecker, P.; Kümpfel, T.; Kallmann, B.; Gottschalk, M.; Grauer, O.; Rieckmann, P.; Trenkwalder, C.; Toyka, K.V. Fatigue in multiple sclerosis: A comparison of different rating scales and correlation to clinical parameters. *Mult. Scler.* **2002**, *8*, 523–526. [CrossRef]
27. Barak, Y.; Achiron, A. Cognitive fatigue in multiple sclerosis: Findings from a two-wave screening project. *J. Neurol. Sci.* **2006**, *245*, 73–76. [CrossRef]
28. Lerdal, A.; Celius, E.G.; Moum, T. Fatigue and its association with sociodemographic variables among multiple sclerosis patients. *Mult. Scler.* **2003**, *9*, 509–514. [CrossRef]
29. Harrison, A.M.; das Nair, R.; Moss-Morris, R. Operationalising cognitive fatigability in multiple sclerosis: A Gordian knot that can be cut? *Mult. Scler.* **2017**, *23*, 1682–1696. [CrossRef]
30. Fisk, J.D.; Pontefract, A.; Ritvo, P.G.; Archibald, C.J.; Murray, T.J. The impact of fatigue on patients with multiple sclerosis. *Can. J. Neurol. Sci.* **1994**, *21*, 9–14. [CrossRef]
31. Genova, H.M.; Rajagopalan, V.; Deluca, J.; Das, A.; Binder, A.; Arjunan, A.; Chiaravalloti, N.; Wylie, G. Examination of cognitive fatigue in multiple sclerosis using functional magnetic resonance imaging and diffusion tensor imaging. *PLoS ONE* **2013**, *8*, e78811. [CrossRef]
32. Chaudhuri, A.; Behan, P.O. Fatigue in neurological disorders. *Lancet* **2004**, *363*, 978–988. [CrossRef]
33. Lindqvist, D.; Kaufman, E.; Brundin, L.; Hall, S.; Surova, Y.; Hansson, O. Non-motor symptoms in patients with Parkinson's disease—Correlations with inflammatory cytokines in serum. *PLoS ONE* **2012**, *7*, e47387. [CrossRef]
34. Fabbrini, G.; Latorre, A.; Suppa, A.; Bloise, M.; Frontoni, M.; Berardelli, A. Fatigue in Parkinson's disease: Motor or non-motor symptom? *Parkinsonism Relat. Disord.* **2013**, *19*, 148–152. [CrossRef]
35. Pavese, N.; Metta, V.; Bose, S.K.; Chaudhuri, K.R.; Brooks, D.J. Fatigue in Parkinson's disease is linked to striatal and limbic serotonergic dysfunction. *Brain* **2010**, *133*, 3434–3443. [CrossRef] [PubMed]
36. Friedman, J.H.; Brown, R.G.; Comella, C.; Garber, C.E.; Krupp, L.B.; Lou, J.S.; Marsh, L.; Nail, L.; Shulman, L.; Taylor, C.B. Fatigue in Parkinson's disease: A review. *Mov. Disord.* **2007**, *22*, 297–308. [CrossRef] [PubMed]
37. Sáez-Francàs, N.; Hernández-Vara, J.; Corominas Roso, M.; Alegre Martín, J.; Casas Brugué, M. The association of apathy with central fatigue perception in patients with Parkinson's disease. *Behav. Neurosci.* **2013**, *127*, 237–244. [CrossRef]
38. Friedman, J.H.; Alves, G.; Hagell, P.; Marinus, J.; Marsh, L.; Martinez-Martin, P.; Goetz, C.G.; Poewe, W.; Rascol, O.; Sampaio, C.; et al. Fatigue rating scales critique and recommendations by the Movement Disorders Society task force on rating scales for Parkinson's disease. *Mov. Disord.* **2010**, *25*, 805–822. [CrossRef] [PubMed]
39. Kostić, V.S.; Tomić, A.; Ječmenica-Lukić, M. The Pathophysiology of Fatigue in Parkinson's Disease and its Pragmatic Management. *Mov. Disord. Clin. Pract.* **2016**, *3*, 323–330. [CrossRef]
40. Alves, G.; Wentzel-Larsen, T.; Larsen, J.P. Is fatigue an independent and persistent symptom in patients with Parkinson disease? *Neurology* **2004**, *63*, 1908–1911. [CrossRef]
41. Schrag, A.; Horsfall, L.; Walters, K.; Noyce, A.; Petersen, I. Prediagnostic presentations of Parkinson's disease in primary care: A case-control study. *Lancet Neurol.* **2015**, *14*, 57–64. [CrossRef]
42. Pont-Sunyer, C.; Hotter, A.; Gaig, C.; Seppi, K.; Compta, Y.; Katzenschlager, R.; Mas, N.; Hofeneder, D.; Brücke, T.; Bayés, A.; et al. The onset of nonmotor symptoms in Parkinson's disease (the ONSET PD study). *Mov. Disord.* **2015**, *30*, 229–237. [CrossRef] [PubMed]

43. Kluger, B.M.; Parra, V.; Jacobson, C.; Garvan, C.W.; Rodriguez, R.L.; Fernandez, H.H.; Fogel, A.; Skoblar, B.M.; Bowers, D.; Okun, M.S. The prevalence of fatigue following deep brain stimulation surgery in Parkinson's disease and association with quality of life. *Parkinsons Disord.* **2012**, *2012*, 769506. [CrossRef]
44. Franssen, M.; Winward, C.; Collett, J.; Wade, D.; Dawes, H. Interventions for fatigue in Parkinson's disease: A systematic review and meta-analysis. *Mov. Disord.* **2014**, *29*, 1675–1678. [CrossRef] [PubMed]
45. Goldman, J.G.; Stebbins, G.T.; Leung, V.; Tilley, B.C.; Goetz, C.G. Relationships among cognitive impairment, sleep, and fatigue in Parkinson's disease using the MDS-UPDRS. *Parkinsonism Relat. Dis.* **2014**, *20*, 1135–1139. [CrossRef]
46. Chou, K.L.; Kotagal, V.; Bohnen, N.I. Neuroimaging and clinical predictors of fatigue in Parkinson disease. *Parkinsonism Relat. Disord.* **2016**, *23*, 45–49. [CrossRef] [PubMed]
47. Tessitore, A.; Giordano, A.; De Micco, R.; Caiazzo, G.; Russo, A.; Cirillo, M.; Esposito, F.; Tedeschi, G. Functional connectivity underpinnings of fatigue in "Drug-Naïve" patients with Parkinson's disease. *Mov. Disord.* **2016**, *31*, 1497–1505. [CrossRef]
48. Clayton, M.S.; Yeung, N.; Cohen Kadosh, R. The roles of cortical oscillations in sustained attention. *Trends Cogn. Sci.* **2015**, *19*, 188–195. [CrossRef]
49. Seppi, K.; Ray Chaudhuri, K.; Coelho, M.; Fox, S.H.; Katzenschlager, R.; Perez Lloret, S.; Weintraub, D.; Sampaio, C. Update on treatments for nonmotor symptoms of Parkinson's disease-an evidence-based medicine review. *Mov. Disord.* **2019**, *34*, 180–198. [CrossRef]
50. Lazcano-Ocampo, C.; Wan, Y.M.; van Wamelen, D.J.; Batzu, L.; Boura, I.; Titova, N.; Leta, V.; Qamar, M.; Martinez-Martin, P.; Ray Chaudhuri, K. Identifying and responding to fatigue and apathy in Parkinson's disease: A review of current practice. *Expert Rev. Neurother.* **2020**, *20*, 477–495. [CrossRef]
51. Dafsari, H.S.; Martinez-Martin, P.; Rizos, A.; Trost, M.; Dos Santos Ghilardi, M.G.; Reddy, P.; Sauerbier, A.; Petry-Schmelzer, J.N.; Kramberger, M.; Borgemeester, R.W.K.; et al. EuroInf 2: Subthalamic stimulation, apomorphine, and levodopa infusion in Parkinson's disease. *Mov. Disord.* **2019**, *34*, 353–365. [CrossRef]
52. Dafsari, H.S.; Reddy, P.; Herchenbach, C.; Wawro, S.; Petry-Schmelzer, J.N.; Visser-Vandewalle, V.; Rizos, A.; Silverdale, M.; Ashkan, K.; Samuel, M.; et al. Beneficial Effects of Bilateral Subthalamic Stimulation on Non-Motor Symptoms in Parkinson's Disease. *Brain Stimul.* **2016**, *9*, 78–85. [CrossRef] [PubMed]
53. Chou, K.L.; Persad, C.C.; Patil, P.G. Change in fatigue after bilateral subthalamic nucleus deep brain stimulation for Parkinson's disease. *Parkinsonism Relat. Disord.* **2012**, *18*, 510–513. [CrossRef]
54. Yavari, F.; Jamil, A.; Mosayebi Samani, M.; Vidor, L.P.; Nitsche, M.A. Basic and functional effects of transcranial Electrical Stimulation (tES)—An introduction. *Neurosci. Biobehav. Rev.* **2018**, *85*, 81–92. [CrossRef]
55. Sale, M.V.; Mattingley, J.B.; Zalesky, A.; Cocchi, L. Imaging human brain networks to improve the clinical efficacy of non-invasive brain stimulation. *Neurosci. Biobehav. Rev.* **2015**, *57*, 187–198. [CrossRef] [PubMed]
56. Elyamany, O.; Leicht, G.; Herrmann, C.S.; Mulert, C. Transcranial alternating current stimulation (tACS): From basic mechanisms towards first applications in psychiatry. *Eur. Arch. Psychiatry Clin. Neurosci.* **2021**, *271*, 135–156. [CrossRef] [PubMed]
57. Bikson, M.; Grossman, P.; Thomas, C.; Zannou, A.L.; Jiang, J.; Adnan, T.; Mourdoukoutas, A.P.; Kronberg, G.; Truong, D.; Boggio, P.; et al. Safety of Transcranial Direct Current Stimulation: Evidence Based Update 2016. *Brain Stimul.* **2016**, *9*, 641–661. [CrossRef] [PubMed]
58. Neuling, T.; Wagner, S.; Wolters, C.H.; Zaehle, T.; Herrmann, C.S. Finite-Element Model Predicts Current Density Distribution for Clinical Applications of tDCS and tACS. *Front. Psychiatry* **2012**, *3*, 83. [CrossRef] [PubMed]
59. Heimrath, K.; Fiene, M.; Rufener, K.S.; Zaehle, T. Modulating Human Auditory Processing by Transcranial Electrical Stimulation. *Front. Cell. Neurosci.* **2016**, *10*, 53. [CrossRef]
60. Bindman, L.J.; Lippold, O.C.; Redfearn, J.W. Long-lasting changes in the level of the electrical activity of the cerebral cortex produced bypolarizing currents. *Nature* **1962**, *196*, 584–585. [CrossRef]
61. Nitsche, M.A.; Paulus, W. Excitability changes induced in the human motor cortex by weak transcranial direct current stimulation. *J. Physiol.* **2000**, *527 Pt 3*, 633–639. [CrossRef]
62. Fox, D. Neuroscience: Brain buzz. *Nature* **2011**, *472*, 156–158. [CrossRef]
63. D'Anselmo, A.; Prete, G.; Tommasi, L.; Brancucci, A. The Dichotic Right Ear Advantage Does not Change with Transcranial Direct Current Stimulation (tDCS). *Brain Stimul.* **2015**, *8*, 1238–1240. [CrossRef]
64. Kunzelmann, K.; Meier, L.; Grieder, M.; Morishima, Y.; Dierks, T. No Effect of Transcranial Direct Current Stimulation of the Auditory Cortex on Auditory-Evoked Potentials. *Front. Neurosci.* **2018**, *12*, 880. [CrossRef]
65. Chen, J.C.; Hämmerer, D.; Strigaro, G.; Liou, L.M.; Tsai, C.H.; Rothwell, J.C.; Edwards, M.J. Domain-specific suppression of auditory mismatch negativity with transcranial direct current stimulation. *Clin. Neurophysiol.* **2014**, *125*, 585–592. [CrossRef]
66. Hanenberg, C.; Getzmann, S.; Lewald, J. Transcranial direct current stimulation of posterior temporal cortex modulates electrophysiological correlates of auditory selective spatial attention in posterior parietal cortex. *Neuropsychologia* **2019**, *131*, 160–170. [CrossRef] [PubMed]
67. Zaehle, T.; Beretta, M.; Jäncke, L.; Herrmann, C.S.; Sandmann, P. Excitability changes induced in the human auditory cortex by transcranial direct current stimulation: Direct electrophysiological evidence. *Exp. Brain Res.* **2011**, *215*, 135–140. [CrossRef] [PubMed]
68. Jackson, M.P.; Rahman, A.; Lafon, B.; Kronberg, G.; Ling, D.; Parra, L.C.; Bikson, M. Animal models of transcranial direct current stimulation: Methods and mechanisms. *Clin. Neurophysiol.* **2016**, *127*, 3425–3454. [CrossRef] [PubMed]

69. Rahman, A.; Reato, D.; Arlotti, M.; Gasca, F.; Datta, A.; Parra, L.C.; Bikson, M. Cellular effects of acute direct current stimulation: Somatic and synaptic terminal effects. *J. Physiol.* **2013**, *591*, 2563–2578. [CrossRef]
70. Bikson, M.; Inoue, M.; Akiyama, H.; Deans, J.K.; Fox, J.E.; Miyakawa, H.; Jefferys, J.G. Effects of uniform extracellular DC electric fields on excitability in rat hippocampal slices in vitro. *J. Physiol.* **2004**, *557 Pt 1*, 175–190. [CrossRef]
71. Marquardt, L.; Kusztrits, I.; Craven, A.R.; Hugdahl, K.; Specht, K.; Hirnstein, M. A multimodal study of the effects of tDCS on dorsolateral prefrontal and temporo-parietal areas during dichotic listening. *Eur. J. Neurosci.* **2021**, *53*, 449–459. [CrossRef] [PubMed]
72. Sala, G.; Bocci, T.; Borzì, V.; Parazzini, M.; Priori, A.; Ferrarese, C. Direct current stimulation enhances neuronal alpha-synuclein degradation in vitro. *Sci. Rep.* **2021**, *11*, 2197. [CrossRef]
73. Borragán, G.; Gilson, M.; Atas, A.; Slama, H.; Lysandropoulos, A.; De Schepper, M.; Peigneux, P. Cognitive Fatigue, Sleep and Cortical Activity in Multiple Sclerosis Disease. A Behavioral, Polysomnographic and Functional Near-Infrared Spectroscopy Investigation. *Front. Hum. Neurosci.* **2018**, *12*, 378. [CrossRef] [PubMed]
74. Fiene, M.; Rufener, K.S.; Kuehne, M.; Matzke, M.; Heinze, H.J.; Zaehle, T. Electrophysiological and behavioral effects of frontal transcranial direct current stimulation on cognitive fatigue in multiple sclerosis. *J. Neurol.* **2018**, *265*, 607–617. [CrossRef] [PubMed]
75. McIntire, L.K.; McKinley, R.A.; Nelson, J.M.; Goodyear, C. Transcranial direct current stimulation versus caffeine as a fatigue countermeasure. *Brain Stimul.* **2017**, *10*, 1070–1078. [CrossRef] [PubMed]
76. McIntire, L.K.; McKinley, R.A.; Goodyear, C.; Nelson, J. A comparison of the effects of transcranial direct current stimulation and caffeine on vigilance and cognitive performance during extended wakefulness. *Brain Stimul.* **2014**, *7*, 499–507. [CrossRef] [PubMed]
77. Linnhoff, S.; Wolter-Weging, J.; Zaehle, T. Objective electrophysiological fatigability markers and their modulation through tDCS. *Clin. Neurophysiol.* **2021**. [CrossRef]
78. Mangia, A.L.; Pirini, M.; Cappello, A. Transcranial direct current stimulation and power spectral parameters: A tDCS/EEG co-registration study. *Front. Hum. Neurosci.* **2014**, *8*, 601. [CrossRef]
79. Miller, J.; Berger, B.; Sauseng, P. Anodal transcranial direct current stimulation (tDCS) increases frontal-midline theta activity in the human EEG: A preliminary investigation of non-invasive stimulation. *Neurosci. Lett.* **2015**, *588*, 114–119. [CrossRef]
80. Zaehle, T.; Sandmann, P.; Thorne, J.D.; Jaencke, L.; Herrmann, C.S. Transcranial direct current stimulation of the prefrontal cortex modulates working memory performance: Combined behavioural and electrophysiological evidence. *BMC Neurosci.* **2011**, *12*, 2. [CrossRef]
81. Ferrucci, R.; Vergari, M.; Cogiamanian, F.; Bocci, T.; Ciocca, M.; Tomasini, E.; De Riz, M.; Scarpini, E.; Priori, A. Transcranial direct current stimulation (tDCS) for fatigue in multiple sclerosis. *NeuroRehabilitation* **2014**, *34*, 121–127. [CrossRef]
82. Tecchio, F.; Cancelli, A.; Cottone, C.; Zito, G.; Pasqualetti, P.; Ghazaryan, A.; Rossini, P.M.; Filippi, M.M. Multiple sclerosis fatigue relief by bilateral somatosensory cortex neuromodulation. *J. Neurol.* **2014**, *261*, 1552–1558. [CrossRef] [PubMed]
83. Chalah, M.A.; Riachi, N.; Ahdab, R.; Mhalla, A.; Abdellaoui, M.; Créange, A.; Lefaucheur, J.P.; Ayache, S.S. Effects of left DLPFC versus right PPC tDCS on multiple sclerosis fatigue. *J. Neurol. Sci.* **2017**, *372*, 131–137. [CrossRef] [PubMed]
84. Ayache, S.S.; Lefaucheur, J.P.; Chalah, M.A. Long term effects of prefrontal tDCS on multiple sclerosis fatigue: A case study. *Brain Stimul.* **2017**, *10*, 1001–1002. [CrossRef] [PubMed]
85. Liu, M.; Fan, S.; Xu, Y.; Cui, L. Non-invasive brain stimulation for fatigue in multiple sclerosis patients: A systematic review and meta-analysis. *Mult. Scler. Relat. Disord.* **2019**, *36*, 101375. [CrossRef] [PubMed]
86. Madrid, J.; Benninger, D.H. Non-invasive brain stimulation for Parkinson's disease: Clinical evidence, latest concepts and future goals: A systematic review. *J. Neurosci. Methods* **2021**, *347*, 108957. [CrossRef] [PubMed]
87. Suarez-García, D.M.A.; Grisales-Cárdenas, J.S.; Zimerman, M.; Cardona, J.F. Transcranial Direct Current Stimulation to Enhance Cognitive Impairment in Parkinson's Disease: A Systematic Review and Meta-Analysis. *Front. Neurol.* **2020**, *11*, 597955. [CrossRef]
88. Forogh, B.; Rafiei, M.; Arbabi, A.; Motamed, M.R.; Madani, S.P.; Sajadi, S. Repeated sessions of transcranial direct current stimulation evaluation on fatigue and daytime sleepiness in Parkinson's disease. *Neurol. Sci.* **2017**, *38*. [CrossRef]
89. Dobbs, B.; Pawlak, N.; Biagioni, M.; Agarwal, S.; Shaw, M.; Pilloni, G.; Bikson, M.; Datta, A.; Charvet, L. Generalizing remotely supervised transcranial direct current stimulation (tDCS): Feasibility and benefit in Parkinson's disease. *J. Neuroeng. Rehabil.* **2018**, *15*, 114. [CrossRef]
90. Yuan, T.F.; Li, W.G.; Zhang, C.; Wei, H.; Sun, S.; Xu, N.J.; Liu, J.; Xu, T.L. Targeting neuroplasticity in patients with neurodegenerative diseases using brain stimulation techniques. *Transl. Neurodegener.* **2020**, *9*, 44. [CrossRef]
91. Boggio, P.S.; Fregni, F.; Bermpohl, F.; Mansur, C.G.; Rosa, M.; Rumi, D.O.; Barbosa, E.R.; Odebrecht Rosa, M.; Pascual-Leone, A.; Rigonatti, S.P.; et al. Effect of repetitive TMS and fluoxetine on cognitive function in patients with Parkinson's disease and concurrent depression. *Mov. Disord.* **2005**, *20*, 1178–1184. [CrossRef]
92. Pal, E.; Nagy, F.; Aschermann, Z.; Balazs, E.; Kovacs, N. The impact of left prefrontal repetitive transcranial magnetic stimulation on depression in Parkinson's disease: A randomized, double-blind, placebo-controlled study. *Mov. Disord.* **2010**, *25*, 2311–2317. [CrossRef] [PubMed]

*Article*

# Extradural Motor Cortex Stimulation in Parkinson's Disease: Long-Term Clinical Outcome

**Carla Piano [1], Francesco Bove [1,2,*], Delia Mulas [3], Enrico Di Stasio [4,5], Alfonso Fasano [6,7,8], Anna Rita Bentivoglio [1,2], Antonio Daniele [1,2], Beatrice Cioni [9], Paolo Calabresi [1,2] and Tommaso Tufo [9]**

1 Neurology Unit, Fondazione Policlinico Universitario A. Gemelli IRCCS, 00168 Rome, Italy; carla.piano@policlinicogemelli.it (C.P.); annarita.bentivoglio@policlinicogemelli.it (A.R.B.); antonio.daniele@unicatt.it (A.D.); paolo.calabresi@policlinicogemelli.it (P.C.)
2 Department of Neurosciences, Università Cattolica del Sacro Cuore, 00168 Rome, Italy
3 Institute of Neurology, Mater Olbia Hospital, 07026 Olbia, Italy; delia.mulas@gmail.com
4 Institute of Biochemistry and Clinical Biochemistry, Università Cattolica del Sacro Cuore, 00168 Rome, Italy; enrico.distasio@unicatt.it
5 Cheminstry, Biocheminstry and Clinical Molecular Biology, Fondazione Policlinico Universitario A. Gemelli IRCCS, 00168 Rome, Italy
6 Edmond J. Safra Program in Parkinson's Disease, Morton and Gloria Shulman Movement Disorders Clinic, Toronto Western Hospital and Division of Neurology, University of Toronto, Toronto, ON M5T 2S8, Canada; alfonso.fasano@uhn.ca
7 Krembil Brain Institute, Toronto, ON M5T 1M8, Canada
8 Center for Advancing Neurotechnological Innovation to Application (CRANIA), Toronto, ON M5G 2A2, Canada
9 Neurosurgery Unit, Fondazione Policlinico Universitario A. Gemelli IRCCS, 00168 Rome, Italy; beatrice.cioni@policlinicogemelli.it (B.C.); tommaso.tufo@policlinicogemelli.it (T.T.)
* Correspondence: francescobove86@gmail.com

**Citation:** Piano, C.; Bove, F.; Mulas, D.; Di Stasio, E.; Fasano, A.; Bentivoglio, A.R.; Daniele, A.; Cioni, B.; Calabresi, P.; Tufo, T. Extradural Motor Cortex Stimulation in Parkinson's Disease: Long-Term Clinical Outcome. *Brain Sci.* **2021**, *11*, 416. https://doi.org/10.3390/brainsci11040416

Academic Editor: Ulrich Palm

Received: 26 February 2021
Accepted: 23 March 2021
Published: 26 March 2021

**Publisher's Note:** MDPI stays neutral with regard to jurisdictional claims in published maps and institutional affiliations.



**Abstract:** Previous investigations have reported on the motor benefits and safety of chronic extradural motor cortex stimulation (EMCS) for patients with Parkinson's disease (PD), but studies addressing the long-term clinical outcome are still lacking. In this study, nine consecutive PD patients who underwent EMCS were prospectively recruited, with a mean follow-up time of 5.1 ± 2.5 years. As compared to the preoperatory baseline, the Unified Parkinson's Disease Rating Scale (UPDRS)-III in the off-medication condition significantly decreased by 13.8% at 12 months, 16.1% at 18 months, 18.4% at 24 months, 21% at 36 months, 15.6% at 60 months, and 8.6% at 72 months. The UPDRS-IV decreased by 30.8% at 12 months, 22.1% at 24 months, 25% at 60 months, and 36.5% at 72 months. Dopaminergic therapy showed a progressive reduction, significant at 60 months (11.8%). Quality of life improved by 18.0% at 12 months, and 22.4% at 60 months. No surgical complication, cognitive or behavioral change occurred. The only adverse event reported was an infection of the implantable pulse generator pocket. Even in the long-term follow-up, EMCS was shown to be a safe and effective treatment option in PD patients, resulting in improvements in motor symptoms and quality of life, and reductions in motor complications and dopaminergic therapy.

**Keywords:** motor cortex stimulation; Parkinson's disease; movement disorders; neuromodulation

## 1. Introduction

Chronic motor cortex stimulation by implanted extradural electrodes (EMCS) is a minimally invasive therapy proposed as an alternative surgical treatment for Parkinson's disease (PD) patients who are not eligible for deep brain stimulation (DBS) [1,2]. The pathophysiological rationale for EMCS in PD derives from several clinical and experimental observations. The primary motor cortex (M1) and lateral premotor cortex are hyperactive in advanced parkinsonism, with increased excitability of corticospinal projections at rest, concomitant with or resulting from reduced intracortical inhibition (ICI) [3]. EMCS may restore normal ICI by acting on small inhibitory interneurons within M1 [4]. Indeed, it may act by desynchronizing pathological oscillations in the beta-band between the basal ganglia and cortical neurons, influencing the electrical activity of subcortical structures in primate models [5]. Moreover, functional neuroimaging studies suggest that EMCS might restore the activity of cortical areas that are hypoactive in PD (e.g., the supplementary motor area, SMA) [6].

Initial reports suggesting effectiveness of EMCS in treatment of PD motor symptoms date back almost 20 years [7]. To date, over 100 PD patients have been treated by EMCS, although the available studies concern small samples with a short-term follow-up [2,8–13]. Differences among various centers in terms of patient selection criteria, electrode placement and stimulation parameters may explain the inconsistent findings reported so far [13]. Although several Authors have reported that EMCS is safe and may improve PD motor symptoms, studies addressing the long-term outcomes are still lacking [14,15]. Recently, for selected patients, spinal cord stimulation has emerged as an alternative neuromodulation procedure to DBS, emphasizing the need for less-invasive surgical options for PD and effective treatments for axial symptoms [16].

This prospective, single center, open-label study was aimed at assessing the efficacy and safety of EMCS in PD patients with long-term follow-up, up to eight years after surgery. In particular, the following outcomes were investigated: efficacy of EMCS on PD motor symptoms and motor complications (motor fluctuations and dyskinesias); reduction in dopaminergic therapy after surgery; impact of EMCS on daily living activities (DLA) and quality of life (QoL); cognitive and behavioral safety of the stimulation, peri- or post-operative adverse events (AEs).

## 2. Materials and Methods

Consecutive PD patients who underwent EMCS implantation at Fondazione Policlinico Universitario Agostino Gemelli IRCCS in Rome between 2003 and 2007 were included after local ethics committee approval (#400-A763). All patients signed a detailed informed consent form.

All enrolled patients had completed at least a 24-month follow-up and fulfilled the following inclusion criteria: PD diagnosis according to United Kingdom PD Brain Bank criteria [17]; disease duration longer than 5 years; dopaminergic responsiveness confirmed by a pharmacological test showing at least a 33% decrease in the Unified Parkinson's Disease Rating Scale (UPDRS) motor score [18]; unsatisfactory pharmacological management of fluctuations; lack of eligibility for DBS (i.e., not accepted by patients or contraindicated according to Core Assessment Program for Surgical Interventional Therapies in PD [19]); ability to give informed consent; stable drug regimen and motor condition for at least 3 months preoperatively.

Exclusion criteria included: history of epilepsy or epileptic activity on electroencephalography; alcohol or drug abuse; previous brain surgery; severe psychiatric symptoms (e.g., psychosis, major depression); moderate or severe cognitive impairment (score $< 24$ on the Mini-Mental State Examination [20]); diagnosis of dementia according to the Diagnostic and Statistical Manual of Mental Disorders, Fourth Edition (DSM-IV) [21]; medical condition contraindicating a surgical procedure under general anesthesia.

### 2.1. Surgical Technique

The surgical procedure was performed with patients under total intravenous anesthesia: a quadripolar electrode strip (model Resume; Medtronic Inc, Minneapolis, Minnesota) was epidurally placed over M1 (through a burr hole over, contralateral to the most affected body side in three patients and bilaterally in remaining patients) and connected to a Soletra or Kinetra (Medtronic Inc) implantable pulse generator (IPG) located in the subclavian region. In all patients, contacts were oriented along the craniocaudal axis of the precentral gyrus: contact 3 was 2 to 3 cm from midline, contact 0 was 4 cm more lateral (Figure 1). Implantation site was preoperatively defined (using magnetic resonance imaging and neuronavigation) and verified by means of motor-evoked potentials and by identifying N20-P20 phase reversal of somatosensory evoked potentials obtained from contralateral median nerve stimulation [22]. Patients postoperatively underwent a computed tomography scan to confirm that the electrode paddle was correctly placed and to rule out surgical complications.

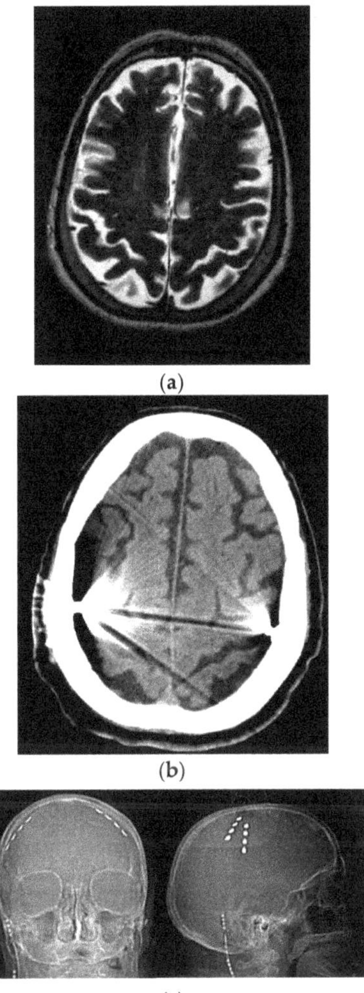

**Figure 1.** Placement of quadripolar electrode strips over the motor cortices. (**a**) Preoperative brain MRI. (**b**) Postoperative brain CT scan. (**c**) Postoperative skull X-ray. R = right; L = left.

### *2.2. Parameter and Medication Adjustments*

Parameter setting was performed in the weeks following surgery. In all patients, stimulation was unilateral (namely, contralateral to the body side with more severe motor impairment) for the first 12 months, and afterwards was bilateral for six patients with electrodes over both hemispheres. Stimulation was continuously delivered through the two most distant contacts of the electrode paddle under the bipolar setting. Stimulation parameters were: biphasic wave of 120 µs duration and 80 Hz frequency; voltage maintained at 50% of threshold intensity for any movement or sensation (between 3 and 5 V). EMCS parameters were stable during the first 24 months of follow-up, to evaluate the chronic effects of stimulation with constant parameters. Additional parameter adjustments (voltage increasing or reduction) were attempted according to clinical response and electrode impedances, after 24-month follow-up. Antiparkinsonian medications were gradually decreased according to individual clinical condition.

### 2.3. Outcome Measures

Patients were evaluated at preoperative baseline and postoperatively (12, 18, 24, 36, 48, 60, 72, 84 and 96 months after implantation). Evaluations were performed in the morning, in the practically defined off-medication condition (off-med, at least 12 h after medication withdrawal) and in the on-med condition, following administration of a standard liquid levodopa at doses 50% higher than the usual morning dose of dopaminergic treatment [19]. Postoperative assessments were performed during the on-stimulation condition since a carryover effect was expected to last several days after switching off pulse generator [6]. Motor evaluation was video-recorded for independent analysis. Outcome measures included the UPDRS subscales I, II, III (in off- and on-med), and IV [18]. Arising from chair, posture, gait, postural stability, body bradykinesia (items 27–31 of UPDRS-III) were also separately scored and a total axial subscore (range 0–20) was calculated by addition of these scores. Disease severity was measured by the UPDRS total score (i.e., the sum of UPDRS I, II, III (only in off-med condition) and IV). At each evaluation, QoL, by means of the PD QoL questionnaire (PDQL) [23], and levodopa equivalent daily dose (LEDD) [24] were assessed.

Cognitive and behavioral assessments were performed at preoperative baseline and postoperatively (at 12, 18, 36, 60 and 96 months after implantation). Cognitive assessment was performed by means of a neuropsychological test battery [25], including Mini-Mental State Examination (MMSE), tasks exploring visuospatial (Corsi Block-Tapping Test) and verbal (digit span forward and backward) working memory, episodic verbal memory (Rey's Auditory Verbal Learning Test, RAVLT), nonverbal abstract reasoning (Raven's Progressive Matrices '47, RPM '47), phonological and semantic verbal fluency, problem-solving and set-shifting abilities (modified Wisconsin Card Sorting Test, mWCST), response inhibition (Stroop test). Tests sensitive to motor speed were not included in neuropsychological tasks, to minimize the bias due to motor impairment. Behavioral assessments included Zung's Self-Rating Depression and Anxiety Scales [26,27], and a clinical interview aimed at detecting behavioral abnormalities or psychiatric disorders. Cognitive and behavioral assessments were performed in the on-med condition.

Stimulation- or device-related AEs were collected at each postoperative evaluation. An electroencephalogram was recorded preoperatively, and postoperatively at 6 and 12 months.

### 2.4. Statistical Analysis

Data were analyzed for normality of distribution using the Kolmogorov–Smirnov test of normality. Continuous data (comparisons between preoperative and postoperative scores at each follow-up visit up to 72 months) were analyzed by means of the Wilcoxon signed-rank test, and are presented as mean ± standard deviation. Comparisons between categorical variables were performed by Fisher's test. Given the explorative nature of our study, the standard non-corrected significance $\alpha$ level of $p < 0.05$ was used to reduce the risk of a type II error. However, considering the risk of a type I error deriving from multiple comparisons, significant values should be interpreted with cautions when levels are only marginally lower than 0.05. All statistical computations were two-sided and relied on Statistical Package for the Social Sciences (SPSS) software, version 15.0 (IBM Co., Armonk, NY, USA).

## 3. Results

Nine PD patients were included (Table 1). Mean age at implantation was 64.0 ± 6.4 years and mean disease duration was 14.6 ± 5.9 years. DBS was refused by four patients and contraindicated in five patients (in two patients because of multi-infarctual encephalopathy, in three patients because of age).

**Table 1.** Patient's demographic and clinical data at baseline.

| Patient | Gender | Age at Surgery | UPDRS III Off-Med | UPDRS III On-Med | Hoehn and Yahr Stage | UPDRS IV | LEDD (mg) | Reason DBS Not Performed |
|---|---|---|---|---|---|---|---|---|
| 1 | F | 57 | 8 | 55 | 24 | 5 | 13 | Patient refusal |
| 2 | M | 55 | 16 | 68 | 19 | 5 | 13 | Brain atrophy |
| 3 | M | 62 | 9 | 62 | 18 | 4 | 11 | Patient refusal |
| 4 | F | 57 | 14 | 64 | 24 | 5 | 15 | Brain atrophy |
| 5 | F | 66 | 28 | 61 | 35 | 5 | 12 | Patient refusal |
| 6 | F | 69 | 17 | 59 | 15 | 5 | 12 | Age and comorbidities |
| 7 | F | 70 | 14 | 62 | 17 | 5 | 9 | Age and comorbidities |
| 8 | M | 72 | 11 | 43 | 16 | 3 | 6 | Age and comorbidities |
| 9 | M | 68 | 14 | 40 | 18 | 3 | 3 | Patient refusal |
| Total | 4 M/5 F | 64.0 ± 6.4 | 14.6 ± 5.9 | 57.1 ± 9.5 | 20.7 ± 6.2 | 4.4 ± 0.9 | 10.4 ± 3.8 | |

Abbreviations: UPDRS = Unified Parkinson's Disease Rating Scale; DBS = deep brain stimulation.

Mean follow-up time was 5.1 ± 2.5 years (range 2–8 years). Two patients died during the follow-up period: one of lung cancer and one of heart failure. Four patients were lost to follow-up for difficulty in reaching our center.

### 3.1. Motor Efficacy

Compared to baseline, the UPDRS-III score in the off-med condition significantly decreased by 13.8% at 12 months ($p$ = 0.01), 16.1% at 18 months ($p$ = 0.04), 18.4% at 24 months ($p$ = 0.01), 21% at 36 months ($p$ = 0.02), 15.6% at 60 months ($p$ = 0.04), and 8.6% at 72 months ($p$ = 0.04). Postoperative motor improvement was mostly related to decreases in the axial subscore: as compared to the baseline, in the off-med condition, the total axial subscore significantly decreased by 26.7% at 18 months ($p$ = 0.02), 28.1% at 24 months ($p$ = 0.03), 5.9% at 60 months ($p$ = 0.04), with a slight persistent improvement up to 96 months (Figure 2).

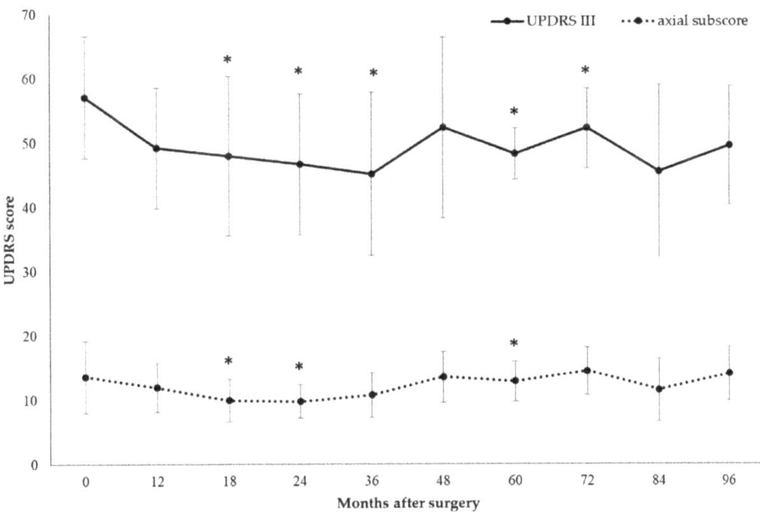


**Figure 2.** Motor efficacy of extradural motor cortex stimulation (EMCS) in the off-med condition in all patients, evaluated at different times after the implant. Error bars represent standard deviation. * $p < 0.05$ at comparisons between preoperative and postoperative scores, analyzed by means of the Wilcoxon signed-rank test.

Compared to the baseline, a progressive motor worsening was observed in the on-med condition, with a significant increase in the UPDRS-III score of 44.9% at 36 months ($p = 0.04$), 68.1% at 48 months ($p = 0.03$), 60.3% at 60 months ($p = 0.04$), 89.4% at 72 months ($p = 0.04$).

Comparisons between subgroups of patients with bilateral ($n = 6$) versus unilateral stimulation ($n = 3$) were not carried out, because of the small sample size. However, at the 24-month follow-up stage, we observed a greater improvement of the UPDRS-III score and total axial subscore in the off-med condition in bilaterally stimulated patients (18.4% decrease in UPDRS III score from 52.1 ± 9.7 at baseline to 42.5 ± 12.3 at 24 months; 29.2% decrease in axial subscore from 12.0 ± 5.7 at baseline to 8.5 ± 2.4 at 24 months), as compared to unilaterally stimulated patients (15.5% decrease in UPDRS-III score from 64.7 ± 3.1 at baseline to 54.7 ± 4.0 at 24 months; 26.9% decrease in axial subscore from 16.3 ± 2.9 at baseline to 12.0 ± 1.0 at 24 months).

Compared to the baseline, the UPDRS-IV score (assessing motor fluctuations and dyskinesias) was significantly decreased by 30.8% at 12 months ($p = 0.03$), 22.1% at 24 months ($p = 0.02$), 25% at 60 months ($p = 0.04$), 36.5% at 72 months ($p = 0.04$). Postoperatively, LEDD showed a progressive reduction, which was significant at the 60-month follow-up stage (11.8%, $p = 0.04$) (Figure 3).

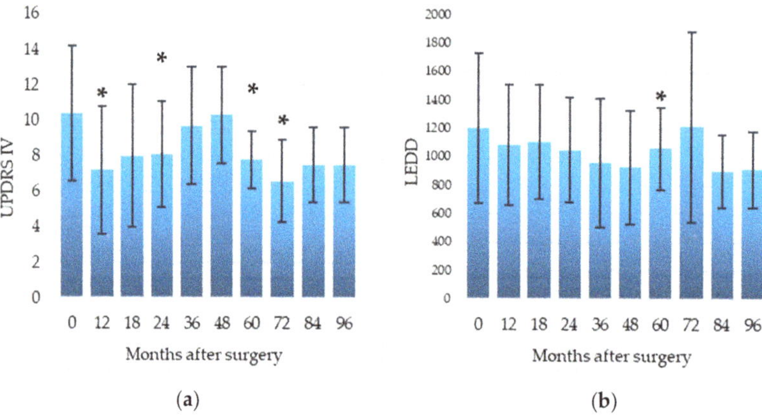


**Figure 3.** Efficacy of EMCS on motor complications (**a**) and reduction in dopaminergic therapy (**b**) in all patients, evaluated at different times after the implant. Error bars represent standard deviation. * $p < 0.05$ at comparisons between preoperative and postoperative scores, analyzed by means of the Wilcoxon signed-rank test.

All data about motor efficacy of EMCS are reported in Table 2.

**Table 2.** Efficacy of stimulation in all patients, evaluated at different times after ECMS implant.

| | **Baseline** ($n = 9$) | **12 m** ($n = 9$) | **18 m** ($n = 9$) | **24 m** ($n = 9$) | **36 m** ($n = 7$) | **48 m** ($n = 6$) | **60 m** ($n = 5$) | **72 m** ($n = 5$) | **84 m** ($n = 2$) | **96 m** ($n = 2$) |
|---|---|---|---|---|---|---|---|---|---|---|
| UPDRS III med-off | 57.1 ± 9.5 | 49.2 ± 9.4 * | 47.9 ± 12.4 * | 46.6 ± 11.0 * | 45.1 ± 12.7 * | 52.3 ± 14.0 | 48.2 ± 4.0 * | 52.2 ± 6.2 * | 45.5 ± 13.4 | 49.5 ± 9.2 |
| UPDRS III axial score med-off | 13.6 ± 5.6 | 11.9 ± 3.8 | 9.9 ± 3.3 * | 9.7 ± 2.6 * | 10.7 ± 3.5 | 13.5 ± 4.0 | 12.8 ± 3.1 * | 14.4 ± 3.7 | 11.5 ± 4.9 | 14.0 ± 4.2 |
| UPDRS III med-on | 20.7 ± 6.2 | 27.1 ± 8.7 | 26.0 ± 6.5 | 26.7 ± 6.2 | 30.0 ± 7.1 * | 34.8 ± 9.9 * | 33.2 ± 5.1 * | 39.2 ± 9.0 * | 37.5 ± 13.4 | 45.0 ± 9.9 |
| UPDRS IV | 10.4 ± 3.8 | 7.2 ± 3.6 * | 8.0 ± 4.0 | 8.1 ± 3.0 * | 9.7 ± 3.3 | 10.3 ± 2.7 | 7.8 ± 1.6 * | 6.6 ± 2.3 * | 7.5 ± 2.1 | 7.5 ± 2.1 |
| LEDD | 1203.0 ± 528.1 | 1085.8 ± 424.5 | 1108.0 ± 402.3 | 1050.8 ± 369.3 | 962.4 ± 455.6 | 929.9 ± 401.0 | 1061.3 ± 286.5 * | 1214.2 ± 670.8 | 905.0 ± 254.6 | 915.0 ± 268.7 |
| UPDRS II | 27.5 ± 10.0 | 22.4 ± 6.9 * | 19.1 ± 6.9 | 24.7 ± 9.0 | 27.5 ± 6.1 | 25.3 ± 8.8 | 22.6 ± 4.0 * | 23.6 ± 3.2 * | 25.0 ± 1.4 | 27.5 ± 2.1 |
| PDQL | 83.5 ± 19.1 | 98.5 ± 20.0 * | 96.0 ± 15.0 | 95.4 ± 15.5 | 92.7 ± 15.7 | 91.8 ± 22.1 | 102.2 ± 19.6 * | 99.2 ± 28.3 | 88.0 ± 15.6 | 102.5 ± 44.5 |
| UPDRS tot | 98.8 ± 24.5 | 82.3 ± 16.9 * | 79.1 ± 21.5 * | 83.7 ± 22.1 * | 89.3 ± 17.5 | 92.3 ± 23.8 | 82.8 ± 5.5 | 88.0 ± 8.5 | 83.5 ± 10.6 | 90.0 ± 8.5 |

Abbreviations: m = months; UPDRS = Unified Parkinson's Disease Rating Scale; LEDD = levodopa equivalent daily dose; PDQL = Parkinson's disease (PD) quality of life (QoL) questionnaire. * $p < 0.05$ at comparisons between preoperative and postoperative scores.

### 3.2. Disease Severity

Compared to the preoperative baseline, the UPDRS total score decreased by 16.7% at 12 months ($p = 0.01$), 19.9% at 18 months ($p = 0.02$), 15.3% at 24 months ($p = 0.03$). Although not significant, the improvement was sustained up to 96 months (8.9%) (Table 2).

### 3.3. DLA and QoL

Compared to the preoperative baseline, difficulties in DLA, assessed by the UPDRS-II, significantly decreased by 18.5% at 12 months ($p = 0.03$), 17.8% at 60 months ($p = 0.04$), 14.2% at 72 months ($p = 0.04$); QoL, assessed by means of the PDQL, significantly improved by 18.0% at 12 months ($p = 0.02$), and 22.4% at 60 months ($p = 0.04$) (Table 2, Figure 4).

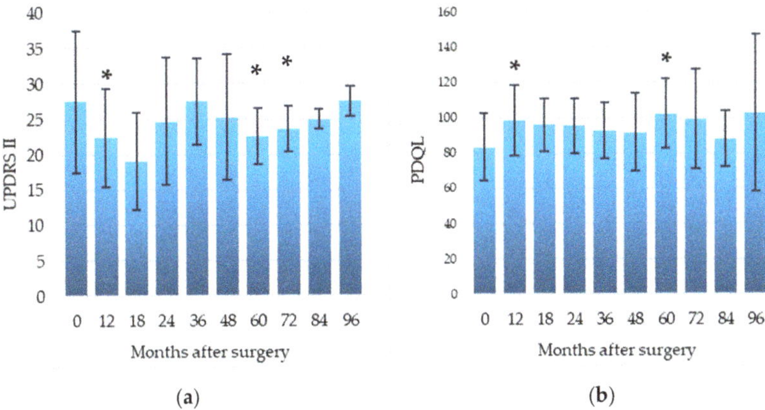


**Figure 4.** Impact of EMCS on daily living activities (**a**) and quality of life (**b**) in all patients, evaluated at different times after the implant. Error bars represent standard deviation. * $p < 0.05$ at comparisons between preoperative and postoperative scores, analyzed by means of the Wilcoxon signed-rank test.

### 3.4. Cognitive and Behvioral Outcome

No patients treated by EMCS developed dementia, as diagnosed according to the DSM-IV diagnostic criteria. No significant postoperative change was observed in terms of the UPDRS-I score. Postoperatively, no significant behavioral change was detected via scales assessing mood and anxiety or by clinical interviews. Improvements in MMSE (from 25.3 ± 2.6 to 26.7 ± 3.2, $p = 0.04$) and two subtests of episodic verbal memory (immediate and delayed recall of RAVLT, from 35.3 ± 6.8 to 45.9 ± 10.9 ($p = 0.03$) and from 7.0 ± 2.2 to 9.7 ± 3.0 ($p = 0.03$), respectively were detected 18 months postoperatively. Interestingly, a slight non-significant improvement was observed in the verbal phonological fluency task (from 17.3 ± 8.3 at baseline to 19.5 ± 11.9 at 12 months, 23.0 ± 13.6 at 18 months, 19.2 ± 11.0 at 36 months, 20.7 ± 9.3 at 60 months, 21.0 ± 15.6 at 96 months). No other change in cognitive performance on neuropsychological tests was detected across postoperative assessments (Table 3). Anticholinergic medications were not used in this population.

**Table 3.** Cognitive and behavioral evaluations in all patients, evaluated at different times after ECMS implant.

| | Baseline ($n$ = 9) | 12 m ($n$ = 9) | 18 m ($n$ = 9) | 36 m ($n$ = 7) | 60 m ($n$ = 5) | 96 m ($n$ = 2) |
|---|---|---|---|---|---|---|
| MMSE | 25.3 ± 2.6 | 25.9 ± 3.0 | 26.7 ± 3.2 * | 26.5 ± 3.9 | 25.0 ± 4.6 | 25.0 ± 2.8 |
| Digit Span forward | 4.6 ± 0.7 | 4.7 ± 1.0 | 4.4 ± 0.5 | 4.0 ± 0.6 | 5.0 ± 0 | 4.5 ± 0.7 |
| Digit Span backward | 3.2 ± 0.7 | 3.0 ± 0.7 | 3.0 ± 0.8 | 3.3 ± 0.8 | 3.0 ± 0 | 3.5 ± 0.7 |
| Corsi's Span forward | 4.8 ± 1.3 | 4.6 ± 1.2 | 4.4 ± 1.5 | 4.2 ± 1.2 | 4.0 ± 0 | 3.5 ± 0.7 |
| Corsi's Span backward | 3.8 ± 1.0 | 3.4 ± 1.2 | 3.1 ± 0.9 | 4.2 ± 1.5 | 3.7 ± 0.6 | 3.0 ± 0 |
| RAVLT immediate recall | 35.3 ± 6.8 | 42.2 ± 12.2 | 45.9 ± 10.9 * | 32.7 ± 5.7 | 43.7 ± 7.0 | 23.5 ± 2.1 |
| RAVLT delayed recall | 7.0 ± 2.2 | 8.5 ± 3.7 | 9.7 ± 3.0 * | 6.2 ± 1.2 | 10.0 ± 4.4 | 4.0 ± 0 |
| RPM '47 | 20.0 ± 7.3 | 21.5 ± 2.7 | 21.4 ± 6.9 | 22.7 ± 8.6 | 24.0 ± 9.2 | 19.5 ± 4.9 |
| Phonological verbal fluency | 17.3 ± 8.3 | 19.5 ± 11.9 | 23.0 ± 13.6 | 19.2 ± 11.0 | 20.7 ± 9.3 | 21.0 ± 15.6 |
| Semantic verbal fluency | 14.1 ± 4.0 | 12.7 ± 4.1 | 13.9 ± 3.4 | 14.0 ± 5.1 | 16.3 ± 6.8 | 15.0 ± 8.5 |
| mWCST criteria | 2.4 ± 1.6 | 2.6 ± 1.7 | 3.7 ± 2.1 | 2.8 ± 1.6 | 2.3 ± 2.3 | 1.5 ± 0.7 |
| mWCST total errors | 23.4 ± 10.4 | 20.5 ± 8.7 | 17.1 ± 11.6 | 19.8 ± 8.3 | 24.0 ± 17.6 | 21.0 ± 1.4 |
| mWCST perseverative errors | 8.1 ± 4.9 | 8.5 ± 5.6 | 7.0 ± 5.4 | 6.5 ± 3.3 | 6.6 ± 7.0 | 6.5 ± 3.5 |
| Stroop interference time | 33.7 ± 32.2 | 35.2 ± 17.8 | 35.0 ± 25.1 | 25.7 ± 14.0 | 48.3 ±55.2 | 37.5 ±10.6 |
| Stroop interference errors | 1.5 ± 3.9 | 2.2 ± 3.6 | 3.1 ± 4.8 | 2.3 ± 2.1 | 3.0 ± 3.3 | 6.0 ± 8.5 |
| Zung Depression Scale | 45.6 ± 13.6 | 46.4 ± 12.5 | 44.6 ± 8.1 | 47.3 ± 8.8 | 47.6 ± 9.3 | 55.0 ± 1.4 |
| Zung Anxiety Scale | 44.9 ± 12.9 | 47.5 ± 10.0 | 44.4 ± 8.0 | 46.8 ± 8.5 | 43.0 ± 11.5 | 44.0 ± 0 |

Abbreviations: m = months; MMSE = Mini-Mental State Examination; RAVLT = Rey's Auditory Verbal Learning Test; RPM = Raven's Progressive Matrices; mWCST = modified Wisconsin Card Sorting Test. * $p < 0.05$ at comparisons between preoperative and postoperative scores. m = months.

### 3.5. Safety

No serious adverse events occurred during the surgical procedure. During follow-up, no stimulation-related adverse events occurred. No patients presented seizures or epileptic discharges on postoperative electroencephalograms. In one patient, the whole implant was removed 36 months after surgery, due to an infection of the IPG pocket. Six battery replacements were performed, without complications.

## 4. Discussion

To our knowledge, the present study reports the longest clinical follow-up in PD patients treated by ECMS. This therapy induced a slight sustained motor improvement, with benefits to axial symptoms, and a reduction in motor complications and dopaminergic therapy. Despite its small size, the improvement was clinically meaningful as patients reported improvements in QoL and DLA. EMCS was safe, without detrimental effects on cognition and behavior.

In this study, we prospectively assessed the long-term efficacy and safety of EMCS in nine PD patients, followed-up for at least two years and up to eight years. EMCS showed beneficial effects on parkinsonian motor symptoms (measured by reductions in the UPDRS III in the off-medication condition) not only in the short-term, but also in the long-term follow-up. These findings, consistent with previous studies with a short-term follow-up [2,8,9,11,13], suggest that beneficial effects on motor symptoms may persist over time. Likewise, we found a progressive motor improvement up to three years after surgery. The sustained effect and progressive postoperative motor improvement may reflect processes of neural plasticity within the motor cortex, modulated by chronic stimulation [9].

Beneficial effects of EMCS were also observed on axial symptoms, which often display a poor response to levodopa and DBS [28]. This therapeutic role may be explained by the electrode position and polarity over M1. As demonstrated in 3D-volume conductor models, in bipolar EMCS, the anode of the dipole gives the largest motor response, exciting neural elements perpendicular to the electrode surface and corticofugal fibers [29]. The orientation of the electrode paddle over M1 (contact 3: 2 to 3 cm from the midline; contact 0: 4 cm more laterally) and electrode polarity (contact 3 was the anode, contact 0 was the cathode) may explain the effects on cortical areas, which are critical for axial motor function. EMCS may also modulate the activities of the cortical areas related to M1, such as the SMA, which plays a role in the pathophysiology of bradykinesia and axial symptoms [30]. Since bilateral activation of the SMA may be induced not only by bilateral but also by unilateral stimulation [6], both unilateral and bilateral EMCS stimulation may improve the axial symptoms [13,31].

As previously described [13], in our study, motor benefits were detectable only in the off-medication condition, similarly to what is seen in DBS patients. This finding rules out a synergic effect of EMCS and anti-parkinsonian medication and suggests that EMCS is able to induce motor improvements. In the on-medication condition, a progressive worsening of the UPDRS motor score was observed over time which is consistent with disease progression [32]. The absence of motor benefits detected in the on-medication condition may be explained by the remarkable effects of levodopa administration on parkinsonian motor symptoms, which might mask the slight beneficial effects of EMCS on such symptoms. Importantly, since the levodopa dose administered at each evaluation was higher than the usual morning levodopa dose, the UPDRS-III score at each evaluation might not reflect the usual on-medication condition.

In this study, EMCS was also effective in the management of PD motor complications (motor fluctuations and dyskinesias), as shown by a significant reduction in the postoperative UPDRS-IV score up to 72 months. The mean daily dose of dopaminergic drugs (measured by LEDD) showed a slight persistent decrease over time, which became significant postoperatively after 60 months. The reduction in motor complications might be explained by the slight postoperative decrease in LEDD and by a direct effect of EMCS on cortical plasticity, as suggested by transcranial magnetic stimulation studies [33,34]. Indeed, prospective EMCS studies suggested a direct effect of EMCS on dyskinesias [9–11]. In recent studies, the same electrodes have been used on M1 to detect cortical signs that are useful to close the loop for adaptive DBS, as a cortical narrowband gamma oscillation related to dyskinesias [35]. A direct effect of EMCS on dyskinesias may be postulated by the desynchronization of cortical narrowband gamma oscillations between the basal ganglia and cortical neurons.

EMCS also resulted in a sustained improvement of disease severity, measured by the total UPDRS score. The reduction in this score was significant up to 24 months, but despite the underlying disease progression, persisted up to 96 months of follow-up.

Not surprisingly, such beneficial effects of EMCS resulted in significant postoperative improvement of both activities of DLA (UPDRS-II score decrease) and QoL (PDQL score increase), which persisted in the long-term follow-up.

We found no detrimental effects on cognition or behavior over the long-term follow-up. We observed a significant improvement in MMSE and two subtests of episodic verbal memory only at the 18-month follow-up, which may be at least partially explained by a practice effect, and a slight (although not significant) improvement in the phonological fluency task over time, up to eight years after surgery. This latter finding is consistent with a previous observation of improvements in the fluency task in nine PD patients treated by EMCS with a 12-month follow-up. In particular, in two patients with unilateral stimulation of the left cerebral hemisphere, who showed an improvement in such fluency task at 3- and 12-month postoperative assessments [13]. In contrast, phonological fluency is consistently impaired after DBS of the subthalamic nucleus [36]. This interesting finding and other effects of EMCS on cognition should be assessed in future studies. Since EMCS induced beneficial effects on language in cases of pure akinesia with gait freezing [31] and improvements in tasks of episodic verbal memory and working memory in two patients with PD [37], it is possible that EMCS might induce beneficial effects on various cognitive processes across distinct disorders.

This study also confirms that EMCS is safe in the long-term follow-up, since adverse events occur very rarely [15]. The only adverse event in our sample was an infection of the IPG pocket, which required the removal of the whole implant three years postoperatively.

*Limitations*

This study has several limitations. The first is the open-label study design, which does not allow us to rule out a placebo effect, which is nevertheless unlikely given the persistence of beneficial effects over the long-term follow-up. In fact, while it is possible that a placebo effect would occur in the first few months after the intervention, we found a sustained motor effect of EMCS beyond five years after the intervention. Moreover, this was also the case in patients with neurodegenerative disease, which usually leads to motor aggravation over the years. Indeed, beneficial effects on specific motor tasks (such as capability of rising from a chair), that did not occur immediately, but rather several months after the stimulation being switched on, reduce any potential biases deriving from the expectations of patients and physicians. Other limitations include the small number of enrolled patients and the lack of a control group. Few patients were enrolled because this treatment is intended for a highly selected population of PD patients who are not eligible for other advanced therapies. As most of the significant $p$ values were marginally lower than 0.05, due to the small number of cases, they should be interpreted with caution, considering the risk of a type I error deriving from multiple comparisons. Furthermore, the availability of other effective treatments for the motor complications of PD, such as DBS or levodopa/carbidopa intestinal gel, limits the possibility to obtain a control group for standard medical treatments. Finally, the high percentage of patients lost at follow-up is an inevitable limit of long-term studies with advanced PD patients, as reported in many DBS studies. In fact, the high degree of motor disability, which characterizes this stage of disease, often leads to difficulties in reaching the center, institutionalization, and death.

## 5. Conclusions

EMCS may be a useful treatment option for advanced PD patients to reduce motor complications and dopaminergic therapy. It provides a small but sustained effect on motor symptoms (both appendicular and axial symptoms), with improvements in DLA and QoL.

Although EMCS seems to be less effective on motor symptoms than other advanced therapies (DBS or levodopa/carbidopa intestinal gel), it may be a safe alternative when these options are contraindicated or refused by patients.

Prospective controlled studies in larger samples of PD patients, possibly with prominent axial symptoms, evaluated by specific scales for axial motor symptoms and dyskinesias, are needed to further define clinical indications for EMCS in PD.

**Author Contributions:** Conceptualization, C.P., F.B., A.F., A.R.B., A.D., B.C., P.C. and T.T.; methodology, C.P., A.F., A.D. and B.C.; formal analysis, F.B. and E.D.S.; investigation, C.P., A.F., A.D. and B.C.; data curation, C.P., F.B., E.D.S., A.D. and B.C.; writing—original draft preparation, F.B. and D.M.; writing—review and editing, C.P., E.D.S., A.F., A.R.B., A.D., B.C., P.C. and T.T.; supervision, A.R.B., P.C., T.T. All authors have read and agreed to the published version of the manuscript.

**Funding:** This research received no external funding.

**Institutional Review Board Statement:** The study was conducted according to the guidelines of the Declaration of Helsinki, and approved by the Ethics Committee of Fondazione Policlinico Universitario Agostino Gemelli IRCCS in Rome (protocol code #400-A763).

**Informed Consent Statement:** Informed consent was obtained from all subjects involved in the study.

**Data Availability Statement:** Anonymized data will be shared with qualified external researchers, after approval of their requests.

**Conflicts of Interest:** A.F. received honoraria from Medtronic. All other authors declare no conflict of interest.

## References

1. Canavero, S. Invasive Cortical Stimulation for Parkinson's Disease (PD): Why, where and how. In *Textbook of Therapeutic Cortical Stimulation*; Nova Science: New York, NY, USA, 2009; pp. 217–227.
2. Pagni, C.A.; Altibrandi, M.G.; Bentivoglio, A.; Caruso, G.; Cioni, B.; Fiorella, C.; Insola, A.; Lavano, A.; Maina, R.; Mazzone, P.; et al. Extradural Motor Cortex Stimulation (EMCS) for Parkinson's disease. History and first results by the study group of the Italian neurosurgical society. *Funct. Rehabil. Neurosurg. Neurotraumatol.* **2005**, *93*, 113–119. [CrossRef]
3. Priori, A.; Lefaucheur, J.-P. Chronic epidural motor cortical stimulation for movement disorders. *Lancet Neurol.* **2007**, *6*, 279–286. [CrossRef]
4. Cioni, B. Motor cortex stimulation for Parkinson's disease. *Oper. Neuromodul.* **2007**, *97*, 233–238. [CrossRef]

5. Drouot, X.; Oshino, S.; Jarraya, B.; Besret, L.; Kishima, H.; Remy, P.; Dauguet, J.; Lefaucheur, J.P.; Dollé, F.; Condé, F.; et al. Functional Recovery in a Primate Model of Parkinson's Disease following Motor Cortex Stimulation. *Neuron* **2004**, *44*, 769–778. [CrossRef] [PubMed]
6. Fasano, A.; Piano, C.; De Simone, C.; Cioni, B.; Di Giuda, D.; Zinno, M.; Daniele, A.; Meglio, M.; Giordano, A.; Bentivoglio, A.R. High frequency extradural motor cortex stimulation transiently improves axial symptoms in a patient with Parkinson's disease. *Mov. Disord.* **2008**, *23*, 1916–1919. [CrossRef] [PubMed]
7. Canavero, S.; Paolotti, R. Extradural motor cortex stimulation for advanced Parkinson's disease: Case report. *Mov. Disord.* **2000**, *15*, 169–171. [CrossRef]
8. Canavero, S.; Paolotti, R.; Bonicalzi, V.; Castellano, G.; Greco-Crasto, S.; Rizzo, L.; Davini, O.; Zenga, F.; Ragazzi, P. Extradural motor cortex stimulation for advanced Parkinson disease. *J. Neurosurg.* **2002**, *97*, 1208–1211. [CrossRef] [PubMed]
9. Cilia, R.; Landi, A.; Vergani, F.; Sganzerla, E.; Pezzoli, G.; Antonini, A. Extradural motor cortex stimulation in Parkinson's disease. *Mov. Disord.* **2007**, *22*, 111–114. [CrossRef]
10. Arle, J.E.; Apetauerova, D.; Zani, J.; Deletis, D.V.; Penney, D.L.; Hoit, D.; Gould, C.; Shils, J.L. Motor cortex stimulation in patients with Parkinson disease: 12-month follow-up in 4 patients. *J. Neurosurg.* **2008**, *109*, 133–139. [CrossRef]
11. Gutiérrez, J.C.; Seijo, F.J.; Vega, M.A.Á.; Gonzalez, F.F.; Aragoneses, B.L.; Blázquez, M. Therapeutic extradural cortical stimulation for Parkinson's Disease: Report of six cases and review of the literature. *Clin. Neurol. Neurosurg.* **2009**, *111*, 703–707. [CrossRef]
12. Moro, E.; Schwalb, J.M.; Piboolnurak, P.; Poon, Y.-Y.W.; Hamani, C.; Hung, S.W.; Arenovich, T.; Lang, A.E.; Chen, R.; Lozano, A.M. Unilateral subdural motor cortex stimulation improves essential tremor but not Parkinson's disease. *Brain* **2011**, *134*, 2096–2105. [CrossRef] [PubMed]
13. Bentivoglio, A.R.; Fasano, A.; Piano, C.; Soleti, F.; Daniele, A.; Zinno, M.; Piccininni, C.; De Simone, C.; Policicchio, D.; Tufo, T.; et al. Unilateral Extradural Motor Cortex Stimulation Is Safe and Improves Parkinson Disease at 1 Year. *Neurosurgery* **2012**, *71*, 815–825. [CrossRef] [PubMed]
14. Cioni, B.; Tufo, T.; Bentivoglio, A.; Trevisi, G.; Piano, C. Motor cortex stimulation for movement disorders. *J. Neurosurg. Sci.* **2016**, *60*, 230–241.
15. Lavano, A.; Guzzi, G.; De Rose, M.; Romano, M.; Della Torre, A.; Vescio, G.; Deodato, F.; Lavano, F.; Volpentesta, G. Minimally invasive motor cortex stimulation for Parkinson's disease. *J. Neurosurg. Sci.* **2017**, *61*, 77–87.
16. Samotus, O.; Parrent, A.; Jog, M. Spinal Cord Stimulation Therapy for Gait Dysfunction in Advanced Parkinson's Disease Patients. *Mov. Disord.* **2018**, *33*, 783–792. [CrossRef] [PubMed]
17. Hughes, A.J.; Daniel, S.E.; Kilford, L.; Lees, A.J. Accuracy of clinical diagnosis of idiopathic Parkinson's disease: A clini-co-pathological study of 100 cases. *J. Neurol. Neurosurg. Psychiatry* **1992**, *55*, 181–184. [CrossRef] [PubMed]
18. Fahn, S.; Elton, R.; Members of the UPDRS Development Committee. The Unified Parkinson's Disease Rating Scale. In *Recent Developments in Parkinson's Disease*; Macmillan Healthcare Information: Florham Park, NJ, USA, 1987; Volume 2, pp. 153–163.
19. Defer, G.L.; Widner, H.; Marie, R.M.; Remy, P.; Levivier, M. Core assessment program for surgical interventional therapies in Parkinson's disease (CAPSIT-PD). *Mov. Disord.* **1999**, *14*, 572–584. [CrossRef]
20. Folstein, M.F.; Folstein, S.E.; McHugh, P.R. "Mini-mental state". A practical method for grading the cognitive state of patients for the clinician. *J. Psychiatr. Res.* **1975**, *12*, 189–198. [CrossRef]
21. American Psychiatric Association. *Diagnostic and Statistical Manual of Mental Disorders*, 5th ed.; DSM-5; American Psychiatric Association: Washington, DC, USA, 2013; p. 5.
22. Cioni, B.; Meglio, M.; Perotti, V.; De Bonis, P.; Montano, N. Neurophysiological aspects of chronic motor cortex stimulation. *Neurophysiol. Clin. Neurophysiol.* **2007**, *37*, 441–447. [CrossRef]
23. De Boer, A.G.; Wijker, W.; Speelman, J.D.; De Haes, J.C. Quality of life in patients with Parkinson's disease: Development of a questionnaire. *J. Neurol. Neurosurg. Psychiatry* **1996**, *61*, 70–74. [CrossRef]
24. Tomlinson, C.L.; Stowe, R.; Patel, S.; Rick, C.; Gray, R.; Clarke, C.E. Systematic review of levodopa dose equivalency report-ing in Parkinson's disease. *Mov. Disord.* **2010**, *25*, 2649–2653. [CrossRef]
25. Daniele, A.; Albanese, A.; Contarino, M.F.; Zinzi, P.; Barbier, A.; Gasparini, F.; Romito, L.M.A.; Bentivoglio, A.R.; Scerrati, M. Cognitive and behavioural effects of chronic stimulation of the subthalamic nucleus in patients with Parkinson's disease. *J. Neurol. Neurosurg. Psychiatry* **2003**, *74*, 175–182. [CrossRef]
26. Zung, W.W.K. A Self-Rating Depression Scale. *Arch. Gen. Psychiatry* **1965**, *12*, 63–70. [CrossRef]
27. Zung, W.W. A Rating Instrument for Anxiety Disorders. *J. Psychosom. Res.* **1971**, *12*, 371–379. [CrossRef]
28. Fasano, A.; Aquino, C.C.; Krauss, J.K.; Honey, C.R.; Bloem, B.R. Axial disability and deep brain stimulation in patients with Parkinson disease. *Nat. Rev. Neurol.* **2015**, *11*, 98–110. [CrossRef] [PubMed]
29. Manola, L.; Holsheimer, J.; Veltink, P.; Buitenweg, J.R. Anodal vs cathodal stimulation of motor cortex: A modeling study. *Clin. Neurophysiol.* **2007**, *118*, 464–474. [CrossRef]
30. Shirota, Y.; Ohtsu, H.; Hamada, M.; Enomoto, H.; Ugawa, Y.; For the Research Committee on rTMS Treatment of Parkinson's Disease. Supplementary motor area stimulation for Parkinson disease: A randomized controlled study. *Neurology* **2013**, *80*, 1400–1405. [CrossRef]
31. Piano, C.; Fasano, A.; Daniele, A.; Di Giuda, D.; Ciavarro, M.; Tufo, T.; Zinno, M.; Bentivoglio, A.R.; Cioni, B. Extradural motor cortex stimulation improves gait, speech, and language in a patient with pure akinesia. *Brain Stimul.* **2018**, *11*, 1192–1194. [CrossRef] [PubMed]

32. Schrag, A.; Dodel, R.; Spottke, A.; Bornschein, B.; Siebert, U.; Quinn, N.P. Rate of clinical progression in Parkinson's disease. A prospective study. *Mov. Disord.* **2007**, *22*, 938–945. [CrossRef] [PubMed]
33. Koch, G. rTMS effects on levodopa induced dyskinesias in Parkinson's disease patients: Searching for effective cortical targets. *Restor. Neurol. Neurosci.* **2010**, *28*, 561–568. [CrossRef]
34. Calabresi, P.; Ghiglieri, V.; Mazzocchetti, P.; Corbelli, I.; Picconi, B. Levodopa-induced plasticity: A double-edged sword in Parkinson's disease? *Philos. Trans. R. Soc. B Biol. Sci.* **2015**, *370*, 20140184. [CrossRef] [PubMed]
35. Swann, N.C.; De Hemptinne, C.; Thompson, M.C.; Miocinovic, S.; Miller, A.M.; Gilron, R.; Ostrem, J.L.; Chizeck, H.J.; Starr, P.A. Adaptive deep brain stimulation for Parkinson's disease using motor cortex sensing. *J. Neural Eng.* **2018**, *15*, 046006. [CrossRef] [PubMed]
36. Fasano, A.; Romito, L.M.; Daniele, A.; Piano, C.; Zinno, M.; Bentivoglio, A.R.; Albanese, A. Motor and cognitive outcome in patients with Parkinson's disease 8 years after subthalamic implants. *Brain* **2010**, *133*, 2664–2676. [CrossRef] [PubMed]
37. Piano, C.; Ciavarro, M.; Bove, F.; Di Giuda, D.; Cocciolillo, F.; Bentivoglio, A.R.; Cioni, B.; Tufo, T.; Calabresi, P.; Daniele, A. Extradural Motor Cortex Stimulation might improve episodic and working memory in patients with Parkinson's disease. *NPJ Park. Dis.* **2020**, *6*, 26. [CrossRef] [PubMed]

*Article*

# The Effects of 10 Hz and 20 Hz tACS in Network Integration and Segregation in Chronic Stroke: A Graph Theoretical fMRI Study

**Cheng Chen [1], Kai Yuan [1], Winnie Chiu-wing Chu [2] and Raymond Kai-yu Tong [1,*]**

[1] Department of Biomedical Engineering, The Chinese University of Hong Kong, Hong Kong 999077, China; chen_cheng@link.cuhk.edu.hk (C.C.); kaiyuan@link.cuhk.edu.hk (K.Y.)
[2] Department of Imaging and Interventional Radiology, The Chinese University of Hong Kong, Hong Kong 999077, China; winniechu@cuhk.edu.hk
* Correspondence: kytong@cuhk.edu.hk

**Abstract:** Transcranial alternating current stimulation (tACS) has emerged as a promising technique to non-invasively modulate the endogenous oscillations in the human brain. Despite its clinical potential to be applied in routine rehabilitation therapies, the underlying modulation mechanism has not been thoroughly understood, especially for patients with neurological disorders, including stroke. In this study, we aimed to investigate the frequency-specific stimulation effect of tACS in chronic stroke. Thirteen chronic stroke patients underwent tACS intervention, while resting-state functional magnetic resonance imaging (fMRI) data were collected under various frequencies (sham, 10 Hz and 20 Hz). The graph theoretical analysis indicated that 20 Hz tACS might facilitate local segregation in motor-related regions and global integration at the whole-brain level. However, 10 Hz was only observed to increase the segregation from whole-brain level. Additionally, it is also observed that, for the network in motor-related regions, the nodal clustering characteristic was decreased after 10 Hz tACS, but increased after 20 Hz tACS . Taken together, our results suggested that tACS in various frequencies might induce heterogeneous modulation effects in lesioned brains. Specifically, 20 Hz tACS might induce more modulation effects, especially in motor-related regions, and they have the potential to be applied in rehabilitation therapies to facilitate neuromodulation. Our findings might shed light on the mechanism of neural responses to tACS and facilitate effectively designing stimulation protocols with tACS in stroke in the future.

**Keywords:** transcranial alternating current stimulation; chronic stroke; functional magnetic resonance imaging; graph theory; segregation and integration of brain networks

**Citation:** Chen, C.; Yuan, K.; Chu, W.C.-w.; Tong, R.K.-y. The Effects of 10 Hz and 20 Hz tACS in Network Integration and Segregation in Chronic Stroke: A Graph Theoretical fMRI Study. *Brain Sci.* **2021**, *11*, 377. https://doi.org/10.3390/brainsci11030377

Academic Editor: Ulrich Palm

Received: 13 February 2021
Accepted: 13 March 2021
Published: 16 March 2021

**Publisher's Note:** MDPI stays neutral with regard to jurisdictional claims in published maps and institutional affiliations.



## 1. Introduction

Nowadays, stroke is the leading cause of death worldwide, and survivors undergo different dysfunctions, especially in the motor aspect [1]. Hence, it is essential for stroke subjects to restore functional abilities in order to diminish the inconvenience in daily-living activities. The existence of neuroplasticity, which is an intrinsic property of the human brain to change its function and reorganize after a lesion forms, makes this possible [2]. Meanwhile, there have been various rehabilitation strategies proposed, including conventional physical therapies as well as advanced robot-assisted methods [3,4]. Except for these therapies that influence brain reorganization in a round-about way, the transcranial current stimulation (tCS), which non-invasively modulates the activity of the brain, has attracted increasing attention [5].

Among many available tCS techniques, transcranial direct current stimulation (tDCS) and transcranial alternating current stimulation (tACS) are two typical methods that have intrigued the researchers in the field of neuroscience. The core difference between these two simulations is the form of the currents elicited. In tDCS, a direct current flows from anodal to cathodal electrodes. The effect of tDCS is often related to membrane depolarization, which leads to an increase of the excitation in neurons underneath anodal electrode, but the inhibition of neurons under cathodal electrode [6,7]. When compared with tDCS, tACS has not been thoroughly investigated due to its potentially complicated mechanism and interfering with inherent frequency-specific oscillations in the human brain, let alone the effects in patients with neurological disorders, including stroke. While the underlying neurophysiological mechanism is unknown, the stimulation effect is often attributed to the manipulation and entrainment of intrinsic oscillations in the brain [8]. In the human brain, the communication within and between brain regions was facilitated by synchronized oscillatory activities and the components with various frequencies playing different roles in functioning [9]. This implied that the stimulation effect of tACS might differ, depending on the eliciting frequency of the alternating current. At the same time, it has been indicated that the activity of alpha (8–12 Hz) and beta (13–30 Hz) frequency is prominent in the sensorimotor cortex in the resting-state [10]. Therefore, in our study, because the tACS was imposed on the primary motor area (M1), 10 Hz and 20 Hz as representative alpha and beta stimulating frequencies, respectively, were adopted. Some previous studies have investigated the stimulation effect of 10 Hz and 20 Hz on M1 in healthy subject. It was observed that beta-tACS could be used to induce neurophysiologically detectable state-dependent enhancement effects [11]. 10 Hz and 20 Hz tACS could both facilitate motor sequence learning during a serial reaction time task (SRTT). Additionally, 20 Hz tACS could further stabilize motor control to retain the initial learning rate under interference [12]. Because of the existence of frequency difference, it is reasonable to expect differential effects and a recent study has suggested that this effect exists not only in motor behavior, but also in M1 excitability [13]. However, the tACS studies on stroke subjects were quite scarce, which implied that further investigation is needed.

It is observed that the effects of tCS are not restricted to the stimulated sites, and it also has an impact on the brain network [14]. Besides, for tACS, if the stimulation frequency matches the endogenous oscillation frequency, more pronounced oscillatory effect could be found at the cortical network level [15]. Hence, it is meaningful to explore the stimulation effect of tACS from the brain network perspective. To investigate the brain networks, functional magnetic resonance imaging (fMRI), especially resting-state fMRI, which measures the blood oxygen level (BOLD) signal of different regions in the resting-state, has been widely utilized [16,17]. On the other hand, the graph theory approach could provide an efficient perspective to model and understand the information of integration and segregation property, as well as regional communication in complex networks. Hereby, graph theory could be applied to brain network that is derived from fMRI to provide a framework to evaluate the properties of the constructed network, which has been generally adopted in human neuroscience [18].

Similar to tDCS, which has been applied in rehabilitation therapies [19], tACS could also be a powerful auxiliary tool added to existing therapies. However, before widely utilizing tACS, it is of considerale significance to understand the underlying mechanism of how tACS influences the patterns of patient's brain. The current study aims to thoroughly explore the frequency-specific stimulation effect of tACS on chronic stroke subjects using graph theory analysis in resting-state fMRI and investigated the modulation effect in motor-related cortical regions. At the same time, the integration and segregation characteristics of the network at the whole-brain level were also investigated. We hypothesized the potential differential effect of 10 Hz and 20 Hz tACS as well as the resulting modulation difference in brain networks in motor-related regions and at the whole-brain level.

## 2. Materials and Methods

### 2.1. Subjects

Thirteen chronic stroke patients (eight males, mean age = 61 ± 10 years) with the right ($n$ = 7) or left ($n$ = 6) hemisphere impairment were recruited from the local community. The inclusion criteria were: (1) first-ever stroke, (2) sufficient cognitive function to understand instructions (Montreal Cognitive Assessment, Moca score ≥ 22), (3) a single unilateral brain lesion, and (4) more than six months before the experiment. The exclusion criteria were: (1) history of alcohol, drug abuse, or epilepsy, (2) severe cognitive deficits, and (3) any contraindication to tACS or MRI. Fugl-Meyer Assessment for upper-extremity (FMA) and Action Research Arm Test (ARAT) were utilized to assess the motor function of the paretic upper limbs for all stroke patients. The lesion map, detailing demographics and clinical properties of the participants, could be found in the Supplementary Materials (Figure S1 and Table S1). This study was approved by the Joint Chinese University of Hong Kong-New Territories East Cluster Clinical Research Ethics Committee. This study was registered at https://clinicaltrials.gov (accessed on 13 February 2021) (NCT04638192). All of the subjects gave written consent before the intervention.

### 2.2. tACS Intervention

According to the international 10–20 system, one electrode ($5 \times 5$ cm$^2$) was positioned over the ipsilesional M1, while the return one was placed over the contralesional supraorbital ridge (Figure 1B). Both of the electrodes were fixed to the patient's scalp with straps before MRI scanning. For 10 Hz and 20 Hz tACS, an MRI compatible DC-stimulator (NeuroConn GmbH, Ilmenau, Germany) was utilized to deliver the current with 1-mA peak-to-peak intensity for 20 min. The 30-s ramp-up and ramp-down periods at the beginning and end of stimulation, respectively, were adopted. For the sham group, the stimulator was switched off after the 30-s ramp-up period to induce typical tingling sensation [20] (Figure 1C). Each subject would undergo these three stimulation protocols in a randomized order. Meanwhile, the stimulation conditions were performed with a wash-off period of at least one week between each other [21].

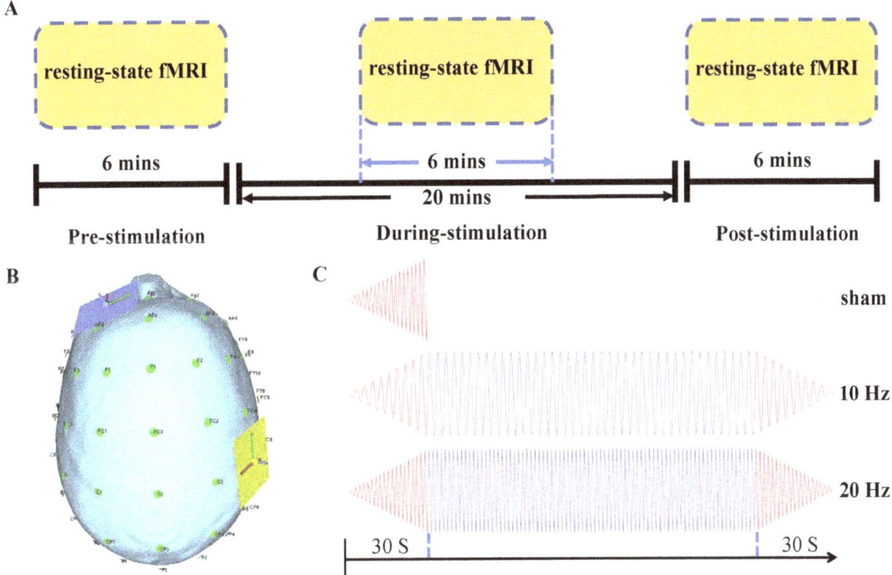


**Figure 1.** (**A**) The protocol of MRI acquisition and transcranial alternating current stimulation (tACS) intervention. (**B**) The montage of stimulation electrodes (drawn by SimNIBS [22]). The yellow one was put on the ipsilesional primary motor cortex, and the blue one is on the contralesional supraorbital ridge. (**C**) The currents of sham, 10 Hz, and 20 Hz.

### 2.3. Image Acquisition and Preprocessing

A 3T Philips MR scanner (Achieva TX, Philips Medical System, Best, The Netherlands) with an eight-channel head coil was used to acquire high resolution T1-weighted anatomical images (TR/TE = 7.47/3.45 ms, flip angle = 8°, 308 slices, voxel size = $0.6 \times 1.042 \times$

1.042 mm$^3$) while using a T1-TFE sequence (ultrafast spoiled gradient echo pulse sequence), and BOLD fMRI images (TR/TE = 2000/30 ms, flip angle = 70°, 37 slices/volume, voxel size = 2.8 × 2.8 × 3.5 mm$^3$) using a GE-EPI sequence (gradient-echo echo-planar-imaging sequence). Resting-state fMRI data were acquired before, during, and immediately after stimulation. Each run lasted for 6 min. with 180 volumes for each fMRI image (Figure 1A). During acquisition, the patient was instructed to keep awake while focusing on a white cross presented in black background.

The fMRI data were mainly preprocessed using DPARSF toolbox [23]. The first four volumes were removed to assure the remaining volumes of fMRI were at magnetization steady state. Subsequently, the remaining volumes were corrected with slice timing and realigned to correct head motion. Nuisance variables, including white matter, cerebrospinal fluid (CSF), global mean signal, and Friston 24 head motion parameters, were then regressed out [24]. To further control for head motion, the scrubbing process were performed for the volumes with framewise displacement (FD) value exceeding 0.3 [25]. If over 25% of all the volumes exceed the threshold, the corresponding data would be discarded, and no data were discarded in our study. Afterward, the fMRI data were aligned to anatomical images. To remove higher frequency physiological noise and lower frequency scanner drift, detrending, and the 0.01–0.1 Hz band-pass temporal filtering was performed [26]. Subsequently, the functional images were normalized to the Montreal Neurological Institute (MNI) template, resliced to 2 × 2 × 2 mm$^3$ voxels, and then spatially smoothed with a Gaussian kernel with a full-width at half-maximum (FWHM) of 6 mm. The fMRI data of subjects who had left-hemispheric lesions were flipped along the midsagittal plane using MRIcron (www.mccauslandcenter.sc.edu/mricro/mricron) for group statistical analysis, so that the lesions of all subjects were in the right hemisphere.

### 2.4. Graph Theory Analysis

#### 2.4.1. Construction of Brain Functional Networks

In order to investigate the modification of brain functioning induced by tACS located at the lesioned motor area, the whole brain was first parcelled into 116 regions based on Automated Anatomical Labeling (AAL) atlas [27] to construct the network at the whole-brain level, and 20 regions of interest (ROIs) related to motor function based on previous studies [28] were extracted to constitute the nodes of the network in motor-related regions (Listed in the Supplementary Materials Table S2). The mean time series of each ROI was averaged. The temporal correlation matrix for each subject under each condition was obtained by calculating the Pearson correlation coefficients between the time courses of each pair of regions.

In graph theory, an adjacency matrix was often adopted to characterize the structure of the graph. In the present study, we would threshold the fMRI temporal correlation matrix to acquire group adjacency matrix as well as individual adjacency matrix for each time point under each stimulation condition. First of all, a Fisher's r-to-z transform was utilized to map correlation r value to z score value for all individual correlation matrices to improve normality [29]. A two-tailed one-sample t-test was then used to test the significance of the correlation different from zero for each possible pair of nodes across subjects. A significant level of $p < 0.01$ with Bonferroni correction was adopted to threshold the temporal correlation matrices to obtain the binarized group adjacency matrices. The ratio of the number of existing edges and the maximum number of all edges derived from the resulting group adjacency matrix were used to binarize individual temporal correlation matrix in a proportional-threshold way [30]. Therefore, the inherent structural property of individual adjacency matrix could be consistent with the group adjacency matrix to maximally reduce the bias that is caused by selecting a priori thresholding parameter [31].

#### 2.4.2. Graph Theoretical Measures

After constructions of brain networks, several measures that characterize the property of modular organization and nodes were evaluated. All graph theory analysis was conducted while using Brain Connectivity Toolbox [32] thta was implemented in MATLAB (The MathWorks Inc., Natick, MA, USA).

Modularity. It is often assumed that a brain network always works with several well-partitioned modules or communities, and each community is responsible for specialized functional processing. Modularity is to measure such goodness of graph partitioning, which is defined as [33]:

$$Q = \sum_{u \in M} \left[ e_{uu} - \left( \sum_{v \in M} e_{uv} \right)^2 \right] \quad (1)$$

where $u$ and $v$ represent the specific modules in the set of all subdivided non-overlapping modules $M$, $e_{uv}$ represents the proportion of all links connecting nodes in module $u$ and $v$, respectively. $Q$ is normally treated as an objective function to the maximize the number of within-module links and minimize the number of inter-module links to optimally subdivide the graph into communities.

Within-module degree z-score. Based on the community assignment of all nodes, the role of a specific node could be determined with respect to its own community as well as other communities. The within-module degree z-score is a classical measure to characterize how 'well-connected' a specific node is to other nodes that belong to the same community. Normally speaking, a high value of within-module z-score indicates dense within-module linking [34]. It is defined as:

$$z_i = \frac{k_i(m_i) - \bar{k}(m_i)}{\sigma^{k(m_i)}} \quad (2)$$

where $k_i$ represents the degree of node $i$, which is equal to the number of links connected to node $i$ in the whole network, $k_i(m_i)$ represents the number of links between node i and other nodes in the same module, and $\bar{k}(m_i)$ and $\sigma^{k(m_i)}$ represent the mean and standard deviation of degree distribution in the same module, respectively.

Participation coefficient. Some of the nodes might not merely connect with nodes near them within the same community, but also have connections with nodes in other communities. The participation coefficient is used to evaluate such diversity of inter-modular interconnection for an individual node. Complementary to within-module degree z-score, the participation coefficient characterizes 'how-distributed' the links of a specific node among various communities [35], which is defined as:

$$P_i = 1 - \sum_{m \in M} \left( \frac{k_i(m)}{k_i} \right)^2 \quad (3)$$

where $k_i(m)$ represents the number of links connecting node $i$ and all other nodes in module $m$. It is noted that, if almost all links of a node are restricted within its own community, the participation coefficient of this node is close to 0. Otherwise, the participation coefficient of the node with almost uniformly distributed links tends to be 1.

Clustering coefficient. The clustering coefficient is a kind of measure of segregation that is related to the number of triangles in the network. The nodal clustering coefficient is equivalent to the fraction of the node's neighbors that are also neighbors of each other, which is defined as [36]:

$$C_i = \frac{2t_i}{k_i(k_i - 1)} \quad (4)$$

where $t_i = \frac{1}{2} \sum_{j,h \in N} a_{ij} a_{ih} a_{jh}$ ($a_{ij}$ indicates the link between node $i$ and node $j$, and $N$ means the set of all nodes in the network) represents the number of triangles surrounding node $i$.

Local efficiency. Local efficiency is a nodal measure to characterize the efficiency of local information transmission and mainly focus on the property of communication among neighbors for a specific node [37]. It is defined as:

$$LE_i = \frac{\sum_{j,h \in N, j \neq h} a_{ij} a_{ih} \left[ d_{jh}(N_i) \right]^{-1}}{k_i(k_i - 1)} \tag{5}$$

where $d_{jh}(N_i)$ indicates the shortest path length of node $j$ and node $h$, which contains the only neighbor of node $i$.

Specifically, since we would like to investigate the modulation effect of tACS in motor-related brain regions, we mainly focused on the analysis of the distribution of nodal metrics. The corresponding measure for a node would be calculated by averaging all values across all subjects. The nodal measures during and after stimulation were baseline corrected by subtracting the corresponding values that were derived from the pre-stimulation session to characterize the modulation effect. Besides, the integration and segregation characteristics of the network at the whole-brain level were also investigated.

### 2.5. Statistical Analysis

The statistical tests were conducted using SPSS 25 (IBM SPSS, NY, USA). A two-way repeated-measure Analysis of Variance (ANOVA) with factors of stimulation (sham, 10 Hz, and 20 Hz) and time (during and post) was carried out in order to investigate the change of distributions of graph theoretically nodal measures, including within-module degree z-score, participation coefficient, clustering coefficient, and local efficiency. The Greenhouse–Geisser correction would be adopted if Mauchly's test of sphericity was significant. Paired $t$-tests were applied as *post-hoc* tests to examine whether there exists significant difference in different combinations of three stimulation conditions for each time point. The significance level was set at $p < 0.05$. Bonferroni correction was used to counteract the problem of multiple comparisons.

## 3. Results

### 3.1. Community Structure

First of all, we investigated the modulation effect of different stimulation protocols that were imposed on the structure of the network in motor-related regions, which is related to the community assignment and affiliation.

When no stimulation (sham) applied to the brain, it could be observed that the community structure and node affiliation to specific functional modules did not change significantly along with time (Figure 2). Different from sham stimulation, there existed evidence showing that 10 Hz stimulation tended to uniformly distribute the nodes to different communities. In some specific regions, the nodes that belong to various communities became more miscellaneous (Figure 3). Interestingly, opposite to 10 Hz stimulation, 20 Hz stimulation showed the ability to merge sub-modules into a larger community. All of the nodes in the same community dominantly located in one specific region and the space encompassed by nodes of different communities scarcely overlap with each other after 20 Hz stimulation (Figure 4).

### 3.2. Graph Theoretically Nodal Measures

It is often assumed that the role of a node could be determined by its position in the *P-z* parameter plane, which is called *P-z* plot [34]. Hence, we investigated the change of distributions of within-module degree z score and participation coefficient, respectively. From the *P-z* plots illustrated in Figure 2–4, it could be observed that, for within-module degree z-score, the distributions did not show significant fluctuations, and the mean values of all conditions were located around zero. The repeated measure ANOVA also did not show any significant effect in within-module degree z-score. $Post-hoc$ tests also indicated that no significant difference was observed for pairwise comparison.

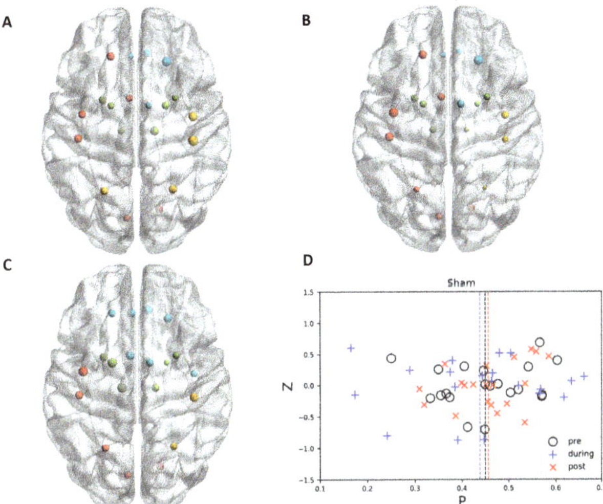


**Figure 2.** The node topology and community structure of sham stimulation in motor-related regions at (**A**) pre, (**B**) during, and (**C**) post time points (drawn by BrainNet Viewer [38].). Left orientation represents left side of the brain. Node size was determined by local efficiency. The nodes with the same color belonged to the same community in each subplot. (**D**) illustrated *P-z* plot for sham stimulation and the dashed line represents the mean *P* value for each time point.

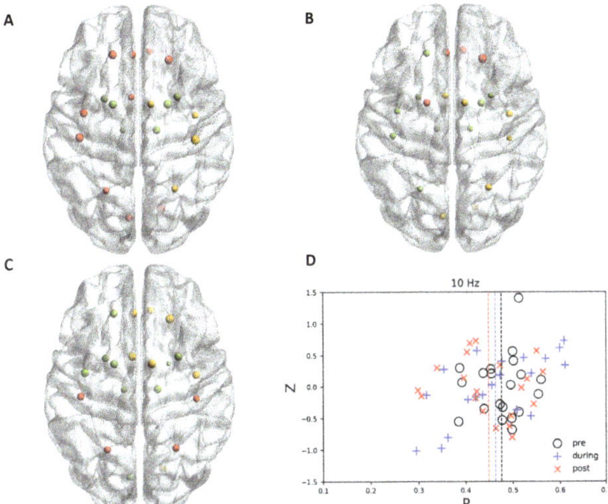


**Figure 3.** The node topology and community structure of 10 Hz stimulation in motor-related regions at (**A**) pre, (**B**) during, and (**C**) post time points. Left orientation represents left side of the brain. Panel (**D**) illustrated *P-z* plot for 10 Hz stimulation.

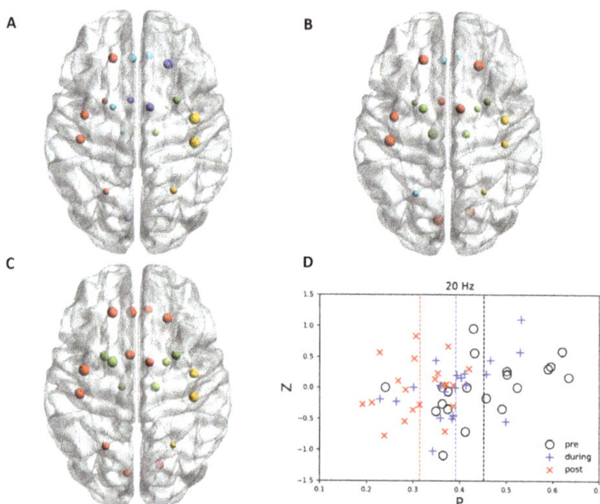


**Figure 4.** The node topology and community structure of 20 Hz stimulation in motor-related regions at (**A**) pre, (**B**) during, and (**C**) post time points. Left orientation represents left side of the brain. Panel (**D**) illustrated *P-z* plot for 20 Hz stimulation.

However, it is observed that, in 20 Hz stimulation, the distribution of participation coefficient shifted to a small value along with time. The repeated measure ANOVA of participation coefficient change indicated significant *stimulation* − *time* interaction effect ($F(2, 38) = 5.527$, $p < 0.008$). *Post* − *hoc* tests indicated a significant difference between *sham* and 20 Hz ($p < 0.001$, Bonferroni corrected) as well as 10 Hz and 20 Hz ($p < 0.020$, Bonferroni corrected) after stimulation (Figure 5A).

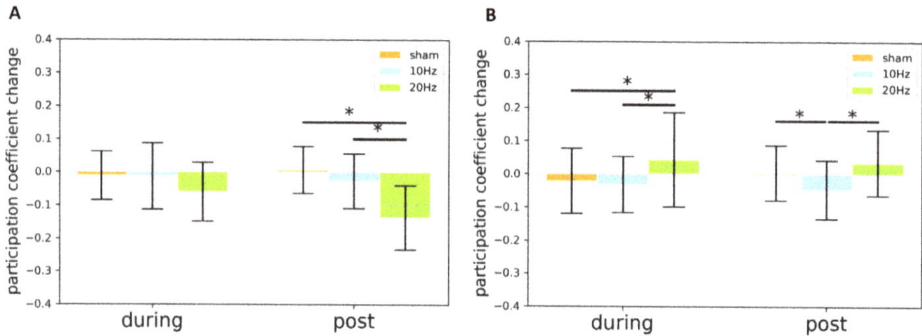


**Figure 5.** The bar chart of participation coefficient change of brain networks (**A**) in motor-related regions as well as (**B**) at the whole-brain level for various conditions during and after stimulation. Error bar stands for the standard error. Asterisk (*) indicates that a significant difference was observed at $p < 0.05$.

For clustering coefficient change, the repeated measure ANOVA showed no significant *stimulation* − *time* interaction effect ($F(2, 38) = 0.092$, $p < 0.912$), but indicated a significant *stimulation* main effect ($F(2, 38) = 10.294$, $p < 0.001$). During stimulation, *Post* − *hoc* tests indicated a significant difference between 10 Hz and 20 Hz ($p < 0.013$, Bonferroni corrected). After stimulation, *Post* − *hoc* tests indicated a significant difference between *sham* and 20 Hz ($p < 0.047$, Bonferroni corrected) as well as between 10 Hz and 20 Hz ($p < 0.027$, Bonferroni corrected) (Figure 6A). Similarly, for local efficiency change, the repeated measure ANOVA also showed no significant *stimulation* − *time* interaction effect ($F(2, 38) = 0.081$, $p < 0.923$), but indicated significant *stimulation* main effect ($F(2, 38) = 14.174$, $p < 0.001$). During stimulation, the *Post − hoc* tests indicated a significant difference between 10 Hz and 20 Hz ($p < 0.005$, Bonferroni corrected). After stimulation, the *Post − hoc* tests indicated a significant difference between *sham* and 10 Hz ($p < 0.007$, Bonferroni corrected), as well as between 10 Hz and 20 Hz ($p < 0.005$, Bonferroni corrected) (Figure 6B).

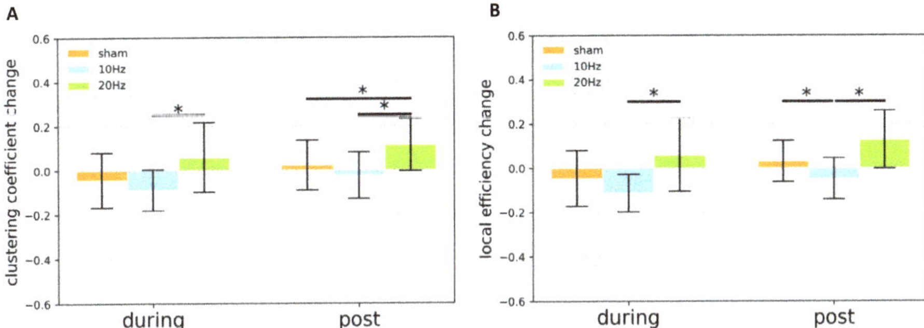


**Figure 6.** The bar chart of (**A**) clustering coefficient change and (**B**) local efficiency change of the network in motor-related regions for various conditions during and after stimulation. Error bar stands for the standard error. Asterisk (*) indicates that a significant difference was observed at $p < 0.05$.

Additionally, the modulation of the modular organization from the whole-brain network was also investigated. The repeated measure ANOVA of participation coefficient change indicated a significant *stimulation − time* interaction effect ($F(2, 230) = 3.536$, $p < 0.035$). During stimulation, *Post − hoc* tests indicated a significant difference between *sham* and 20 Hz ($p < 0.004$, Bonferroni corrected) as well as 10 Hz and 20 Hz ($p < 0.001$, Bonferroni corrected). After stimulation, *Post − hoc* tests indicated a significant difference between *sham* and 10 Hz ($p < 0.001$, Bonferroni corrected), as well as between 10 Hz and 20 Hz ($p < 0.001$, Bonferroni corrected) (Figure 5B). For a comprehensive understanding, the results of clustering coefficient and local efficiency analysis at the whole-brain level were also provided in the Supplementary Materials (Figure S2).

## 4. Discussion

This study aimed to investigate the tACS stimulation effect with 10 Hz and 20 Hz frequencies being applied in chronic stroke subjects using resting-state fMRI from a graph theoretical perspective. The results showed differential modulations induced by tACS with various frequencies. Meanwhile, the difference was also observed from brain networks in motor-related regions and whole-brain level, respectively. Our findings might facilitate effectively designing stimulation protocols with tACS in chronic stroke.

Evidence has accumulated that brain oscillations play an essential role in normal functioning through modulating the timing of neuronal spiking at the microscale and synchronizing distributed related cortical regions at the macroscale [39,40]. Specifically, the oscillations in alpha and beta bands were important and widely investigated by researchers. During relaxed alert states, alpha oscillations are supposed to be the most pronounced across most of the brain regions [41], and its functions were speculated to be involved in some aspects of attention and sensory processing [42,43]. Beta oscillations, especially those in sensorimotor brain regions, were usually motor-related and linked to activities, including motor observation, imagery, and execution [44]. In this context, non-invasive tACS has emerged as a powerful tool to modulate the brain activities and the internal brain states via entrainment of intrinsic frequency-specific oscillations [6,7]. In our study, the significant decrease of participation coefficient that indicated a higher segregation of communities was observed in motor-related regions after 20 Hz tACS, but no such effect was induced by 10 Hz tACS. In an age-related study, it has been exhibited that the participation coefficient was increased in older as compared with younger participants in the somatomotor networks probably due to less efficient use of neural resources [45]. This implied that 20 Hz tACS might be able to improve the cost efficiency of neural resources and make functional modules more differential and specific in the motor system. It has also been proposed that beta-band activity might correspond to an idling rhythm in the motor system [46] and allow for more efficient processing of feedback [47]. Hereby, the entrainment of beta oscillation after 20 Hz tACS might facilitate such processing by assembling sub-modules with higher functional coupling in motor-related regions.

It has been suggested that brain oscillations in different frequency ranges might enable regional interactions at different spatial scales [48]. Previous modeling studies implied that alpha and beta oscillations might support functional coupling over long distances [49]. Hence, the stimulation effect of tACS was also expected to be observed from the whole-brain perspective, although the stimulation site was located at the primary motor cortex. Our results illustrated that, at the whole-brain level, 10 Hz tACS facilitated segregation and 20 Hz facilitated the integration of communities. It has been revealed that the presence of alpha generators existed across all cortical layers [50] and similar alpha physiology was found across the whole brain, which implied an integrative function of alpha wave, especially under the resting condition [51]. Different from alpha oscillation, the function of beta frequency in the whole-brain level was not explicit. In our study, it seemed that, the communication of whole-brain communities, was enhanced after 20 Hz tACS, characterized by an increase in the participation coefficient. This could be partially explained by the findings in previous studies, which exhibited that beta oscillations are ideally suited for communicating across long conduction delays [49,52,53].

Additionally, since the stimulation site was located over the primary motor cortex, we were also interested in the modulation effect of nodal properties in the motor system. Hereby, the nodal measures of clustering coefficient and local efficiency, which shared similar meaning, were adopted. For both local measures, the increasing trend was observed after 20 Hz stimulation in motor-related regions, but the only decreasing trend was observed after 10 Hz stimulation. This implied that 20 Hz tACS might improve the efficiency of information transmission within specific modules and such an effect could maintain after stimulation [18]. Opposite with 20 Hz tACS, 10 Hz tCAS led to decreased local efficiency which might indicate a pruning of task-irrelevant connections [54]. Together, it suggested that the motor system might show frequency-specific responses to extrinsic stimulation and be prone to be more sensitive to entrainment of beta-band oscillations.

It is also worth noting that, although some previous studies have exhibited the modulation effect of tACS with different frequencies on healthy participants, few studies have investigated the influence of tACS in stroke patients [12,20,55,56]. One study claimed that tACS might facilitate lesioned brain self-regulation during neurofeedback intervention [57]. Our study tried to uncover the modulation effect of frequency-specific tACS in the chronic stroke from the view of brain networks with fMRI data at the same time. On the other hand, it has been proposed that the human brain always seeks a balance between the local segregation of function and the global integration of information [58]. Based on our findings, 20 Hz might have the potential to assist the lesioned brain to reach such an optimal state and it could be a promising tool applied in routine rehabilitation therapies. However, randomized controlled trials were needed in order to verify this point and determine the frequency that can maximally accelerate the recovery process for stroke patients in the future. In the current study, the investigation of the effect on motor function and motor learning after the single-session tACS could not be evidenced by the results directly and was quite limited due to the lack of behavioral assessment. Some previous studies have indicated that a single-session tDCS could help chronic stroke subjects to shorten the response time of tasks and improved pinch force in the paretic hand [59,60]. Lefebvre et al. suggested that a single-session dual-tDCS could enhance online motor skill learning and facilitate precision grip as well as dexterity for chronic stroke patients [61,62]. Different from tDCS, the acute effect of a single-session tACS was merely investigated. Hence, the experimental design could be improved by collecting some behavioral data before and after the stimulation to better understand this point in the future.

Several limitations should be noted in our study. The network status of the stroke subjects may be different due to the various lesions in sizes and locations. In the present study, we did not take the lesion information into account, because most subjects had relatively homogeneous lesions in sizes and locations. Meanwhile, we adopted the repeated measures design (each subject underwent all stimulation conditions) which could control for factors that cause variability between subjects. In this way, the influence of network status variations that are caused by lesions could be reduced to some extent. However, to make the results more precise, in the future subjects with sufficiently homogeneous neural injury should be recruited and more advanced analytical methods that take the lesion formation into account should be utilized. On the other hand, to make the findings more valid, it is better to check the difference in the node topography between stroke patients and healthy adults. Caution should be taken when interpreting our findings due to the lack of such comparison in the present study. Besides, it is really inevitable that there are multiple co-existing states when resting as well as diversity among stroke subjects in different experimental sessions, which might introduce some variations in the pre-stimulation community structures. Therefore, due to the existing variability of community structures in the baseline, precaution should be taken while qualitatively interpreting these observations. Although we proposed to mainly stimulate the primary motor area, the current montage with huge stimulation pads might lead to the diffusion of stimulation effect. Hence, a high-definition tACS with the centering montage [56] could be adopted to be more specific in the future. Besides, the sample size was not large, which might limit the generalization power to some extent. More patients should be recruited to validate and extend the findings of the current study.

## 5. Conclusions

In summary, we investigated the frequency-specific stimulation effect of tACS in chronic stroke while using resting-state fMRI data. The graph theoretical analysis mainly indicated the differential modulation effect of network integration and segregation properties in motor-related regions as well as at the whole-brain level after 10 Hz and 20 Hz tACS intervention. This study might facilitate designing neurorehabilitation protocols with tACS for stroke survivors in the future.

**Supplementary Materials:** The following are available online at https://www.mdpi.com/2076-3425/11/3/377/s1, Figure S1: Lesion distribution of stroke subjects, Figure S2: The bar chart of clustering coefficient change and local efficiency change of the network at the whole-brain level for various conditions during and after stimulation, Table S1: Demographics and clinical properties of the participants, Table S2: AAL ROIs in motor-related regions.

**Author Contributions:** C.C. conducted the experiments, performed signal processing, and prepared the manuscripts. K.Y. conducted the experiments, performed signal processing, and prepared the manuscripts. W.C.-w.C. gave valuable suggestions to polish the manuscripts. R.K.-y.T. oversaw and proposed this study, assisted to conduct the experiments, and prepare the manuscripts. All authors have read and agreed to the published version of the manuscript.

**Funding:** This study is supported by General Research Fund (Reference No. 14208118), Research Grant Council of Hong Kong.

**Institutional Review Board Statement:** The study was conducted according to the guidelines of the Declaration of Helsinki and approved by the Joint Chinese University of Hong Kong-New Territories East Cluster Clinical Research Ethics Committee.

**Informed Consent Statement:** Informed consent was obtained from all subjects involved in the study.

**Data Availability Statement:** The data presented in this study are available on proper request from the corresponding author.

**Acknowledgments:** The authors would like to thank all stroke subjects in this study.

**Conflicts of Interest:** The authors declare that they have no competing interests.

## References

1. Stinear, C.; Lang, C.; Zeiler, S.; Byblow, W. Advances and challenges in stroke rehabilitation. *Lancet Neurol.* **2020**, *19*. [CrossRef]
2. Jäncke, L. The plastic human brain. *Restor. Neurol. Neurosci.* **2009**, *27*, 521–538. [CrossRef] [PubMed]
3. Norouzi-Gheidari, N.; Archambault, P.; Fung, J. Effects of robot-assisted therapy on stroke rehabilitation in upper limbs: Systematic review and meta-analysis of the literature. *J. Rehabil. Res. Dev.* **2012**, *49*, 479–496. [CrossRef]
4. Duret, C.; Grosmaire, A.G.; Krebs, H. Robot-Assisted Therapy in Upper Extremity Hemiparesis: Overview of an Evidence-Based Approach. *Front. Neurol.* **2019**, *10*. [CrossRef] [PubMed]
5. Bao, S.C.; Khan, A.; Song, R.; Tong, R.K.Y. Rewiring the Lesioned Brain: Electrical Stimulation for Post-Stroke Motor Restoration. *J. Stroke* **2020**, *22*, 47–63. [CrossRef] [PubMed]
6. Purpura, D.; McMurtry, J. Intacellular activities and evoked potential changes during polarization of motor cortex. *J. Neurophysiol.* **1965**, *28*, 166–185. [CrossRef]
7. Nitsche, M.; Paulus, W. Excitability changes induced in the human motor cortex by weak transcranial direct current stimulation. *J. Physiol.* **2000**, *527 Pt 3*, 633–639. [CrossRef]
8. Thut, G.; Schyns, P.; Gross, J. Entrainment of Perceptually Relevant Brain Oscillations by Non-Invasive Rhythmic Stimulation of the Human Brain. *Front. Psychol.* **2011**, *2*, 170. [CrossRef]
9. Schnitzler, A.; Gross, J. Normal and pathological oscillatory communication in the brain. *Nat. Rev. Neurosci.* **2005**, *6*, 285–296. [CrossRef]
10. Salmelin, R.; Hari, R. Characterization of spontaneous MEG rhythms in healthy adults. *Electroencephalogr. Clin. Neurophysiol.* **1994**, *91*, 237–248. [CrossRef]
11. Feurra, M.; Blagoveshchensky, E.; Nikulin, V.; Nazarova, M.; Lebedeva, A.; Pozdeeva, D.; Yurevich, M.; Rossi, S. State-Dependent Effects of Transcranial Oscillatory Currents on the Motor System during Action Observation. *Sci. Rep.* **2019**, *9*, 1–11. [CrossRef]
12. Pollok, B.; Boysen, A.C.; Krause, V. The effect of transcranial alternating current stimulation (tACS) at alpha and beta frequency on motor learning. *Behav. Brain Res.* **2015**, *293*. [CrossRef]
13. Meier, A.; Krause, V.; Pollok, B. Early motor memory consolidation: Effects of 10 Hz and 20 Hz transcranial alternating current stimulation (tACS) over the left primary motor cortex (M1). *Klin. Neurophysiol.* **2014**, *45*. [CrossRef]
14. Kwon, Y.H.; Ko, M.H.; Ahn, S.; Kim, Y.H.; Song, J.; Lee, C.H.; Chang, M.; Jang, S. Primary motor cortex activation by transcranial direct current stimulation in the human brain. *Neurosci. Lett.* **2008**, *435*, 56–59. [CrossRef]
15. Ali, M.; Sellers, K.; Frohlich, F. Transcranial Alternating Current Stimulation Modulates Large-Scale Cortical Network Activity by Network Resonance. *J. Neurosci. Off. J. Soc. Neurosci.* **2013**, *33*, 11262–11275. [CrossRef]
16. Kimberley, T.; Khandekar, G.; Borich, M. fMRI reliability in subjects with stroke. *Exp. Brain Res.* **2008**, *186*, 183–190. [CrossRef]
17. Van den Heuvel, M.P.; Hulshoff Pol, H.E. Exploring the brain network: A review on resting-state fMRI functional connectivity. *Eur. Neuropsychopharmacol.* **2010**, *20*, 519–534. [CrossRef] [PubMed]
18. Farahani, F.; Karwowski, W.; Lighthall, N. Application of Graph Theory for Identifying Connectivity Patterns in Human Brain Networks: A Systematic Review. *Front. Neurosci.* **2019**, *13*, 585. [CrossRef] [PubMed]
19. Solomons, C. A Review of Transcranial Electrical Stimulation Methods in Stroke Rehabilitation. *Neurol. India* **2019**, *67*, 417. [CrossRef] [PubMed]
20. Wach, C.; Krause, V.; Moliadze, V.; Paulus, W.; Schnitzler, A.; Pollok, B. Effects of 10 Hz and 20 Hz transcranial alternating current stimulation (tACS) on motor functions and motor cortical excitability. *Behav. Brain Res.* **2012**. [CrossRef]
21. Wang, Y.; Shi, L.; Dong, G.; Zhang, Z.; Chen, R. Effects of Transcranial Electrical Stimulation on Human Auditory Processing and Behavior—A Review. *Brain Sci.* **2020**, *10*, 531. [CrossRef]
22. Thielscher, A.; Antunes, A.; Saturnino, G. Field modeling for transcranial magnetic stimulation: A useful tool to understand the physiological effects of TMS? In Proceedings of the 2015 37th Annual International Conference of the IEEE Engineering in Medicine and Biology Society (EMBC), Milan, Italy, 25–29 August 2015; Volume 2015, pp. 222–225. [CrossRef]
23. Yan, C.G.; Zang, Y.F. DPARSF: A MatLab toolbox for "pipeline" data analysis of resting-state fMRI. *Front. Syst. Neurosci.* **2010**, *4*, 13. [CrossRef]
24. Friston, K.; Williams, S.; Howard, R.; Frackowiak, R.; Turner, R. Movement-Related effects in fMRI time-series. *Magn. Reson. Med.* **1996**, *35*, 346–355. [CrossRef]
25. Power, J.; Barnes, K.; Snyder, A.; Schlaggar, B.; Petersen, S. Spurious but systematic conditions in functional connectivity MRI networks arise from subject motion. *Neuroimage* **2012**, *59*, 2141–2154. [CrossRef]
26. Zuo, X.N.; Di Martino, A.; Kelly, C.; Shehzad, Z.; Gee, D.; Klein, D.; Castellanos, F.; Biswal, B.; Milham, M. The Oscillating Brain: Complex and Reliable. *NeuroImage* **2009**, *49*, 1432–45. [CrossRef]
27. Tzourio-Mazoyer, N.; Landeau, B.; Papathanassiou, D.; Crivello, F.; Etard, O.; Delcroix, N.; Mazoyer, B.; Joliot, M. Automated Anatomical Labeling of Activations in SPM Using a Macroscopic Anatomical Parcellation of the MNI MRI Single-Subject Brain. *NeuroImage* **2002**, *15*, 273–289. [CrossRef]
28. Bear, M.; Connors, B.; Paradiso, M. *Neuroscience: Exploring the Brain*, 4th ed.; Jones & Bartlett Learning, LLC: Sudbury, MA, USA, 2015; pp. 1–975.

29. Cambridge, U.; Cohen, P. Applied Multiple Regression/Correlation Analysis for The Behavioral Sciences. *Am. J. Cardiol.* **1983**, *51*, 187.
30. Fornito, A.; Zalesky, A.; Bullmore, E. Network Scaling Effects in Graph Analytic Studies of Human Resting-State fMRI Data. *Front. Syst. Neurosci.* **2010**, *4*, 22. [CrossRef] [PubMed]
31. Garrison, K.; Scheinost, D.; Finn, E.; Shen, X.; Constable, R. The (in)stability of functional brain network measures across thresholds. *NeuroImage* **2015**, *118*. [CrossRef] [PubMed]
32. Rubinov, M.; Sporns, O. Complex network measures of brain connectivity: Uses and interpretations. *NeuroImage* **2010**, *52*, 1059–1069. [CrossRef] [PubMed]
33. Newman, M. Fast algorithm for detecting community structure in networks. *Phys. Rev. E Stat. Nonlinear Soft Matter Phys.* **2004**, *69*, 066133. [CrossRef] [PubMed]
34. Guimerà, R.; Amaral, L. Functional Cartography of Complex Metabolic Networks. *Nature* **2005**, *23*, 22–231. [CrossRef] [PubMed]
35. Guimerà, R.; Amaral, L. Cartography of complex networks: Modules and universal roles. *J. Stat. Mech.* **2005**, *2005*, nihpa35573. [CrossRef] [PubMed]
36. Watts, D.; Strogatz, S. Collective dynamics of 'small-world' networks. *Nature* **2011**. [CrossRef]
37. Latora, V.; Marchiori, M. Efficient Behavior of Small-World Networks. *Phys. Rev. Lett.* **2001**, *87*, 198701. [CrossRef]
38. Xia, M.; Wang, J.; He, Y. BrainNet Viewer: A Network Visualization Tool for Human Brain Connectomics. *PLoS ONE* **2013**, *8*, e68910. [CrossRef] [PubMed]
39. Jacobs, J.; Kahana, M.; Ekstrom, A.; Fried, I. Brain Oscillations Control Timing of Single-Neuron Activity in Humans. *J. Neurosci. Off. J. Soc. Neurosci.* **2007**, *27*, 3839–3844. [CrossRef]
40. Fries, P. A mechanism for cognitive dynamics: Neuronal communication through neuronal coherence. *Trends Cogn. Sci.* **2001**, *9*, 356–362. [CrossRef]
41. Srinivasan, R.; Winter, W.; Nunez, P. Source analysis of EEG oscillations using high-resolution EEG and MEG. *Prog. Brain Res.* **2006**, *159*, 29–42. [CrossRef]
42. Voytek, B.; Canolty, R.; Shestyuk, A.; Crone, N.; Parvizi, J.; Knight, R. Shifts in Gamma Phase–Amplitude Coupling Frequency from Theta to Alpha Over Posterior Cortex During Visual Tasks. *Front. Hum. Neurosci.* **2010**, *4*, 191. [CrossRef]
43. Jensen, O.; Mazaheri, A. Shaping Functional Architecture by Oscillatory Alpha Activity: Gating by Inhibition. *Front. Hum. Neurosci.* **2010**, *4*, 186. [CrossRef]
44. Honaga, E.; Ishii, R.; Kurimoto, R.; Ikezawa, K.; Takahashi, H.; Nakahachi, T.; Iwase, M.; Mizuta, I.; Yoshimine, T.; Takeda, M. Post-movement beta rebound abnormality as indicator of mirror neuron system dysfunction in autistic spectrum disorder: An MEG study. *Neurosci. Lett.* **2010**, *478*, 141–145. [CrossRef]
45. Geerligs, L.; Renken, R.; Saliasi, E.; Maurits, N.; Lorist, M. A Brain-Wide Study of Age-Related Changes in Functional Connectivity. *Cereb. Cortex* **2014**, *25*. [CrossRef]
46. Pfurtscheller, G.; Stancak, A.; Neuper, C. Post-movement beta synchronization. A correlate of an idle motor area? *Electroencephalogr. Clin. Neurophysiol.* **1996**, *98*, 281–293. [CrossRef]
47. Baker, S. Oscillatory interactions between sensorimotor cortex and the periphery. *Curr. Opin. Neurobiol.* **2008**, *17*, 649–655. [CrossRef] [PubMed]
48. Rosanova, M.; Casali, A.; Bellina, V.; Resta, F.; Mariotti, M.; Massimini, M. Natural Frequencies of Human Corticothalamic Circuits. *J. Neurosci. Off. J. Soc. Neurosci.* **2009**, *29*, 7679–7685. [CrossRef]
49. Kopell, N.; Ermentrout, B.; Whittington, M.; Traub, R. Gamma rhythms and beta rhythms have different synchronization properties. *Proc. Natl. Acad. Sci. USA* **2000**, *97*, 1867–1872. [CrossRef] [PubMed]
50. Haegens, S.; Barczak, A.; Musacchia, G.; Lipton, M.; Mehta, A.; Lakatos, P.; Schroeder, C. Laminar Profile and Physiology of the Rhythm in Primary Visual, Auditory, and Somatosensory Regions of Neocortex. *J. Neurosci.* **2015**, *35*, 14341–14352. [CrossRef] [PubMed]
51. Halgren, M.; Ulbert, I.; Bastuji, H.; Fabó, D.; Eross, L.; Rey, M.; Devinsky, O.; Doyle, W.; Mak-McCully, R.; Halgren, E.; et al. The generation and propagation of the human alpha rhythm. *Proc. Natl. Acad. Sci. USA* **2019**, *116*, 201913092. [CrossRef] [PubMed]
52. Groppe, D.; Bickel, S.; Keller, C.; Jain, S.; Hwang, S.; Harden, C.; Mehta, A. Dominant frequencies of resting human brain activity as measured by the electrocorticogram. *NeuroImage* **2013**, *79*. [CrossRef]
53. Bibbig, A.; Traub, R.; Whittington, M. Long-range synchronization of $\gamma$ and $\beta$ oscillations and the plasticity of excitatory and inhibitory synapses: A network model. *J. Neurophysiol.* **2002**, *88*, 1634–1654. [CrossRef]
54. Cohen, J.; D'Esposito, M. The Segregation and Integration of Distinct Brain Networks and Their Relationship to Cognition. *J. Neurosci.* **2016**, *36*, 12083–12094. [CrossRef]
55. Feurra, M.; Pasqualetti, P.; Bianco, G.; Santarnecchi, E.; Rossi, A.; Rossi, S. State-Dependent Effects of Transcranial Oscillatory Currents on the Motor System: What You Think Matters. *J. Neurosci. Off. J. Soc. Neurosci.* **2013**, *33*, 17483–17489. [CrossRef] [PubMed]
56. Heise, K.F.; Kortzorg, N.; Saturnino, G.; Fujiyama, H.; Cuypers, K.; Thielscher, A.; Swinnen, S. Evaluation of a Modified High-Definition Electrode Montage for Transcranial Alternating Current Stimulation (tACS) of Pre-Central Areas. *Brain Stimul.* **2016**, *9*. [CrossRef] [PubMed]
57. Naros, G.; Gharabaghi, A. Physiological and behavioral effects of $\beta$-tACS on brain self-regulation in chronic stroke. *Brain Stimul.* **2017**, *10*, 251–259. [CrossRef] [PubMed]

58. Lord, L.D.; Stevner, A.B.; Deco, G.; Kringelbach, M.L. Understanding principles of integration and segregation using whole-brain computational connectomics: implications for neuropsychiatric disorders. *Philos. Trans. R. Soc. A Math. Phys. Eng. Sci.* **2017**, *375*, 20160283. [CrossRef]
59. Hummel, F.; Voller, B.; Celnik, P.; Floel, A.; Giraux, P.; Gerloff, C.; Cohen, L. Effects of brain polarization on reaction times and pinch force in chronic stroke. *BMC Neurosci.* **2006**, *7*, 73. [CrossRef]
60. Stagg, C.; Bachtiar, V.; O'Shea, J.; Allman, C.; Bosnell, R.; Kischka, U.; Matthews, P.; Johansen-Berg, H. Cortical activation changes underlying stimulation induced behavioral gains in chronic stroke. *Brain J. Neurol.* **2011**, *135*, 276–284. [CrossRef] [PubMed]
61. Lefebvre, S.; Laloux, P.; Peeters, A.; Desfontaines, P.; Jamart, J.; Vandermeeren, Y. Dual-tDCS Enhances Online Motor Skill Learning and Long-Term Retention in Chronic Stroke Patients. *Front. Hum. Neurosci.* **2013**. [CrossRef]
62. Lefebvre, S.; Thonnard, J.L.; Laloux, P.; Peeters, A.; Jamart, J.; Vandermeeren, Y. Single Session of Dual-tDCS Transiently Improves Precision Grip and Dexterity of the Paretic Hand After Stroke. *Neurorehabilit. Neural Repair* **2013**, *28*, 100–110. [CrossRef]

*Article*

# Visual Cortex Transcranial Direct Current Stimulation for Proliferative Diabetic Retinopathy Patients: A Double-Blinded Randomized Exploratory Trial

**Angelito Braulio F. de Venecia III [1,2] and Shane M. Fresnoza [3,4,5,*]**

1 Eye Center, Region 1 Medical Center, Dagupan 2400, Pangasinan, Philippines; abfdv777@gmail.com
2 Nazareth General Hospital, Dagupan 2400, Pangasinan, Philippines
3 Institute of Psychology, University of Graz, 8010 Graz, Austria
4 BioTechMed, 8010 Graz, Austria
5 College of Medicine, De La Salle Health Sciences Institute, Dasmarinas 4114, Cavite, Philippines
* Correspondence: shane.fresnoza@uni-graz.at

**Abstract:** Proliferative diabetic retinopathy (PDR) is a severe complication of diabetes. PDR-related retinal hemorrhages often lead to severe vision loss. The main goals of management are to prevent visual impairment progression and improve residual vision. We explored the potential of transcranial direct current stimulation (tDCS) to enhance residual vision. tDCS applied to the primary visual cortex (V1) may improve visual input processing from PDR patients' retinas. Eleven PDR patients received cathodal tDCS stimulation of V1 (1 mA for 10 min), and another eleven patients received sham stimulation (1 mA for 30 s). Visual acuity (logarithm of the minimum angle of resolution (LogMAR) scores) and number acuity (reaction times (RTs) and accuracy rates (ARs)) were measured before and immediately after stimulation. The LogMAR scores and the RTs of patients who received cathodal tDCS decreased significantly after stimulation. Cathodal tDCS has no significant effect on ARs. There were no significant changes in the LogMAR scores, RTs, and ARs of PDR patients who received sham stimulation. The results are compatible with our proposal that neuronal noise aggravates impaired visual function in PDR. The therapeutic effect indicates the potential of tDCS as a safe and effective vision rehabilitation tool for PDR patients.

**Keywords:** transcranial direct current stimulation; visual acuity; numerical discrimination; diabetes; diabetic neuropathy; visual cortex

**Citation:** de Venecia, A.B.F., III; Fresnoza, S.M. Visual Cortex Transcranial Direct Current Stimulation for Proliferative Diabetic Retinopathy Patients: A Double-Blinded Randomized Exploratory Trial. *Brain Sci.* **2021**, *11*, 270. https://doi.org/10.3390/brainsci11020270

Academic Editors: Ulrich Palm, Moussa Antoine Chalah and Samar S. Ayache

Received: 22 January 2021
Accepted: 19 February 2021
Published: 21 February 2021



## 1. Introduction

Diabetic retinopathy (DR) is a grave ocular complication of diabetes mellitus and the leading cause of preventable blindness. The incidence of DR is projected to increase because the number of diabetic patients is expected to rise from 171 million in 2000 to 366 million by 2030 worldwide [1]. The economic burden of DR comes from direct disease management costs and lost worker productivity because it affects working-age populations [2]. Clinically, in the early asymptomatic stage of DR (non-proliferative diabetic retinopathy (NPDR)), microaneurysms, hemorrhages, and hard exudates are already present in the retina. Significant visual impairment occurs in the advanced stage (proliferative diabetic retinopathy (PDR)) secondary to neovascularization that causes severe bleeding and retinal detachment [3]. The primary treatment goal is to prevent further visual loss with intensive pharmacotherapeutic control of blood glucose level and management of microvascular complications using intravitreal pharmacological agents, laser photocoagulation, and vitreous surgery [3]. Therapeutics to improve residual vision, such as anti-vascular endothelial growth factor (VEGF) therapy, also gained popularity in recent years. However, their cost-effectiveness and complications such as traumatic intraocular injuries and tractional retinal detachment still outweigh the benefits. Therefore, it remains a challenge to improve residual vision in PDR patients with the available interventions.

According to the "residual vision activation theory", strengthening of synaptic transmission and synchronization of partially damaged structures (within-systems plasticity) and downstream neuronal networks (network plasticity) is a promising alternative to reactivate and restore visual functions [4]. One way of exploring this possibility is the use of non-invasive brain stimulation (NIBS) techniques such as transcranial direct current stimulation (tDCS) and transcranial alternating current stimulation (tACS). TDCS involves the application of low-intensity (1–2 mA) direct electrical current to stimulate cortical areas through the intact head. During stimulation, tDCS modulates neuronal activity by alterations of neuronal membrane potentials caused by the opening or closing of voltage-gated ion channels [5]. As shown in the motor cortex, this effect is polarity-dependent because anodal (positive current) and cathodal (negative current) stimulation brings the resting membrane potential closer to depolarization (increase likelihood of neuronal firing) and hyperpolarization (decrease likelihood of neuronal firing), respectively [5]. The effect of tDCS on cortical excitability persists after stimulation and is attributed to prolonged synaptic efficacy changes such as long-term potentiation (LTP) and long-term depression (LTD). In general, anodal stimulation increases and cathodal stimulation decreases cortical activity [6,7].

In healthy individuals, tDCS elicited transient excitability changes in the primary visual cortex (V1) as inferred from the modulation of the N70 component of visual evoked potential (VEP) amplitude and occurrence of transcranial magnetic stimulation (TMS)-induced phosphenes [8–11]. Behaviorally, in the color discrimination task, anodal tDCS improved the blue-yellow range threshold (with no impact in the red-green range) mediated by the koniocellular pathways, while cathodal tDCS impaired discrimination within the red-green range mediated by the parvocellular pathways and at the same time increased koniocellular-driven discrimination [12]. Moreover, anodal tDCS significantly decreased cell discrimination threshold only from the most eccentric regions (peripheral) of the visual field [13]. TDCS is also beneficial for patients with visual field loss due to cortical damage. For instance, in stroke patients, tDCS improved contrast discrimination, motion and object detection, object recognition, attention or visual awareness, as well as performance in complex perceptual tasks such as face recognition and visual search [4,14]. In amblyopic patients, 15 min of anodal tDCS improved contrast sensitivity and increased VEPs of the amblyopic eye, while cathodal tDCS decreased both measures [15,16]. Cathodal tDCS over V1 contralateral to the amblyopic eye improved visual acuity and inhibited VEPs amplitude in the targeted site, while it facilitated VEPs ipsilateral to the amblyopic eye [17]. On the other hand, patients with vision loss secondary to optic neuropathies reported visual field improvement after receiving repetitive transorbital tACS. Evidence suggests that the therapeutic effect of tACS is due to re-synchronization of brain networks, which were desynchronized by vision loss [18]. However, despite the promising results, there is currently no attempt to explore the potential of tDCS as a vision rehabilitation tool for visual impairment due to retinal disease.

Retinal diseases, including the early stage of DR (NDR), are characterized by high internal noise within the visual pathways, further aggravating impaired visual functions [19–21]. In theory, neuronal noise can be more severe during the advanced stage of DR. For instance, resting-state functional connectivity is significantly increased between V1 and the frontal lobe in PDR patients [22]. The increase in functional connectivity can be considered compensatory but may also constitute an aberrant neural network that can interfere when a task recruits these brain regions. Therefore, we hypothesized that the hyperpolarizing effect of cathodal tDCS on neuronal membrane potentials could reduce neural noise and improve the processing of visual inputs from damaged retinas of PDR patients. We applied cathodal tDCS over V1 and measured visual acuity before and after stimulation using the Early Treatment Diabetic Retinopathy Study (ETDRS) chart. We also hypothesized that if visual acuity improves, we can observe changes in task performance associated with vision-related processes upstream from V1. For that purpose, we measured the patients' "number acuity", that is, the ability to discriminate the more numerous of two sets of non-verbal stimuli (e.g., dots) using a numerical discrimination task [23,24]. Number acuity is an established measure of the parietal lobe-based approximate number system (ANS) [23,25], but substantial recent evidence also suggests direct perceptual processing of numerosity in V1 [26–30]. The results of the present study will help us understand the impact of DR at the cortical level and develop a new vision rehabilitation tool.

## 2. Materials and Methods

### 2.1. Participants

Twenty-two clinically diagnosed PDR patients volunteered for the study. The number of patients surpassed our a priori calculated (G*Power 3.1.9.2) sample size (14) required to achieve a statistical power (1-β) of 90% at an alpha level of 0.05 and a moderate effect size (0.50) for a study with between-within subject design. The study was conducted on February 20, 2020, at the Nazareth General Hospital in Dagupan City, Pangasinan, Philippines. The patients are all native Filipino speakers and private patients of Dr de Venecia. They were allocated into either the tDCS group (5 females, 6 males; mean age = 57.73 years, SD = 9.45 years) or the sham group (4 females, 7 males; mean age = 52.73, SD = 12.92 years) using permuted block randomization (Figure 1). They all have corrected-to-normal vision and were right-handed except for 2 patients in the tDCS group and 1 patient in the sham group. The diabetes duration (mean ± SD) was 11.36 ± 6.32 and 13.27 ± 5.85 years for the tDCS group and sham group, respectively. Presented in Table 1 are the demographic and clinical characteristics of each patient. Except for retinopathy, there were no reports of neurological (for example, stroke and epilepsy) and psychiatric disorders (for example, depression), as well as brain injury and surgery. We also cross-checked the patients for contraindications to NIBS methods such as metallic or electrical implants in the head, neck, and chest. The experimental protocols complied with the Helsinki Declaration's guidelines for human studies and approved by the chief of clinics representing the hospital board of Nazareth General Hospital. The study is registered at the International Standard Randomised Controlled Trial Number (ISRCTN) registry (Registration number: ISRCTN70877737). All patients signed written informed consent before the experiment.

### 2.2. Experimental Design and Procedure

The study has a double-blinded, randomized, sham-controlled design. For reliable blinding, another team member set up the stimulator, and neither the patients nor the experimenter knew the stimulation condition. We conducted the experiments in a quiet and well-lighted room. Initially, patients were brief about the study purpose and given detailed instructions (in Filipino or local dialect) to ensure they understood the task. Informed consent was acquired after all questions had been answered and thoroughly clarified. The experiment starts with the measurements of the right and left eyes' visual acuity. Subsequently, we localized V1 following the international 10–20 electroencephalography (EEG) guidelines. The distance between the nasion and inion, as well as between the right and left pre-auricular were measured using a tape measure (cm). The two lines intersection was marked using a washable color marker and designated as the vertex. The vertex's location served as a reference in positioning the EEG cap on the patient's head. We used three elastic EEG caps (size 54, 56, and 58 cm) to accommodate different head sizes (EASYCAP GmbH, Germany). During the mounting of the EEG cap, the vertex always matched the location of the Cz electrode. The scalp area underlying the Oz electrode was marked and designated as the location of V1. After removing the EEG cap, the stimulating (cathode) electrode was placed on V1 and secured using an elastic rubber bandage. The reference (anode) electrode was placed on the right shoulder using adhesive medical tape. After setting up the tDCS electrodes, patients were allowed to perform practice trials. Subsequently, they performed the numerical discrimination task, which was immediately followed by the stimulation (cathodal tDCS or sham). As instructed, the patients closed their eyes and relaxed during stimulation. The patients repeated the numerical discrimination task and underwent visual acuity testing immediately after the stimulation. Each experimental session, including the preparations, lasted for 40 min.

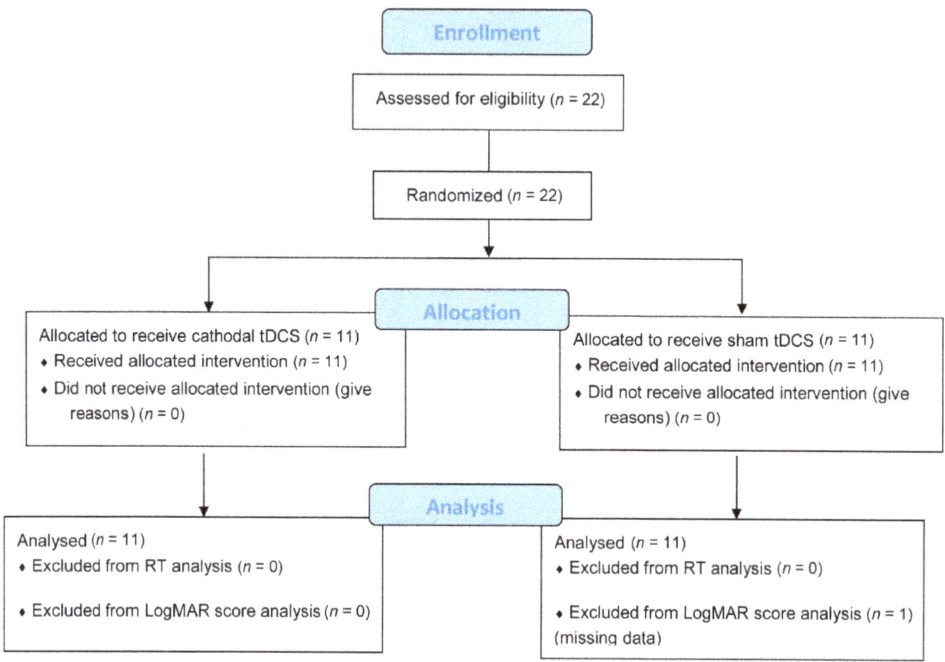


**Figure 1.** Flowchart of patient inclusion (CONSORT 2010).

### 2.3. *Transcranial Direct Current Stimulation (tDCS)*

Delivery of tDCS was done using a pair of saline-soaked (0.9%-NaCl) surface sponge electrodes connected to a battery-driven, constant-current stimulator (DC-STIMULATOR PLUS, NeuroConn Gmbh, Ilmenau, Germany). The cathode electrode was placed over V1 using the 10–20 EEG coordinates. To achieve focal stimulation of V1 and avoid the effect on other cortical areas, the reference/return (anode) electrode was placed on the right shoulder. For cathodal tDCS stimulation, the current intensity was 1 mA and was delivered continuously for 10 min. The rectangular electrodes measured 5 × 7 cm (surface area: 35 $cm^2$) in diameter; therefore, the current density underneath them was 0.029 $mA/cm^2$ during stimulation. The current was slowly ramped-up and ramped-down for 10 s at the start and end of stimulation, respectively. We kept the impedance during stimulation below 10 kΩ to minimize tingling skin sensation. To ensure the patients experience a similar initial skin sensation of tDCS, for sham stimulation, the current was applied for 30 s with an additional 10 s ramped-up from 0 to 1 mA, and 10 s ramped-down to 0 mA. A stimulation duration of 30 s is not known to induce after-effects [7]. All stimulation parameters conformed to the safety guidelines for tDCS [31]. After the experiment, the patients answered a standard tDCS questionnaire to document stimulation-related adverse effects.

**Table 1.** Demographic and clinical characteristics of the study population.

| | Gender | Age | DM Duration (Years) | Comorbid Conditions | Medications |
|---|---|---|---|---|---|
| tDCS group | | | | | |
| 1 | M | 42 | 18 | Diabetic nephropathy | Insulin |
| 2 | F | 65 | 10 | Diabetic nephropathy | Insulin |
| 3 | F | 54 | 5 | | Insulin, Gliclazide, Keto-analogs + essential amino acids |
| 4 | M | 60 | 5 | Hypertension | Metformin, Simvastatin, Amlodipine |
| 5 | F | 71 | 10 | Hypertension | Gliclazide, Pioglitazone, Losartan, Rosuvastatin |
| 6 | M | 56 | 22 | | Metformin, Pioglitazone |
| 7 | F | 59 | 8 | Hypertension | Insulin, Linagliptin, Aspirin, Amlodipine, Losartan |
| 8 | M | 69 | 5 | Hypertension | Metformin, Losartan |
| 9 | F | 63 | 15 | Hypertension | Metformin, Gliclazide, Amlodipine, Vitamins B-complex |
| 10 | M | 53 | 7 | Cardiac disease | Gliclazide, Aspirin, Losartan, Amlodipine |
| 11 | M | 43 | 20 | Hypertension | Metformin, Amlodipine, Ferrous sulfate |
| Sham group | | | | | |
| 1 | M | 35 | 14 | | Metformin, Atorvastatin |
| 2 | M | 41 | 15 | | Insulin |
| 3 | M | 48 | 25 | Hypertension, asthma, diabetic nephropathy | Amlodipine, Valsartan, Calcium carbonate, Ferrous sulfate + Vitamin B-complex + folate |
| 4 | F | 59 | 10 | | Metformin, Aspirin, Simvastatin |
| 5 | M | 63 | 5 | Prostatic hyperplasia, hypertension | Telmisartan, Glimepiride, Finasteride, Keto-analogs + essential amino acids |
| 6 | F | 59 | 15 | Hypertension | Insulin, Empagliflozin |
| 7 | M | 46 | 10 | | Metformin |
| 8 | F | 35 | 10 | Cardiac disease, hypertension | Furosemide, Insulin, Warfarin, Carvedilol, Valsartan, Teneligliptin, Spironolactone, Rosuvastatin, Vitamin B12, Linagliptin, Keto-analogs |
| 9 | M | 76 | 22 | Hypertension | Losartan, Amlodipine, Insulin |
| 10 | M | 55 | 10 | Hypertension | Metformin |
| 11 | F | 63 | 10 | | Insulin |

DM, diabetes mellitus; PDR, proliferative diabetic retinopathy; tDCS, transcranial direct current stimulation; M, male; F, female.

### 2.4. Visual Acuity Assessment

A certified ophthalmologist (Dr de Venecia) conducted the visual acuity testing procedure using the standard ETDRS chart before and immediately after stimulation. The ETDRS chart has similar numbers (5) of Sloan letters per line, equal spacing between lines and letters on a log scale (0.02 logarithm of the minimum angle of resolution (LogMAR), and balanced letter difficulty in the individual lines [17]. During the test (with dimmed room light), the chart was positioned 4 m away from the patients and lit at a standard lighting level (85 cd/m$^2$). Uncorrected visual acuity was measured first in the right eye, while the left eye was occluded. To prevent memorization, we used two different charts for testing the right (Chart 1) and left eye (Chart 2). The patients read the letters on the chart slowly from top to bottom, letter-by-letter, and beginning with the first letter on the top row. If a patient misread >2 letters on a line, we aborted the procedure and added 0.02 log units (for every misread letter) to the LogMAR score of that line. A high LogMAR score is indicative of worsening vision. Patients who cannot read the letters were given a LogMAR score of 1.9 for the ability to count fingers, 2.3 for detecting hand motion, 2.7 for light perception, and 3.0 for the absence of light perception [32].

### 2.5. Numerical Discrimination Task

During the task, patients sat 50 cm away from a 16-inch computer screen with a refresh rate of 60 Hz and a resolution of 1920 × 1080 (HP ENVY dv7-7388sz Notebook PC). Stimuli consisted of 60 intermixed, non-overlapping white and black dots presented on a gray background with a luminance of 36.2 cd/m$^2$ (Figure 2). The atients were instructed to indicate whether there were more black or white dots. Their left and right index fingers rested on the "B" and "N" computer keys, respectively. To familiarise the patients with task rules, they were given a short training period (10 practice trials) before the main task. The practice trials were similar to the test trials except that they only contained 40 dots. After the training, a key press starts the test trials, and the dots appeared in the middle part of the computer screen. In previous studies involving healthy individuals [23,24], the presentation time of the dots was 200 ms, which is within the known duration of rapid information processing (100–400 ms) through the visual system before response output [33]. Considering the patients' visual impairment, we set the stimulus presentation duration to 500 ms to ensure sufficient time for stimulus processing and limit the chances of guessing. This duration, however, is insufficient for the patients to count the exact number of dots serially. The initial colors of the stimulus were yellow and blue, however, during practice trials, patients also had difficulty perceiving them; hence we replaced the color with white and black.

To isolate the effect of stimulation on numerosity, we kept other perceptual confounding stimulus variables (for example, individual dot diameter (8 mm), surface area covered by the dots (113.09 cm$^2$), and inter-dot spacing or sparsity) constant across the trials [34]. Therefore, only the number (ratio) of the black and white dots changes per trial. Patients were made aware that no trial contains equal numbers of black and white dots. Once the dots disappear, an instruction appears on the screen instructing the patients to indicate their answer by pressing the "B" key for black and "N" key for white. The next trial appears after a button press. Although there was no restricted time window for responding, we instructed the patients to indicate their answers as quickly and accurately as possible. They answered two sets of 30 different test trials before and after stimulation. The appearance of test trials containing more black or more white dots was randomized. We drew the ratio of the smaller to the larger set on each trial from one of four ratio bins: 1:2, 3:4, 5:6, or 7:8 [24]. There were seven trials with a 1:2 ratio, eight trials with a 3:4 ratio, seven trials with a 5:6 ratio, and eight trials with a 7:8 ratio. The patients performed the practice and test trials binocularly without optical corrections to rule out corrective lenses' influence in the stimulation-specific effect on residual vision. Stimulus presentation and response recording was made possible by a Psychopy-based program (Psychopy Software in Python, University of Nottingham, Nottingham, United Kingdom). The program recorded reaction time (RT) and accuracy for each trial. RT is defined as the time (seconds) from the dots' disappearance until the patient presses a response key. The task (practice and test trials) and stimulation period were about 20 min.

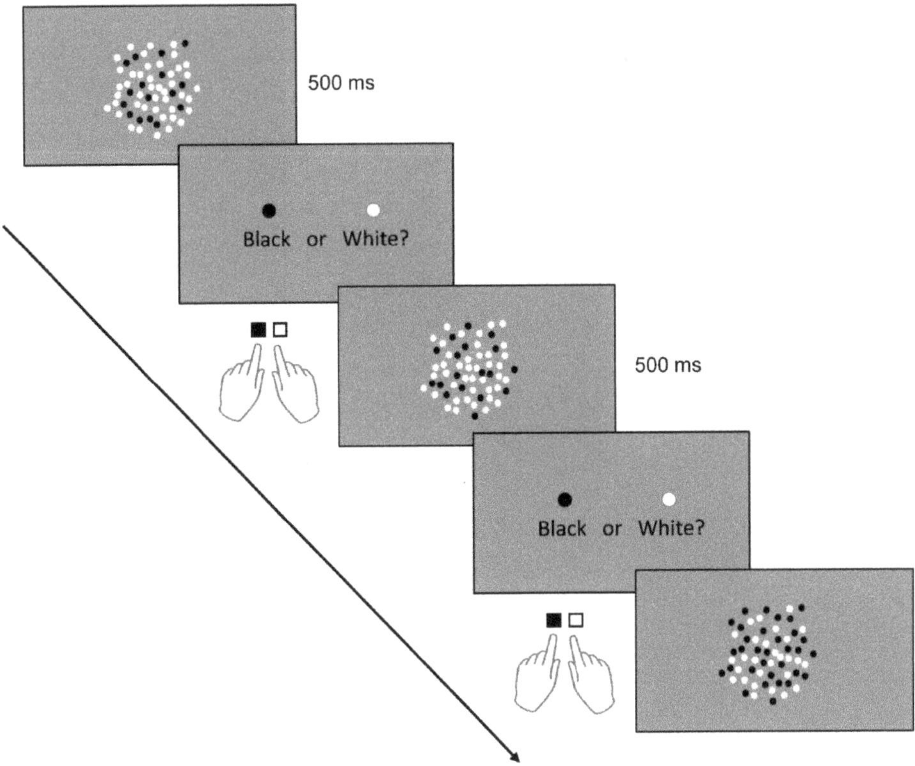


**Figure 2.** Numerical discrimination task. Stimuli (dots) were presented visually for 500 ms, and patients had to decide via button press whether there were more white or black dots.

### 2.6. Statistical Analysis

All data were analyzed using SPSS 26 software (IBM Corp., Armonk, NY, USA). Shapiro–Wilk and Levene's test assessed the normality of the data distribution and homogeneity of variances, respectively. In instances of normality violation (Shapiro–Wilk test: $p \leq 0.05$), the data are logarithmically transformed. The LogMAR scores were analyzed using a three-way (2 × 2 × 2) mixed analysis of variance (ANOVA) with a between-subject factor "stimulation" (cathodal tDCS and sham) and within-subject factors "time" (before and after stimulation) and "eye" (right and left). One patient from the sham group had missing visual acuity data; therefore, we only analyzed LogMAR scores from 11 patients in the tDCS group and 10 patients in the sham group. For the numerosity discrimination task, we calculated the average RTs before and after stimulation. Discarded RT (excluded from the analysis) are those beyond 2 standard deviations (SDs) away from the mean (outliers) and those from incorrect trials. To determine the accuracy rate (AR%), we divided the number of correct trials from the total number of trials and multiplied by 100. The RTs and ARs were analyzed using a two-way (2 × 2) mixed ANOVA with a between-subject factor "stimulation" (cathodal tDCS and sham) and a within-subject factor "time" (before and after stimulation). Effect sizes were reported as partial eta squared ($\eta_p^2$) values (small: 0.01, moderate: 0.06 and large: >0.13). If the ANOVA yielded significant results, we explored it with a Bonferroni corrected post-hoc *t*-test. In all statistical analyses, we set a threshold significance level at a *p*-value of $\leq$0.05. Unless otherwise stated, all reported values are mean $\pm$ SD.

## 3. Results

The tDCS and sham group have comparable mean age and diabetes duration. All patients completed the experimental procedure well. One patient in the sham group reported mild headache, neck fatigue, and increased heart rate after stimulation. He was able to go home after the symptoms alleviated. Patients in the tDCS group did not complain of any side effects. Before the stimulation, the mean LogMAR score on the right eye (sham group: 1.96 $\pm$ 0.33 log units, tDCS group: 0.99 $\pm$ 0.25 log units) significantly differed between the groups ($p$ = 0.029). In contrast, the mean LogMAR score on the left eye (sham group: 1.04 $\pm$ 0.24 log units, tDCS group: 1.25 $\pm$ 0.26 log units) was comparable between the groups ($p$ = 0.835). The numerosity discrimination performance was comparable between the groups as indicated by the mean RT (sham group: 1.22 $\pm$ 0.20 s, tDCS group: 2.47 $\pm$ 0.53 s) ($p$ = 0.056) and mean AR (sham group: 74.24 $\pm$ 3.65%, tDCS group: 74.24 $\pm$ 4.94%) ($p$ = 0.898) before the stimulation (Table S1).

### 3.1. Visual Acuity

The data used for the final analysis was normally distributed after logarithmic transformation and have equal group variances (all $p > 0.05$). The ANOVA conducted on the log-transformed LogMAR scores revealed a significant main effect of time ($F$ (1,19) = 14.97, $p$ = 0.001, $\eta_\rho{}^2$ = 0.441) such that the overall post-stimulation score (1.15 $\pm$ 0.82) was significantly lower than before stimulation (1.42 $\pm$ 0.94). The eye and stimulation interaction was significant ($F$ (1,19) = 4.83, $p$ = 0.041, $\eta_\rho{}^2$ = 0.203). Bonferroni corrected post hoc comparisons indicate that the interaction was mainly driven by the significantly lower overall score of the right eye in the tDCS group (0.99 $\pm$ 0.84) than the right eye in the sham group (1.96 $\pm$ 0.33) ($p$ = 0.016) (Figure 3A,B). In contrast, the overall score of the left eye in the tDCS group (1.25 $\pm$ 0.26) and sham group (1.04 $\pm$ 0.24) were comparable ($p$ = 0.965).

The time and stimulation interaction was also significant ($F$ (1,19) = 8.92, $p$ = 0.008, $\eta_\rho{}^2$ = 0.319). Bonferroni corrected post hoc comparisons showed that the overall score in the tDCS group (before stimulation: 1.13 $\pm$ 0.94, after stimulation: 1.00 $\pm$ 0.95) significantly decreased after stimulation ($p \leq 0.001$), whereas the decrease in the overall score in the sham group (before stimulation: 1.49 $\pm$ 0.98, after stimulation: 1.45 $\pm$ 0.99) was not significant ($p$ = 0.549). The three-way interactions of stimulation, time, and eye was not significant (Table 2). Planned exploratory post hoc comparisons revealed that this result was mainly driven by the absence of significant differences in the sham group's pre- and post-stimulation scores of the right (before stimulation: 1.96 $\pm$ 0.33, after stimulation: 1.91 $\pm$ 0.32; $p$ = 0.794) and left eye (before stimulation: 1.04 $\pm$ 0.24, after stimulation: 1.00 $\pm$ 0.24; $p$ = 0.542) (Figure 3B); and second, the absence of significant differences in the left eye's pre-stimulation (tDCS: 1.25 $\pm$ 0.86, sham: 1.04 $\pm$ 0.75; $p$ = 0.835) and post-stimulation scores (tDCS: 1.16 $\pm$ 0.90, sham: 1.00 $\pm$ 0.75; $p$ = 0.909) between stimulation conditions. However, post hoc comparisons in the tDCS group revealed a significant decrease in the right (before stimulation: 0.99 $\pm$ 0.25, after stimulation: 0.85 $\pm$ 0.27; $p$ = 0.001) and left eye's scores (before stimulation: 1.25 $\pm$ 0.26, after stimulation: 1.16 $\pm$ 0.27; $p$ = 0.020) (Figure 3A). The right eye's post-stimulation scores also differed significantly between stimulation conditions, with the score in the tDCS group (0.85 $\pm$ 0.88) being lower than in the sham group (1.91 $\pm$ 1.03) ($p$ = 0.011). Figure 4 shows the difference between the pre- and post-stimulation LogMAR scores of individual patients in the tDCS and sham group.

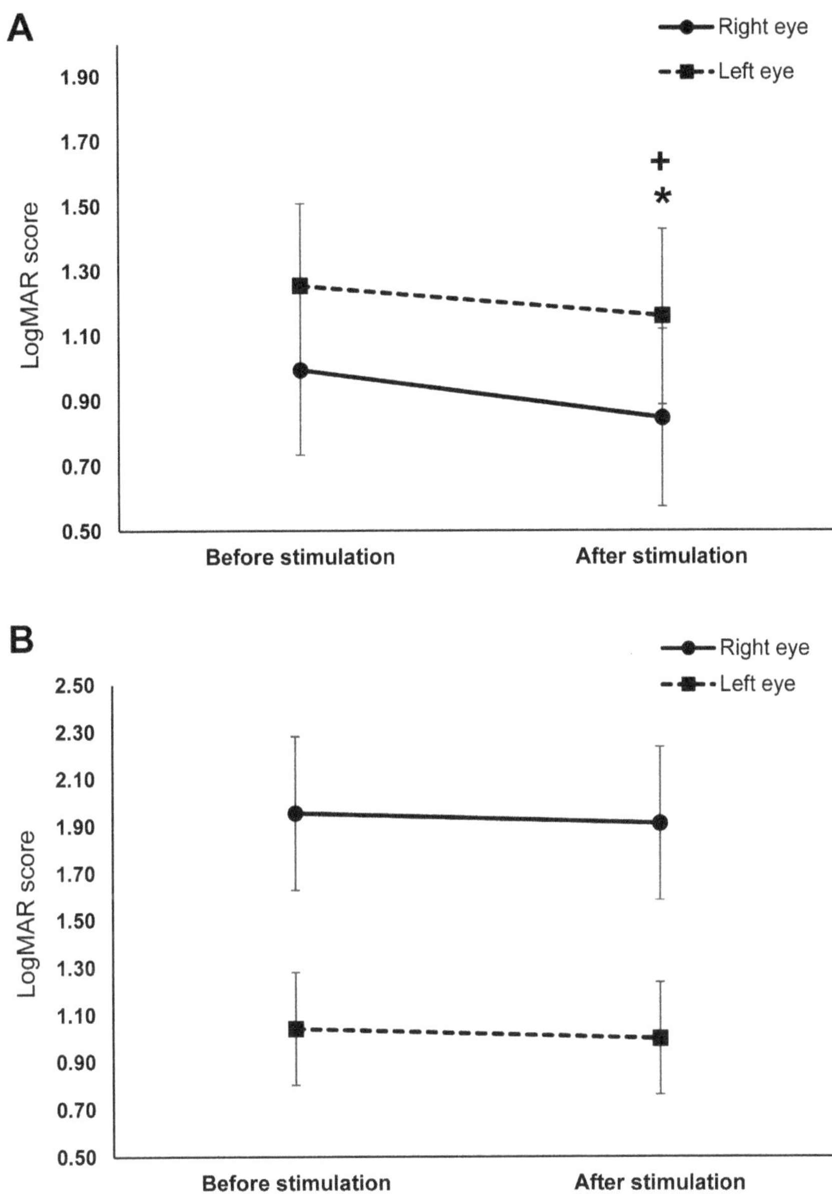


**Figure 3.** The effects of cathodal transcranial direct current stimulation (tDCS) and sham stimulation on visual acuity. The $y$-axis displays the mean LogMAR scores. The $x$-axis displays the time points of the LogMAR score measurement. (**A**) Visual acuity before and after cathodal tDCS stimulation. The right (straight line) and left (dashed line) eye's LogMAR scores significantly decreased after stimulation. (**B**) Visual acuity before and after sham stimulation. There were no significant changes in the LogMAR scores of both eyes after sham stimulation. + = significant differences between the pre- and post-stimulation LogMAR score of the right eye ($p \leq 0.05$), * = significant differences between the pre- and post-stimulation LogMAR score of the left eye ($p \leq 0.05$). Presented data are mean values$\pm$standard error of the mean (SEM).

**Table 2.** Results of the analyses of variances (ANOVAs) performed on the logarithm of the minimum angle of resolution (LogMAR) score and accuracy rate.

| | Numerator *df* | Denominator *df* | *F*-Value | *p*-Value | $\eta_\rho^2$ |
|---|---|---|---|---|---|
| LogMAR score | | | | | |
| Stimulation | 1 | 19 | 2.19 | 0.104 | 0.133 |
| Time | 1 | 19 | 14.97 | 0.001 * | 0.441 |
| Eye | 1 | 19 | 0.51 | 0.482 | 0.026 |
| Time × stimulation | 1 | 19 | 8.92 | 0.008 * | 0.319 |
| Time × eye | 1 | 19 | 1.13 | 0.301 | 0.056 |
| Eye × stimulation | 1 | 19 | 4.83 | 0.041 * | 0.203 |
| Stimulation × time × eye | 1 | 19 | 1.56 | 0.227 | 0.076 |
| Accuracy rate | | | | | |
| Stimulation | 1 | 20 | 0.53 | 0.821 | 0.003 |
| Time | 1 | 20 | 0.18 | 0.677 | 0.009 |
| Time × stimulation | 1 | 20 | 0.12 | 0.732 | 0.006 |

* = $p < 0.05$, *df* = Degrees of freedom.

### 3.2. Number Acuity

For the RTs, the final dataset included 976 trials (73.94% of 1320 trials). RTs from incorrect trials (331) and outliers (13) accounted for 25.08% and 0.98% of the full dataset, respectively, and were excluded in the final analysis. The ANOVA for the log-transformed RTs yielded a significant main effect of time ($F$ (1,20) = 22.02, $p \leq 0.001$, $\eta_\rho^2 = 0.524$) as indicated by the shorter post-stimulation overall RT (before stimulation: 1.85 ± 1.03 s, after stimulation: 1.19 ± 0.05 s). The main effect of stimulation was not significant ($F$ (1,20) = 2.665, $p = 0.118$, $\eta_\rho^2 = 0.118$), indicating comparable overall RT between tDCS (1.89 ± 1.33 s) and sham (1.15 ± 1.33 s) group. However, the effect of time and stimulation interaction was significant ($F$ (1,20) = 5.85, $p = 0.025$, $\eta_\rho^2 = 0.226$). Bonferroni corrected post hoc comparisons showed that RTs significantly decreased in the tDCS group (before stimulation: 2.47 ± 1.77 s, after stimulation: 1.32 ± 0.65 s) ($p \leq 0.001$). However, the RTs were comparable before (1.22 ± 0.68 s) and after (1.07 ± 0.65 s) sham stimulation ($p = 0.124$) (Figure 5A). Figure 6 shows the difference between the pre- and post-stimulation RT of individual patients in the tDCS and sham group. The ANOVA for the AR yielded no significant results (Table 2, Figure 5B).

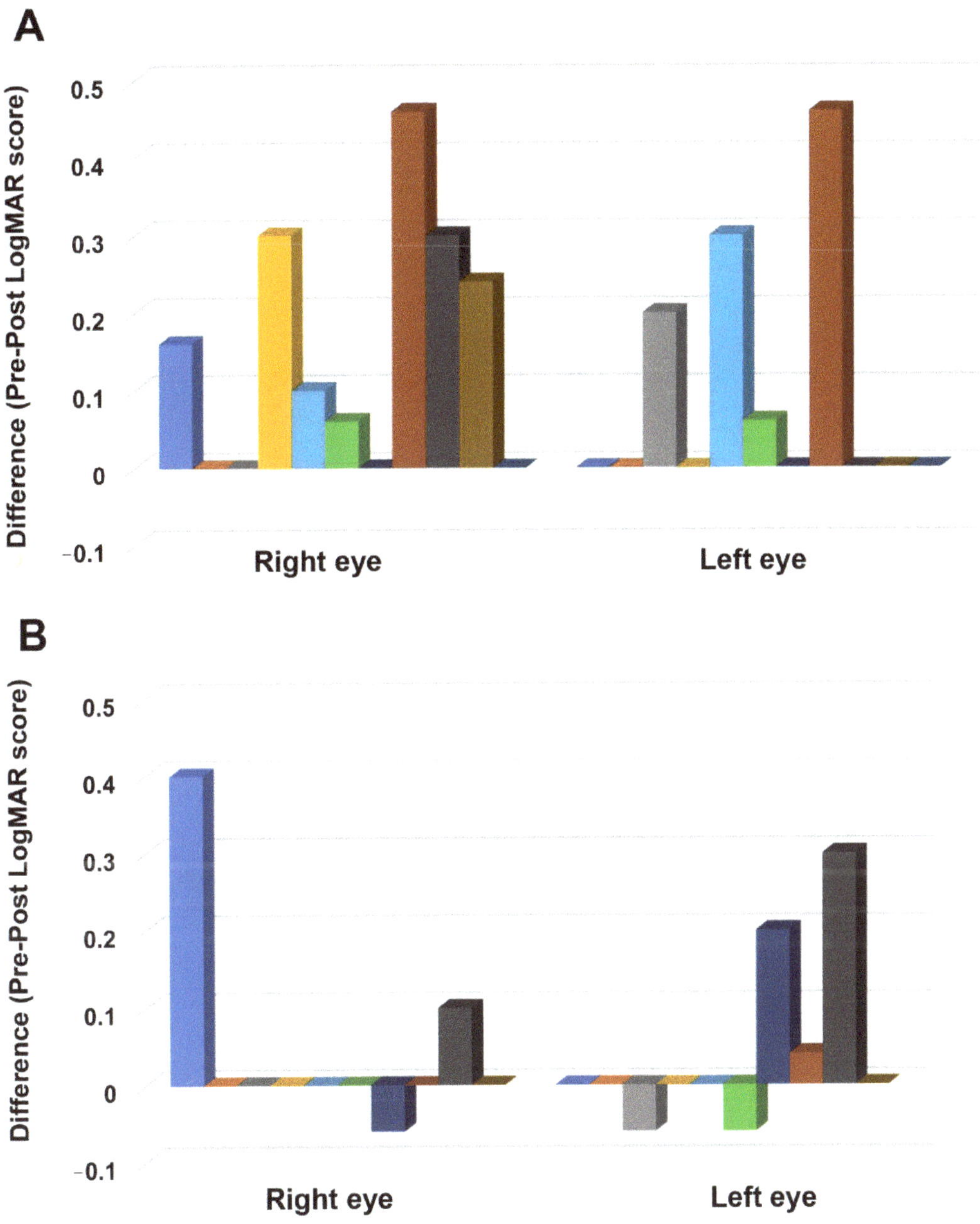


**Figure 4.** The difference between pre- and post-stimulation LogMAR scores. The $y$-axis displays the difference between the before and after stimulation LogMAR scores. The $x$-axis indicates the eye where the measurements are conducted. Each column represents the individual participant's score in the tDCS (**A**) and sham group (**B**). A positive score indicates improvement, and a negative score suggests worsening of visual acuity after stimulation.

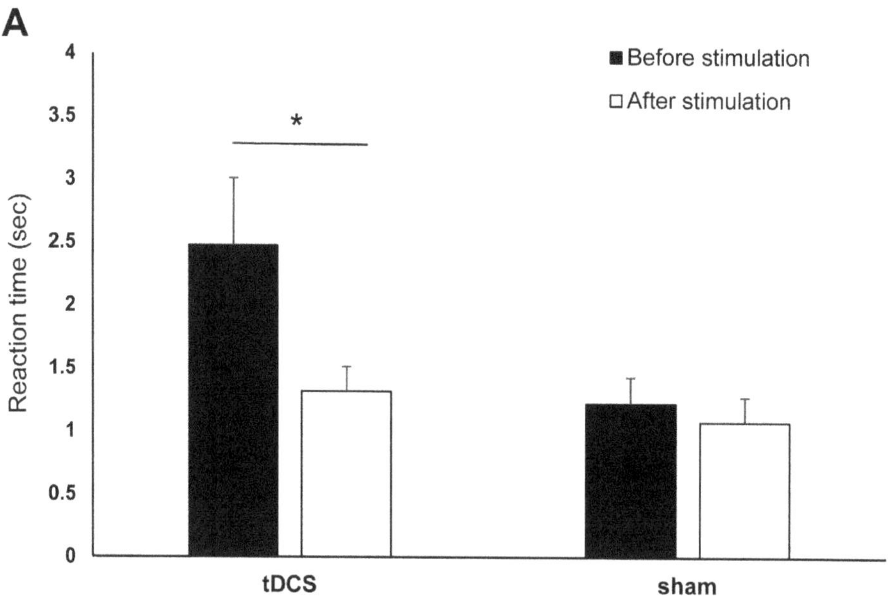


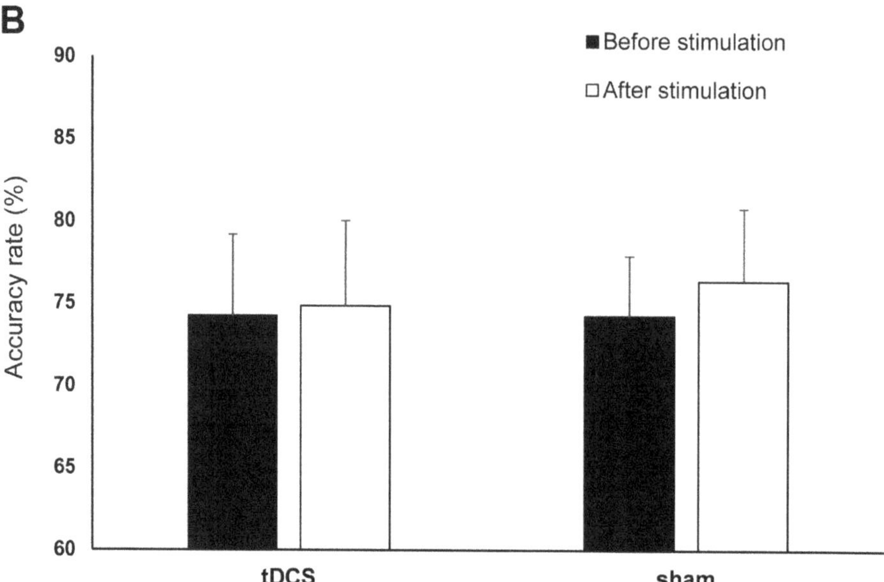


**Figure 5.** The effects of cathodal tDCS and sham stimulation on numerical discrimination. (**A**) Reaction time (RT) before and after cathodal tDCS and sham stimulation. The $y$-axis displays the mean RT (in seconds). The $x$-axis displays the time points of the RTs measurements in the tDCS and sham groups. RT significantly decreased after cathodal tDCS but not after sham. (**B**) Accuracy rate (AR) before and after cathodal tDCS and sham stimulation. The $y$-axis displays the mean AR (%). The $x$-axis displays the time points of the error measurement in the tDCS and sham group. There were no significant changes in AR after tDCS and sham stimulation. tDCS = transcranial direct current stimulation, RT = reaction time, AR = accuracy rate, * = Significant differences between the pre- and post-stimulation measurements ($p \leq 0.05$). Presented data are mean values ± standard error of the mean (SEM).

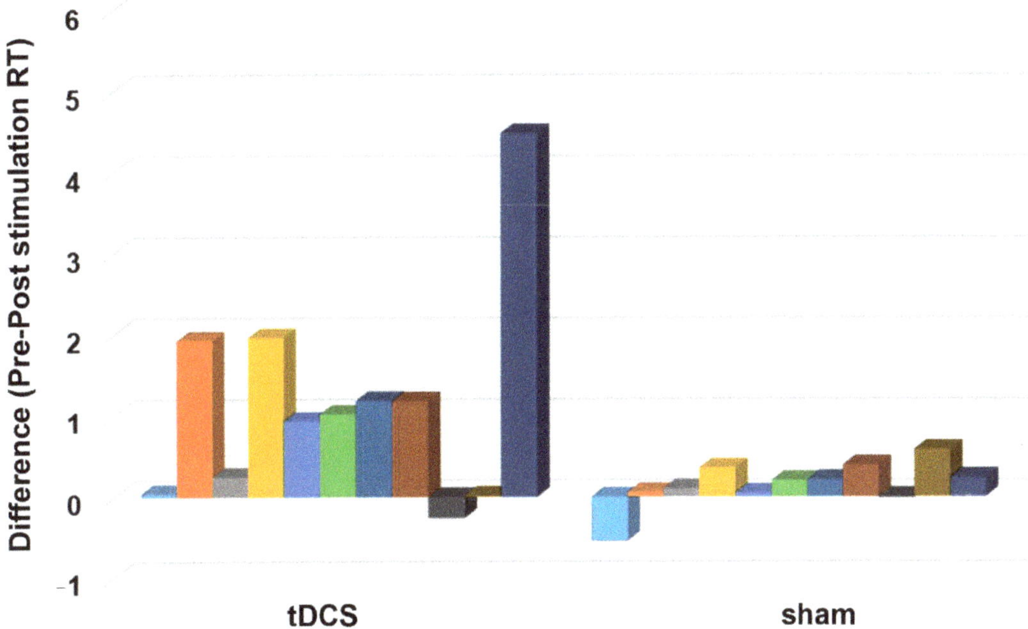


**Figure 6.** The difference between pre- and post-stimulation reaction time (RT). The *y*-axis displays the difference between the before and after stimulation reaction time (s). The *x*-axis indicates the groups (tDCS and sham). Each column represents the individual participant's reaction time difference. A positive and negative score indicates a shorter and longer reaction time after stimulation, respectively.

## 4. Discussion

The present study explored the effect of tDCS stimulation in PDR patients' residual vision. The mean LogMAR score of both eyes and mean RT in numerical discrimination task significantly improved in patients who underwent cathodal tDCS stimulation of V1. Diabetes duration is comparable between the groups and may not have influenced the stimulation's effectiveness (for example, stimulation is less effective for patients with more chronic diabetes). Therefore, the results indicate that cathodal tDCS stimulation of V1 improves the patients' visual and number acuity.

### *4.1. The Effect of Cathodal tDCS on Visual Acuity*

The cathodal tDCS-induced improvement in visual acuity is robust as indicated by the large effect sizes observed on significant interactions involving the factor stimulation (all $\eta_\rho^2 > 0.13$). These results are consistent with our a priori hypothesis that decreasing resting membrane potentials and spontaneous neuronal firing rates in V1 with cathodal tDCS can improve the patients' vision. Concerning the underlying mechanisms, we argue based on the evidence that high internal noise within the visual pathways contributes to vision impairment in retinal diseases [19–21]. Studies suggest that noise-processing neurons act more randomly than neurons that process functional signals; therefore, the inhibitory effect of cathodal tDCS would be more robust over the noise and increase the signal-to-noise ratio in the neuronal network [14,35]. The increased signal-to-noise ratio may boost neural computations in V1 needed for efficient perception and interpretation of sensory signals from eyes with compromised retinal functions. In contrast, modulation of neuronal signals against background noise may not have occurred in the sham group since their visual acuity did not improve after stimulation. Our assumption is in accordance with the findings in amblyopic patients who receive cathodal tDCS over V1 contralateral to the affected eye. In these patients, in addition to a possible stimulation-induced decrease in transcallosal inhibition, a reduction in V1 neuronal excitability as indicated by the reduction of VEPs amplitude may have facilitated visual acuity improvement [17]. Interestingly, cathodal tDCS impaired Vernier acuity, while anodal tDCS improved Vernier and visual acuity (measured with a Landolt gap task) in healthy young individuals [36,37]. Anodal tDCS of V1 also increased VEPs amplitude and improved contrast sensitivity in a group of amblyopic patients who were relatively younger than the participants in the Bocci et al. study [16]. These results theoretically suggest that in cases where neuronal noise is less robust such as in an intact visual system or in patients with less chronic visual system pathology, anodal tDCS may enhance visual acuity by boosting functional neuronal signals. On the other hand, cathodal tDCS can be detrimental in such cases because it may impair both noise and functional neuronal signals. In contrast, in more chronic or advanced visual system disorders, where neuronal noise is more robust, cathodal tDCS can be more beneficial because of its inhibitory effect.

### 4.2. The Effect of Cathodal tDCS on Number Acuity

The decrease in RT after cathodal tDCS was also robust, as indicated by the large effect size of the time and stimulation interaction ($\eta_\rho^2 = 0.226$). In contrast, cathodal tDCS had a limited impact on accuracy because the analysis of ARs revealed no significant results. This is not surprising because task difficultly is low, and the patients' ARs are already high (mean AR: 74.24%) before stimulation. Therefore, the patients could have immediately reached a ceiling effect on task performance. Nonetheless, there is no indication of speed-accuracy trade-off (decrease in RT with an increase AR or vice versa) in either the cathodal tDCS or sham group. Overall, although the between groups' AR is comparable, patients who received cathodal tDCS were faster in discriminating whether there were more black or more white dots than those who received sham stimulation. Here, we argue that this is due to the modulation of the underlying neural mechanism behind the perceptual processing of numerosity, such as discrimination and encoding (individuation of dots) in V1 by cathodal tDCS [27–30,38]. The hyperpolarizing effect of cathodal tDCS may tune or denoise the activity of V1 neurons, particularly those with receptive fields that respond best to outline, contour, and edges defined by the ratio-dependent distribution of black and white dots in space [39–41]. By serving as a "noise filter" in the early visual areas, improvement in visuospatial information processing in other regions of the ventral and dorsal visual streams such as the temporal (object processing), parietal (spatial processing), and frontal (decision making) cortices can be considered a secondary effect of cathodal tDCS stimulation of V1 [25,42]. V1 cathodal stimulation may diminish the robust resting-state functional connectivity between V1 and the frontal lobe, which can interfere with the frontal eye-fields' critical function for saccadic eye movements and perceptual decision-making [22,43]. Furthermore, tDCS may have modulated the ratio-dependent number processing system that responds selectively to nonsymbolic quantities (e.g., dot arrays) of larger ratios (3:1 or 4:1) reported in the subcortical monocular portion of V1 [26]. Interestingly, there are no significant differences in AR or RT between ratio conditions in our study, indicating the absence of cathodal tDCS inhibitory effect on visual attentional skills on which numerosity perception depends. One possible explanation is that our task specifically recruits the ANS, which is relatively "attentional-free" compared to the attentional dependent "subitizing system" (for low numerosities: <4 dots) and the attentional demanding "texture-density mechanism" operating for high dense/numerous stimuli (>60 dots) [44,45] This is because although the ratio of black and white dots is allowed to vary in our task, the total number of dots in the array is fixed (60) as well as other features such as dots sparsity. Modulation of the subcortical monocular portion of V1 with tDCS is possible trans-synaptically [46]. However, this assumption remains to be determined in PDR patients using NIBS techniques with deeper penetration, such as TMS.

### 4.3. Limitations

In the present study, there were some limitations that we should address. First, contrast sensitivity (CS) is not measured because of the unavailability of a standard CS test. We considered this a significant limitation because there is evidence indicating that CS measurement can provide better information about the impact of intrinsic noise in early-stage diabetic patients [21]. In the case of patients with PDR, the effect of neuronal noise is not yet clearly understood. Further investigations should address the effects of tDCS on CS in NDR and PDR patients. Second, for safety reason, the patients' maintenance medications (Table 1) were not discontinued during the experiment. We cannot entirely rule out the influence of these medications on the effect of tDCS, particularly those that can pass the blood-brain-barrier. For future studies, discontinuation of drugs may be possible at least 24 h before the stimulation in patients with stable blood glucose level and those taking fewer medications. Third, we did not identify the dominant eye of each patient. Although the significance of eye dominance is yet to be established [47], particularly in the field of brain stimulation, it is tempting to assume that tDCS effects may differ on the dominant and non-dominant eyes, because in theory, the dominant one provides more input to the visual cortex. Fourth, the patients did not receive anodal stimulation, and therefore the polarity-dependent effect of tDCS is not tested in the present study. Future experiments must apply anodal, cathodal, and sham stimulation in a within-subject design to systematically control for stimulation-specific effects. Finally, our sample size is relatively small, and extrapolating the results to all diabetic patients must be cautioned. Additional studies with a larger sample size are needed further to explore the impact of tDCS in V1 of diabetic patients.

## 5. Conclusions

The present study provides preliminary evidence that non-invasive brain stimulation methods such as tDCS can improve PDR patients' residual vision. The results demonstrated the promising potential of tDCS as a vision rehabilitation tool for patients with other retinal diseases.

**Supplementary Materials:** The following are available online at https://www.mdpi.com/2076-3425/11/2/270/s1, Table S1: LogMAR scores, mean reaction times and mean accuracy rates of individual patient before and after stimulation.

**Author Contributions:** A.B.F.d.V.III and S.M.F. contributed to the study conception and design, material preparation, data collection and analysis. S.M.F. wrote the first draft of the manuscript, and A.B.F.d.V.III commented on previous versions of the manuscript. The authors read and approved the final manuscript. All authors have read and agreed to the published version of the manuscript.

**Funding:** This research received no external funding.

**Institutional Review Board Statement:** All patients gave their informed consent for inclusion before they participated in the study. The study was conducted in accordance with the Declaration of Helsinki, and the protocol was approved by the chief of clinics who represents the hospital board of Nazareth General Hospital.

**Informed Consent Statement:** Informed consent was obtained from all patients involved in the study. Clinical trial registry: ISRCTN. Registration number: ISRCTN70877737.

**Data Availability Statement:** Data is contained within the article or supplementary material.

**Acknowledgments:** Open Access Funding by the University of Graz.

**Conflicts of Interest:** The authors declare no conflict of interest.

## References

1. Wild, S.; Roglic, G.; Green, A.; Sicree, R.; King, H. Global Prevalence of Diabetes. *Diabetes Care.* **2004**, *27*, 1047–1053. [CrossRef]
2. Mansour, S.E.; Browning, D.J.; Wong, K.; Flynn, H.W., Jr.; Bhavsar, A.R. The Evolving Treatment of Diabetic Retinopathy. *Clin. Ophthalmol.* **2020**, *14*, 653–678. [CrossRef]

3. Wang, W.; Lo, A.C.Y. Diabetic Retinopathy: Pathophysiology and Treatments. *Int. J. Mol. Sci.* **2018**, *19*, 1816. [CrossRef]
4. Sabel, B.A.; Thut, G.; Haueisen, J.; Henrich-Noack, P.; Herrmann, C.S.; Hunold, A.; Kammer, T.; Matteo, B.; Sergeeva, E.G.; Waleszczyk, W.; et al. Vision modulation, plasticity and restoration using non-invasive brain stimulation—An IFCN-sponsored review. *Clin. Neurophysiol.* **2020**, *131*, 887–911. [CrossRef] [PubMed]
5. Stagg, C.J.; Nitsche, M.A. Physiological Basis of Transcranial Direct Current Stimulation. *Neuroscience* **2011**, *17*, 37–53. [CrossRef] [PubMed]
6. Nitsche, M.A.; Cohen, L.G.; Wassermann, E.M.; Priori, A.; Lang, N.; Antal, A.; Paulus, W.; Hummel, F.; Boggio, P.S.; Fregni, F.; et al. Transcranial direct current stimulation: State of the art 2008. *Brain Stimul.* **2008**, *1*, 206–223. [CrossRef]
7. Nitsche, M.A.; Paulus, W. Excitability changes induced in the human motor cortex by weak transcranial direct current stimulation. *J. Physiol.* **2000**, *527*, 633–639. [CrossRef] [PubMed]
8. Antal, A.; Kincses, T.Z.; Nitsche, M.A.; Paulus, W. Manipulation of phosphene thresholds by transcranial direct current stimulation in man. *Exp. Brain Res.* **2003**, *150*, 375–378. [CrossRef]
9. Antal, A.; Kincses, T.Z.; Nitsche, M.A.; Paulus, W. Modulation of moving phosphene thresholds by transcranial direct current stimulation of V1 in human. *Neuropsychologia* **2003**, *41*, 1802–1807. [CrossRef]
10. Antal, A.; Kincses, Z.T.; Nitsche, M.; Bartfai, O.; Paulus, W. Excitability changes induced in the human primary visual cortex by transcranial direct current stimulation: Direct electrophyiological evidence. *Investig. Ophthalmol. Vis. Sci.* **2004**, *45*, 702–707. [CrossRef] [PubMed]
11. Wunder, S.; Hunold, A.; Fiedler, P.; Schlegelmilch, F.; Schellhorn, K.; Haueisen, J. Novel bifunctional cap for simultaneous electroencephalography and transcranial electrical stimulation. *Sci. Rep.* **2018**, *8*, 7259. [CrossRef] [PubMed]
12. Costa, T.L.; Nagy, B.V.; Barboni, M.T.S.; Boggio, P.S.; Ventura, D.F. Transcranial direct current stimulation modulates human color discrimination in a pathway-specific manner. *Front. Psychiatry* **2012**, *3*, 78. [CrossRef]
13. Costa, T.L.; Gualtieri, M.; Barboni, M.T.S.; Katayama, R.K.; Boggio, P.S.; Ventura, D.F. Contrasting effects of transcranial direct current stimulation on central and peripheral visual fields. *Exp. Brain Res.* **2015**, *233*, 1391–1397. [CrossRef] [PubMed]
14. Costa, T.L.; Lapenta, O.M.; Boggio, P.S.; Ventura, D.F. Transcranial direct current stimulation as a tool in the study of sensory-perceptual processing. *Atten. Percept. Psychophys.* **2015**, *77*, 1813–1840. [CrossRef] [PubMed]
15. Spiegel, D.P.; Byblow, W.D.; Hess, R.F.; Thompson, B. Anodal Transcranial Direct Current Stimulation Transiently Improves Contrast Sensitivity and Normalises Visual Cortex Activation in Individuals With Amblyopia. *Neurorehabil. Neural Repair* **2013**, *27*, 760–769. [CrossRef]
16. Ding, Z.; Li, J.; Spiegel, D.P.; Chen, Z.; Chan, L.; Luo, G.; Yuan, J.; Deng, D.; Yu, M.; Thompson, B. The effect of transcranial direct current stimulation on contrast sensitivity and visual evoked potential amplitude in adults with amblyopia. *Sci. Rep.* **2016**, *6*, 19280. [CrossRef]
17. Bocci, T.; Nasini, F.; Caleo, M.; Restani, L.; Barloscio, D.; Ardolino, G.; Priori, A.; Maffei, L.; Nardi, M.; Sartucci, F. Unilateral Application of Cathodal tDCS Reduces Transcallosal Inhibition and Improves Visual Acuity in Amblyopic Patients. *Front. Behav. Neurosci.* **2018**, *12*, 109. [CrossRef]
18. Gall, C.; Schmidt, S.; Schittkowski, M.P.; Antal, A.; Ambrus, G.G.; Paulus, W.; Dannhauer, M.; Michalik, R.; Mante, A.; Bola, M.; et al. Alternating Current Stimulation for Vision Restoration after Optic Nerve Damage: A Randomised Clinical Trial. *PLoS ONE* **2016**, *11*, e0156134. [CrossRef]
19. Pelli, D.G.; Levi, D.M.; Chung, S.T.L. Using visual noise to characterize amblyopic letter identification. *J. Vis.* **2004**, *4*, 6. [CrossRef]
20. McAnany, J.J.; Alexander, K.R.; Genead, M.A.; Fishman, G.A. Equivalent Intrinsic Noise, Sampling Efficiency, and Contrast Sensitivity in Patients With Retinitis Pigmentosa. *Investig. Opthalmol. Vis. Sci.* **2013**, *54*, 3857. [CrossRef]
21. McAnany, J.J.; Park, J.C. Reduced Contrast Sensitivity is Associated With Elevated Equivalent Intrinsic Noise in Type 2 Diabetics Who Have Mild or No Retinopathy. *Investig. Opthalmol. Vis. Sci.* **2018**, *59*, 2652. [CrossRef]
22. Yu, Y.; Lan, D.-Y.; Tang, L.-Y.; Su, T.; Li, B.; Jiang, N.; Liang, R.-B.; Ge, Q.-M.; Li, Q.-Y.; Shao, Y. Intrinsic functional connectivity alterations of the primary visual cortex in patients with proliferative diabetic retinopathy: A seed-based resting-state fMRI study. *Ther. Adv. Endocrinol. Metab.* **2020**, *11*, 204201882096029. [CrossRef]
23. Cappelletti, M.; Gessaroli, E.; Hithersay, R.; Mitolo, M.; Didino, D.; Kanai, R.; Kadosh, R.C.; Walsh, V. Transfer of Cognitive Training across Magnitude Dimensions Achieved with Concurrent Brain Stimulation of the Parietal Lobe. *J. Neurosci.* **2013**, *33*, 14899–14907. [CrossRef]
24. Halberda, J.; Mazzocco, M.M.M.; Feigenson, L. Individual differences in non-verbal number acuity correlate with maths achievement. *Nature* **2008**, *455*, 665–668. [CrossRef] [PubMed]
25. Dehaene, S.; Piazza, M.; Pinel, P.; Cohen, L. Three Parietal Circuits for Number Processing. *Cogn. Neuropsychol.* **2003**, *20*, 487–506. [CrossRef] [PubMed]
26. Collins, E.; Park, J.; Behrmann, M. Numerosity representation is encoded in human subcortex. *Proc. Natl. Acad. Sci. USA* **2017**, *114*, E2806–E2815. [CrossRef]
27. Fornaciai, M.; Brannon, E.M.; Woldorff, M.G.; Park, J. Numerosity processing in early visual cortex. *Neuroimage* **2017**, *157*, 429–438. [CrossRef]
28. Guillaume, M.; Mejias, S.; Rossion, B.; Dzhelyova, M.; Schiltz, C. A rapid, objective and implicit measure of visual quantity discrimination. *Neuropsychologia* **2018**, *111*, 180–189. [CrossRef] [PubMed]

29. DeWind, N.K.; Park, J.; Woldorff, M.G.; Brannon, E.M. Numerical encoding in early visual cortex. *Cortex* **2019**, *114*, 76–89. [CrossRef]
30. van Rinsveld, A.; Guillaume, M.; Kohler, P.J.; Schiltz, C.; Gevers, W.; Content, A. The neural signature of numerosity by separating numerical and continuous magnitude extraction in visual cortex with frequency-tagged EEG. *Proc. Natl. Acad. Sci. USA* **2020**, *117*, 5726–5732. [CrossRef] [PubMed]
31. Matsumoto, H.; Ugawa, Y. Adverse events of tDCS and tACS: A review. *Clin. Neurophysiol. Pract.* **2017**, *2*, 19–25. [CrossRef]
32. Lange, C.; Feltgen, N.; Junker, B.; Schulze-Bonsel, K.; Bach, M. Resolving the clinical acuity categories "hand motion" and "counting fingers" using the Freiburg Visual Acuity Test (FrACT). *Graefe's Arch. Clin. Exp. Ophthalmol.* **2009**, *247*, 137–142. [CrossRef] [PubMed]
33. Foxe, J.; Simpson, G. Flow of activation from V1 to frontal cortex in humans. *Exp. Brain Res.* **2002**, *142*, 139–150. [CrossRef] [PubMed]
34. Reinhart, R.M.G.; Zhu, J.; Park, S.; Woodman, G.F. Medial–Frontal Stimulation Enhances Learning in Schizophrenia by Restoring Prediction Error Signaling. *J. Neurosci.* **2015**, *35*, 12232–12240. [CrossRef]
35. Antal, A.; Paulus, W. Transcranial Direct Current Stimulation and Visual Perception. *Perception* **2008**, *37*, 367–374. [CrossRef] [PubMed]
36. Reinhart, R.M.G.; Xiao, W.; McClenahan, L.J.; Woodman, G.F. Electrical Stimulation of Visual Cortex Can Immediately Improve Spatial Vision. *Curr. Biol.* **2016**, *26*, 1867–1872. [CrossRef]
37. Bonder, T.; Gopher, D.; Yeshurun, Y. The Joint Effects of Spatial Cueing and Transcranial Direct Current Stimulation on Visual Acuity. *Front. Psychol.* **2018**, *9*, 159. [CrossRef] [PubMed]
38. Salthouse, T.A. Speed of behavior and its implications for cognition. In *Handbook of the Psychology of Aging*; Birren, J.E., Schaie, K.W., Eds.; Van Nostrand Reinhold: New York, NY, USA, 1985.
39. Koulakov, A.; Chklovskii, D. Orientation Preference Patterns in Mammalian Visual Cortex: A Wire Length Minimisation Approach. *Neuron* **2001**, *29*, 519–527. [CrossRef]
40. Sasaki, Y.; Watanabe, T. The primary visual cortex fills in color. *Proc. Natl. Acad. Sci. USA* **2004**, *101*, 18251–18256. [CrossRef] [PubMed]
41. Seymour, K.; Clifford, C.W.G.; Logothetis, N.K.; Bartels, A. Coding and Binding of Color and Form in Visual Cortex. *Cereb. Cortex.* **2009**, *20*, 1946–1954. [CrossRef]
42. Chan, J.S.; Newell, F.N. Behavioral evidence for task-dependent "what" versus "where" processing within and across modalities. *Percept. Psychophys.* **2008**, *70*, 36–49. [CrossRef]
43. Murd, C.; Moisa, M.; Grueschow, M.; Polania, R.; Ruff, C.C. Causal contributions of human frontal eye fields to distinct aspects of decision formation. *Sci. Rep.* **2020**, *10*, 7317. [CrossRef] [PubMed]
44. Anobile, G.; Tomaiuolo, F.; Campana, S.; Cicchini, G.M. Three-systems for visual numerosity: A single case study. *Neuropsychologia* **2020**, *136*, 107259. [CrossRef]
45. Pomè, A.; Anobile, G.; Cicchini, G.M.; Scabia, A.; Burr, D.C. Higher attentional costs for numerosity estimation at high densities. *Atten. Percept. Psychophys.* **2019**, *81*, 2604–2611. [CrossRef] [PubMed]
46. Nonnekes, J.; Arrogi, A.; Munneke, M.A.M.; van Asseldonk, E.H.F.; Nijhuis, L.B.O.; Geurts, A.C.; Weerdesteyn, V. Subcortical Structures in Humans Can Be Facilitated by Transcranial Direct Current Stimulation. *PLoS ONE* **2014**, *9*, e107731. [CrossRef]
47. AMapp, P.; Ono, H.; Barbeito, R. What does the dominant eye dominate? A brief and somewhat contentious review. *Percept. Psychophys.* **2003**, *65*, 310–317. [CrossRef]

*Review*

# Extremely Low-Frequency Magnetic Field as a Stress Factor—Really Detrimental?—Insight into Literature from the Last Decade

**Angelika Klimek and Justyna Rogalska ***

Department of Animal Physiology and Neurobiology, Faculty of Biological and Veterinary Sciences, Nicolaus Copernicus University, 87-100 Torun, Poland; klimek@doktorant.umk.pl
* Correspondence: rogal@umk.pl; Tel.: +48-56-611-26-31

**Abstract:** Biological effects of extremely low-frequency magnetic field (ELF-MF) and its consequences on human health have become the subject of important and recurrent public debate. ELF-MF evokes cell/organism responses that are characteristic to a general stress reaction, thus it can be regarded as a stress factor. Exposure to ELF-MF "turns on" different intracellular mechanisms into both directions: compensatory or deleterious ones. ELF-MF can provoke morphological and physiological changes in stress-related systems, mainly nervous, hormonal, and immunological ones. This review summarizes the ELF-MF-mediated changes at various levels of the organism organization. Special attention is placed on the review of literature from the last decade. Most studies on ELF-MF effects concentrate on its negative influence, e.g., impairment of behavior towards depressive and anxiety disorders; however, in the last decade there was an increase in the number of research studies showing stimulating impact of ELF-MF on neuroplasticity and neurorehabilitation. In the face of numerous studies on the ELF-MF action, it is necessary to systematize the knowledge for a better understanding of the phenomenon, in order to reduce the risk associated with the exposure to this factor and to recognize the possibility of using it as a therapeutic agent.

**Keywords:** magnetic field; stress; HPA axis; catecholamines; cytokines; hormones; behavior; anxiety; neuroplasticity; cell survival

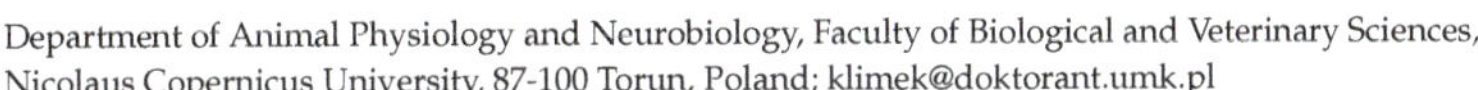

**Citation:** Klimek, A.; Rogalska, J. Extremely Low-Frequency Magnetic Field as a Stress Factor—Really Detrimental?—Insight into Literature from the Last Decade. *Brain Sci.* **2021**, *11*, 174. https://doi.org/10.3390/brainsci11020174

Academic Editor: Ulrich Palm
Received: 22 December 2020
Accepted: 27 January 2021
Published: 31 January 2021



## 1. Introduction

Many studies have suggested an association between extremely low-frequency magnetic field (ELF-MF) exposure and anxiety and/or depression. On the other hand, the ELF-MF-induced improvement of brain function has also been found. The mechanism of these effects is assumed to be a stress response induced by ELF-MF exposure. Extremely low-frequency MF is natural physical phenomenon in our environment. The rapid development of science and technology resulted in the introduction of many new devices and technologies in industry, agriculture, and everyday life. We are continuously exposed in our environment to ELF-MF (range of 0–300 Hz) [1]. MFs are either of natural origin (geomagnetic field, intense solar activity, thunderstorms) or human-made (factories, transmission lines, electric appliances at work and home, magnetic resonance imaging, medical treatment, etc.) [2]. Common used frequencies of electric and magnetic fields of the electric power supply and of electric and magnetic fields generated by electricity power lines and electric/electronic devices are 50 Hz in Europe and 60 Hz in North America [2]. Biological effects of ELF-MF and their consequences on human health have become the subject of important and recurrent public debate. Until now the reported studies are largely contradictory with regard to epidemiologic studies (some of the research studies found a relationship with development of diseases while the others failed to find any [3–8] (Table S1). Whether or not ELF-MF exposure is related to increased health risks, it has led many scientists to examine the potential mechanisms by which ELF-MF might affect human health. Special attention is paid to the adverse impact of both low- and high-frequency MF (radio waves) due to many possible pathological effects and numerous reports on MF-induced carcinogenicity [9]. ELF-MF was proved to be a stress factor and as a consequence, it can provoke morphological and physiological changes in stress-related systems [10]. Some authors argue that ELF-MF evokes cell/organism responses that are characteristic to general stress reaction. ELF-MF exposure "turns on" different intracellular—compensatory or deleterious—mechanisms and modifies stress-related function of nervous, hormonal and immunological systems (Figure 1). ELF-MF influence on living matter can cause a detrimental increase in free radicals levels and radical-evoked damages in macromolecules [11]. Most studies on ELF-MF effects concentrate on its negative influence; however, in the last decade there was an increase in the number of research studies showing stimulating impact of ELF-MF on brain plasticity processes (the production of protective proteins (e.g., Hsp70 or BDNF) or an increase in the activity of antioxidant enzymes) [12]. Furthermore, long-term exposure to ELF-MF can cause permanent changes in behavior (towards depressive and anxiety disorders) that are related to exposure to chronic stress [10,13,14].

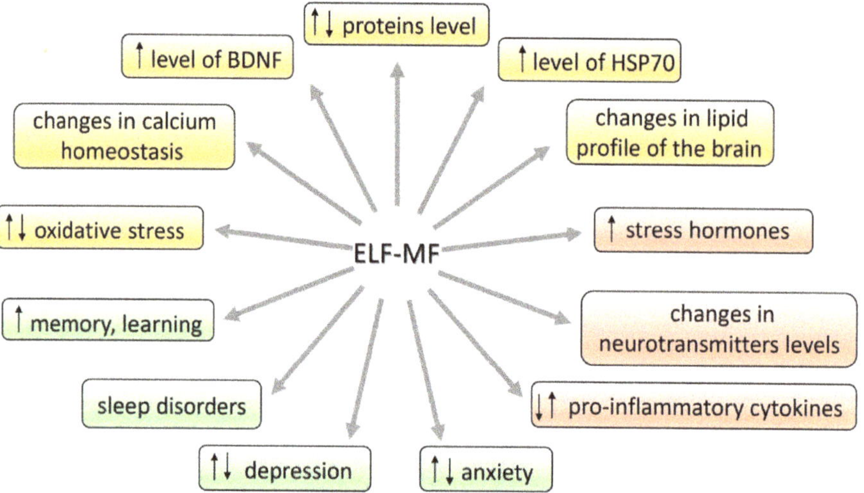


**Figure 1.** Effects of extremely low-frequency magnetic field (ELF-MF) action in the organism.

In the face of numerous studies on the effects of the ELF-MF, it is necessary to systematize the knowledge for a better understanding of this phenomenon, in order to reduce the risk associated with exposure to this factor, but also to recognize the possibility of using it as a therapeutic agent.

## 2. Stress—A Factor Determining the Function of Organism at All Levels of Organization

It is accepted that ELF-MF exposure may count as a mild stress situation [10,15–17] and it could activate a wide spectrum of interacting neuronal, molecular, and neurochemical systems that underpin behavioral and physiological responses. Chronic stress can promote and exacerbate pathophysiology leading to allostatic overload in human body [18]. The brain developed some adaptive mechanisms in the face of changing environments and stress factors imposed on the nervous system. Integrated response to stressful stimuli is an essential component of adaptive processes critical for survival of the organism. Failure of this stress adaptation is considered as one of the primary neuropathological causes of stress-related disorders. A healthy organism is able to turn on or off effectively physiological and psychological responses to stimuli; however, if the stress system response is not adequate—too slow or too high, its mediators will enhance vulnerability to stress-related disease to which the individual is predisposed. Adaptation to repeated stress is associated with a complex cascade of molecular and cellular events, ranging from regulation of gene expression to release of neurotransmitters [19].

The definition of stress is not precise because the process is differently understood by people representing various fields of science. Stress can be discussed in the context of its influence on all levels of an organism's organization: molecular, cellular, physiological, and behavioral as well as psychological. The term "stress" was introduced by Hans Selye [20] and described as a result of disturbed homeostasis in the organism. Seyle [21] stated that stress is a "nonspecific response of the body to any demand". McEwen defined stress as "experiences that are challenging emotionally and physiologically" [18]. Other authors describe stress as a process involving perception, interpretation, response, and adaptation to harmful, threatening, or challenging events [22]. The reaction to a stress event is necessary for the organism to cope with danger [18]. Alarming signals include an internal, psychological, or environmental stimulus—such as ELF-MF. Some authors postulate that the changes turned on under the influence of exposure to ELF-MF are similar to those caused by other stress factors. The consequences of stress can be different and are dependent mainly on the strength of the stimulus. A low dose of stress can drive adaptive processes such as plasticity processes, e.g., the growth of postsynaptic (dendritic) spines, production of stress-resistant proteins, e.g., BDNF (brain-derived neurotrophic factor) and stimulation of neural stem cells to form new neurons that replace or cooperate with the existing ones [23], whereas even one high dose of a given factor may be harmful or even lethal [24].

In the neuroendocrine approach, stress is related to activation of the autonomic nervous system (SAM) and hypothalamo–pituitary–adrenal (HPA) axis. First, the autonomic nervous system is activated causing the release of noradrenaline and adrenaline from the adrenal medulla into the circulation, which—being a hormone—can rapidly regulate the function of peripheral organs [25] as well as the immunological response, which is supposed to adapt the organism to new, stressful conditions [26]. Acute activation of this system leads to release of noradrenaline from an extensive network of neurons throughout the brain, producing an enhanced state of arousal, which is critical for adaptive responses to stress [27]. Somewhat later, the HPA axis is activated, which causes the secretion of corticosteroid hormones from the adrenal cortex [25]. In response to a stressor, corticotropin-releasing hormone (CRH) is secreted in the hypothalamus. CRH is then driven with blood to anterior pituitary, where it causes adrenocorticotropic hormone (ACTH) release. In the next stage, ACTH reaches the adrenal glands and as a consequence glucocorticoids (cortisol and corticosterone) are secreted. The glucocorticoids cause increased arousal that ensures the organism's readiness for action. Thus, the HPA axis system regulates the intensity, dynamics, and termination of the stress response [28,29]. Hippocampus, which role is the inhibition the HPA axis via the negative feedback, is of crucial importance for the dynamic of stress response [30–32]. On the other hand, corticosteroids can also modulate hippocampal function in the opposite directions: causing neuron's dysfunction or plasticity and as a result, hippocampus-related behavioral changes can be observed. The hypothalamic–pituitary–adrenal (HPA) axis is sensitive to a broad spectrum of experimental and environmental events [30] that may result in physiological and behavioral changes in both directions: detrimental and compensatory ones. Sometimes these modifications are very subtle, but in some conditions they can even underlie the stress-related disorders or mediate the reversion of brain damage.

## 3. Molecular Stress Response to ELF-MF

Many studies show that stress induces the disruption in homeostasis [23,33] and as a consequence an overcompensation response is triggered to re-establish homeostasis. It needs gene expression and protein synthesis that progresses over time and leads to establish a new set-point for stress response systems. As ELF-MF is able to change the stress parameters, it is suggested that it can shift the set-point of endocrinological regulations and determine the health status of the organism as a consequence [15]. The effects of exposure to ELF-MF are particularly prevalent in the hippocampal area of the brain [34,35]. As mentioned, the hippocampus is involved in regulating the HPA axis activity, but on the other hand, stress hormones (mainly corticosterone and noradrenaline) are known to modulate hippocampal function and they may determine the plasticity processes in this area; it means an adaptive response to ELF-MF's exposure. Targets for ELF-MF at molecular level include the cell membrane (e.g., its permeability, inorganic ion transport, receptor function), second messengers synthesis, chromosome structural changes and chemical changes in DNA structure, genes expression and protein synthesis (e.g., metabolism-related), free radicals, and neurotrophic factors. Such profound modifications have to be reflected in neurotransmitter activity, hormone release, and metabolism of the brain [36–44]. What is important, the effect of ELF-MF on molecular and/or cellular mechanisms is not obvious—it can be detrimental or protective. However, the research study on these mechanisms can shed some light on the possible metabolic pathway being possibly influenced by ELF-MF. ELF-MF-evoked cellular stress includes the modifications of key substances in cell metabolism—proteins and lipids. The mentioned alternations are mainly related to ELF-MF-induced oxidative stress. The consequence of these processes can be cell death such as apoptosis, necrosis, or autophagy [42,45–48].

### *3.1. Proteins and Lipids*

As shown in in vivo research, ELF-MF can affect levels and function of proteins—crucial for maintenance of cell homeostasis, e.g., proteins anchored in lipid bilayer of the cell membrane functioning as ion channels, enzymes, and receptors, as well as the other proteins of key importance for the response to stress, regulation of apoptosis, and a number of metabolic processes [37,45,47,49,50]. Total protein level as well as its activity (e.g., alanine aminotransferase (ALT), aspartate aminotransferase (AST), alkaline phosphatase (ALP), albumin, bilirubin) was augmented in rats exposed to 1.5 mT ELF-MF [45]. Exposure to both 0.5 and 1 mT ELF-MF altered protein pattern in rat's hippocampus. Gene ontology analysis showed that the most important function of the identified proteins altered after ELF-MF exposure is to ensure the functioning of the brain. Exposure to ELF-MF caused extreme downregulation of two proteins: Sptan1 and Dpysl2. The first is responsible for stabilization of cell membrane and organization of intracellular organelles. The second, Dpysl2, plays a key role in neuronal development and polarity, and additionally in neuron projection morphogenesis. Notably, the increased intensity of ELF-MF may be associated with more alteration in cell protein expression, and subsequent cell morphology and proliferation rate changes [47]. The chromogranin A (CgA) is another protein that should be mentioned as important in stress response and as a new target for electromagnetic radiation. It is a neuroendocrine secretory protein costored and coreleased with catecholamines from adrenal medulla, adrenergic nerve endings, and neuroendocrine cells. CgA is also a marker of sympathoadrenal activity, so its level gives information on the course of stress response [51]. The protein is also involved in maintaining calcium homeostasis in the cell [52]. What is important, its level increases during a depressive mood or stress situation [52]. The serum level of CgA in volunteer subjects chronically exposed to ELF-MF in the range 0.1–0.3 μT did not differ from the level in control group. However, a trend toward lower concentrations of CgA was observed in the group exposed to higher level of ELF-MF (>0.3 μT). Suppressive effects of ELF-MF on CgA level could be recognized as having inhibitory effects on the activity of the sympathetic nervous system [53].

In addition to proteins, the brain lipid profile is also influenced by ELF-MF exposure and taking into account multiple roles for lipids, they can be the medium for the ELF-MF action in the cell. Lipids are structural components of the cell membrane and they are involved in transfer of signals across membranes [17]. Apart from being structural elements, they are also required for axonal elongation and act as precursors for various secondary messengers, including arachidonic acid, docosahexaenoic acid, or 1,2-diacylglycerol [54]. Any changes in brain lipid metabolism lead to disturbances in homeostasis and are responsible for altered functioning at the cell and tissue levels. It was shown that 60 Hz 2.4 mT ELF-MF induces changes in the brain lipid profile and in corticosterone concentration. The level of these changes was similar to that in the positive control group of rats exposed to stress-RS (movement restraint). After 21 days of exposure to ELF-MF or RS or combined model (ELF-MF + RS), a general tendency to the decrease of total lipid level in brain structures was observed in each experimental group. Total cholesterol level was significantly increased in the cortex in the ELF-MF and RS + ELF-MF groups, and in subcortical structures in the RS + ELF-MF group. Inversely, polar lipids level in ELF-MF and RS + ELF-MF groups was decreased both in the cortex and in subcortical structures. Nonesterified fatty acid levels were found to be slightly higher in subcortical structures of the RS + ELF-MF group as compared to the control and RS groups. The analysis of fatty acid methyl esters revealed that the level of polyunsaturated fatty acids in cerebellum of ELF-MF-exposed rats was decreased, whereas their level in subcortical structures in the same group was increased. In addition to the changes in the amount of different kinds of lipids, the ELF-MF-induced lipid oxidative modifications were also noticed. The concentration of thiobarbituric acid reactive substances (TBARS, byproduct of lipid peroxidation) in lipids was higher, especially in the cortex and cerebellum of all treated groups [17]. Previous research has shown that immediate changes in lipid profile and TBARS levels after 2 h of singular exposure were visible only in the RS + ELF-MF group, whereas single exposure to ELF-MF or RS alone did not cause any changes in reduced glutathione and nitric oxide levels [55]. The increased level of lipid peroxidation was also noticed in rats exposed to ELF-MF (100 μT and 500 μT) [56]. The interesting research on ELF-MF-induced (50 Hz, 3 mT) changes in lipid profile (proteomic and transcriptomic profiling) in *Caenorhabditis elegans* was performed by Sun et al. [57]. In the glycerolipids (GLs) group, total triacylglycerols (TGs) content was increased while diacylglycerols (DGs) level was decreased. It should be also noted that among the most enriched proteins evaluated in this research, there were ones involved in lipid transport [57]. These studies indicate that ELF-MF affects the brain's lipid balance in a similar way to physiological stressors.

### 3.2. Oxidative Stress and Antioxidant Status

Stress can be a factor causing an increase of the level of oxidative stress parameters in the brain, including lipid peroxidation and on the other hand, it can activate antioxidant response [58,59]. Oxidative stress is the result of an imbalance between reactive oxygen species (ROS) and antioxidants [29]. Under normal conditions the synthesis of ROS is usually balanced, but when the production of ROS increases they become harmful for organism. The imbalance causes changes at the cellular level, which causes DNA, proteins, and lipids damage. ROS are involved in physiological processes, for instance, in cell signaling and respiratory chain and immune response, but some pathological factors can contribute to their increased level [60]. Overproduction of ROS occurs, inter alia, in response to stress (heat, anoxia, ultraviolet light, injury, environmental pollution, cigarette smoke, psychological trauma, and many others) [61]. It has also been reported that ROS levels increase after ELF-MF exposure and the reason for this phenomenon can be the failure of antioxidant defense.

The disturbance of oxidative homeostasis was proved in in vitro research. Exposure to ELF-MF of 1 mT resulted in free radical increase in mouse macrophages [62] and SH-SY5Y neuroblastoma cells [39]. ELF-MF-induced increased ROS production was also found in K562 human leukemia cell line (50 Hz, 0.025/0.05/0.1 mT) [63,64], and in human osteoarthritic chondrocytes (100 Hz) [65]. The viability decrease and morphological changes of rat hippocampal neurons concomitantly with the increase of MDA (malondialdehyde) and ROS levels and reduction of superoxide dismutase activity were noticed after exposure to ELF-MF (50 Hz, 8 mT) [42]. Similarly, exposure to ELF-MF (50 Hz, 25–200 μT) resulted in increased ROS production and diminished activity of antioxidant enzymes (superoxide dismutase (SOD), glutathione peroxidase (GPx), glutathione reductase (GR)) in the human keratinocyte cell line NCTC 2544 [66].

In vitro results have been confirmed in in vivo research. The shift into oxidative processes, presented as ROS-level elevation and significantly, the total antioxidative capacity (TAC) level decrease, were found in *Caenorhabditis elegans* exposed to ELF-MF (50 Hz, 3 mT) [67]. These results, proving the ELF-MF-induced impairment of antioxidant mechanisms in the organism, were also obtained from research using rodent models. The toxic, increasing oxidative stress level effect of ELF-MF was found mainly in the brain. Akdag et al. [68] demonstrated that the activity of antioxidant enzyme catalase (CAT) was decreased in ELF-MF-exposed animals regardless of ELF-MF intensity (100 and 500 μT). Moreover, in the group exposed to 500 μT, TAC was lower than in the 100 μT group. At the same time, in the 500 μT group the levels of oxidative stress markers, MDA and MPO (myeloperoxidase), and values of total oxidant status (TOS) and oxidative stress index (OSI) were significantly higher. TBARS concentrations increasing concomitantly with decreasing reduced glutathione (GSH), total free-SH group concentrations, and TAC levels were found in rats exposed to ELF-MF (40 Hz, 7 mT and 50 Hz, 12 and 18 kV/m) [69]. The activities of antioxidant enzymes in brain homogenates were also decreased in rats exposed to ELF-MF (50 Hz 10 kV/m, 4.3 pT) [70]. In addition, in mouse brain subjected to ELF-MF (50 Hz, 8 mT), the levels of MDA, ROS, nitric oxide (NO), and nitric oxide synthase (NOS) were increased, whereas activities of SOD, CAT, and GPx were decreased [71,72]. Free radical level (superoxide anion- $O_2\bullet^-$ and $NO_2^-$) was increased in the hypothalamus of rats exposed to ELF-MF (50 Hz, 10 mT) [73]. Acute exposure to ELF-MF (60 Hz 2.4 mT) resulted in the impairment of antioxidant mechanisms in the brain as well as in other tissues: heart, kidney, and plasma (decrease in SOD activity and reduced glutathione level) [55,74]. The disturbance of oxidative status was also found in testes of rats (diabetic model) exposed to ELF-MF (50 Hz, 8.2 mT): the increase in MDA and NO level, and diminished GSH level [75]. Many studies on the effects of ELF-MF have been conducted on people from risk groups, occupationally and residentially (living near high voltage lines) exposed to ELF-MF. El-Helaly and Abu-Hashem [76] carried out their research on a group of 50 electronic equipment installers and repairers. The serum malondialdehyde (MDA) level in the ELF-MF-exposed group was significantly higher than in control, and concomitantly the melatonin level (hormone supporting the antioxidant effect) in this group was lower. Similarly, the increment of oxidative stress and oxidative damage to DNA was also found in other research on power plant workers (occupational exposure, 110–420 kV and 4.09 V/m, 16.27 μT) [40,77,78]. The data suggest that exposure to ELF-MF could cause the failure of the antioxidant response and the collapse of homeostatic capability of the cell, leading to oxidative damage and functional impairment. However, the direct connection to the risk of disease development has not been unequivocally proved.

Subsequent studies shed light on the effects of ELF-MF on the antioxidant mechanisms that can underlie the protection against neurodegeneration. The ELF-MF-induced improvement of antioxidant protection has been evaluated in both in vitro and in vivo research. Exposure of C2C12 cells (myoblasts) to ELF-MF of 1 mT caused a drastic decrease in ROS level while total antioxidant status (TAS) and the activities of CAT and GPx were elevated [79]. Ehnert et al. [80] found ELF-MF-induced (16 Hz 6–282 μT) increase of SOD2, CAT, GPX3, and glutathione-disulfide reductase (GSR) activity concomitant with the reduction of ROS levels in human osteoblasts. Similarly, in the myelogenous leukemia cell line K562 exposed to ELF-MF (50 Hz, 1 mT) and in human blood platelets exposed to different sources of electromagnetic radiation (1 kHz, 0.5 mT; 50 Hz, 10 mT; or 1 kHz 220 V/m) CAT activity was increased [81,82]. Moreover, the exposure IMR-90 human lung fibroblasts for a total of 168 h to 6 mT ELF-MF contributed to decreased ROS level [83]. Exposure of human neuronal cell culture SH-SY5Y to 50 Hz ELF-MF with magnetic field intensity 1 mT resulted in elevated activity of NOS. This enzyme is controlled by proinflammatory cytokines that also activate ROS. After 1, 3, 6, and 24 h of exposure to ELF-MF, the activity of the enzyme was significantly increased. Moreover, the augmented production of $O_2^-$ was also found. However, CAT activity increased as the exposure time increased, possibly indicating a gradual adaptation of cells to the conditions of oxidative stress. On the other hand, these adaptive mechanisms turn out to be insufficient when ELF-MF exposure is combined with additional administration of $H_2O_2$—the oxidative effect is then exacerbated. These data suggest that ELF-MF may to some extent have neuroprotective effect. The combination of ELF-MF exposure and the stressor $H_2O_2$ prevents cells from being effectively defended against ROS [84]. However, when $H_2O_2$-treated cells were exposed to a higher value of ELF-MF (75 Hz, 2 mT), ROS level decreased and MnSOD activity increased [85]. It definitely suggests that the protective effect of ELF-MF depends on its intensity. The results of this in vitro research points out the beneficial effect of ELF-MF as an upregulation of antioxidant pathways, leading to protection against oxidative damage has been noted, reflecting an attempt to stimulate cellular response to neuronal damage.

In addition, ELF-MF as a mild stress factor activates an adaptive response that ensures the oxidative–antioxidant balance in rodent models as well as in humans. In a rat model of Huntington's disease, ELF-MF (60 Hz and 0.7 mT) was found to be able to reverse the process of neuronal degeneration and oxidative stress; it enhanced the antioxidant glutathione content and reduced the oxidative stress markers, 8-hydroxy-2′-deoxyguanosine and oxidized glutathione levels, in the whole-brain tissue [12]. Recent research evaluating the redox state in post-stroke patients demonstrated the beneficial effect of ELF-MF on oxidative status. High magnetic intensity, 5 or 7 mT of 40 Hz ELF-MF, significantly increased enzymatic antioxidant activity as compared to results obtained before treatment. The results were correlated with the improvement in functional and mental status of post-stroke patients [86,87]. These data show that ELF-MF is a factor that may both increase the production of ROS and activate organisms' antioxidant machinery in humans. In consequence, the electromagnetic radiation may drive the mechanisms underlying cell survival and plasticity.

### 3.3. Neuroprotective Proteins: Hsp70 and BDNF

Prosurvival responses include DNA repair processes and the increase in expression of chaperone protein—70-kDa heat shock proteins (Hsp70) and neurotrophin—brain-derived neurotrophic factor (BDNF) [88,89]. The expression of Hsp70 and BDNF appears to be a part of the general stress response and thus it is speculated to be associated with hormonal response to stress [89]. The increase of expression of these proteins would indicate the development of processes adapting neuronal networks in order to optimize circuits responding to the external environment and to integrate the response to challenges [90]. It was shown that stress hormones (mainly corticosterone and noradrenaline) influence via their receptors the plasticity processes in the hippocampus [31,91]. Noradrenaline can even dictate the direction of synaptic strength change in the hippocampus [91]. Under the influence of ELF-MF the expression of stress-response genes increases, resulting in higher levels of molecular chaperones such as Hsp70 [92–94]. The role of Hsp proteins is to stabilize polypeptide chains during their translocation across the cell membranes and to prevent aggregation of proteins with abnormal structure. Moreover, the antiapoptotic properties of Hsp70 and their role in appropriate folding and activation of proteins have also been proved [89,95]. As there are many pathways that could be affected to upregulate Hsp70 expression induced by stress, it is difficult to determine if any specific pathway may be affected.

The protective value of ELF-MF mediated by its influence on Hsp70 level was proved in in vitro research. Perez [96] showed that ELF-MF (50 MHz) leads to higher levels of Hsp70 in human T lymphocytes and fibroblast cell lines when subjected to stress, and that this response was of protective value. It seems to precondition and to enhance the cellular stress response when cells are provoked by toxic stimuli. Moreover, the cell protection was proportional to the levels of Hsp70. Exposure of human leukemia cell line K562 to ELF-MF (less than 0.1 and 1 mT) leads to increased Hsp70 levels [63,64]. More recent in vitro studies on ELF-MF with a density over 1 mT have shown marked effects, including an increase in Hsp70 transcription that results in protection against chronic hypoxia-induced injury [50].

Interesting data were also received in in vivo research on invertebrates. According to Gutzeit [97] exposure to 50 Hz ELF-MF with magnetic flux densities 50–150 μT enhances the response to thermal stress in *C. elegans*. ELF-MF-mediated specific genes activation could enhance transcription of an already activated set of heat shock genes by costressor (heat stress), thus providing an adequate and optimal defense response. Exposure to 60 Hz 8 μT ELF-MF caused regeneration of the heads and tails parts of the Planarian, *Dugesia dorotocethala*. This effect was accompanied by an increase in the level of Hsp70, which is triggered by extracellular signal-regulated kinase (ERK) cascade. It is known that ERK is activated as reaction to injury to promote regeneration [98].

In this approach, ELF-MF appears to be a mild stressor mobilizing the organism to cope with a dangerous situation [98]. The expression of the Hsp70 in response to stress serves to protect against the negative impact of stress. Hsp70 induction and stress systems function were shown to be two important inter-related mechanisms in maintaining the homeostasis under stress conditions [89]. According to the juxtaposition presented, some of beneficial effects of ELF-MF can be due to the protective role of Hsp70.

A substance of high importance for the nervous system is also brain-derived neurotrophic factor (BDNF). This neurotrophin is responsible for differentiation and survival of neurons during development, but it is also important for the adult brain, especially when subjected to stress conditions [90]. In a mature brain, BDNF ensures excitatory and inhibitory synaptic transmission and neuroplasticity [99]. The mechanism of neuroplasticity is crucial for learning and memory processes. It includes enhancement of the long-term potentiation (LTP), and stimulating and controlling neural growth. BDNF has a high affinity to full length tropomyosin receptor kinase B (TrkB) and its truncated isoform, p75 NTR. Through the activation of TrkB, the neurotrophin starts the cascade of signaling pathways, which results in neurogenesis, neuroplasticity, cell survival, and resistance to stress [100]. In vitro research proved that the exposure to pulsed ELF-MF (50 Hz; 1 mT for 2 h) increased the BDNF mRNA expression in cultured dorsal root ganglion neurons [101].

BDNF expression can be modulated by external, physiological, and pathological factors proven mainly in research on both rodents and humans. In the course of some diseases, such as Alzheimer disease, and during aging process or chronic stress, the inhibition of BDNF expression is noted, while exercise, enriched environment, and taking antidepressants are related to the intensified expression of BDNF [99]. ELF-MF is used in physical therapy due to its ability to stimulate BDNF synthesis. Several studies focused on this particular effect of ELF-MF on diseases and pathologies, like Huntington disease or stroke [12,102,103]. In post-stroke patients subjected to ten sessions of 15 min ELF-MF therapy (40 Hz 5 mT), plasma BDNF level was about 200% higher than before the treatment [102]. The study undertaken on a rat model of Huntington disease indicated that exposure to ELF-MF (60 Hz 0.7 mT 2 h in the morning and 2 h in the afternoon for 21 days) significantly elevated BDNF level in the rats with induced Huntington disease. Moreover, changes in the rats' behavior related to Huntington disease were neutralized by ELF-MF [12]. Urnukhsaikhan et al. [103] showed that expression of BDNF, TrkB, and phosphorylated protein kinase B was increased in ELF-MF-stimulated (60 Hz, 10 mT) ischemic mice. In vitro research proved that the exposure to pulsed EMF (50 Hz; 1 mT for 2 h) increased the BDNF mRNA expression in cultured dorsal root ganglion neurons [103]. Thus, there is evidence suggesting that the neuroprotective effect of the exposure to extremely low-frequency MFs may be due to, at least in part, the impact of the fields on neurotrophic factors levels, leading to an increase of cell survival.

### 3.4. Plasticity, Neurogenesis, Proliferation, and Differentiation

The level of activity of voltage-gated $Ca^{2+}$ channels is an important factor determining the synaptic transmission and leading to stimulation of short-term synaptic plasticity [104]. Calcium ions are involved in secretion of neurotransmitters. The influx of $Ca^{2+}$ through presynaptic voltage-gated $Ca^{2+}$ (Cav) channels triggers the release of neurotransmitters from presynaptic part of synapses. Measurements at a large glutamatergic synapse in the mammalian auditory brainstem—the calyx of Held—showed that vesicle endocytosis and synaptic transmission were enhanced in mice (8–10 postnatal days old) kept from birth under the influence of EMF (50 Hz EMF, 1 mT). Moreover, in mice exposed to EMF, the increase in expression of calcium channels at the presynaptic nerve terminal facilitating the influx of calcium was found. The observed mechanism is responsible for increasing endocytosis and synaptic plasticity [105]. In vitro research also evidenced the ELF-MF-induced increase in intracellular $Ca^{2+}$ concentration. This effect was found in C2C12 cells (myoblasts) after 0.1 and 1 mT ELF-MF exposure [79], in human pluripotent stem cells (iPSCs) after 1.5 mT ELF-MF exposure [105], in dorsal root ganglion neurons after 0.1, 1, 10, and 100 mT ELF-MF application [101], and in rat hippocampal neurons exposed to 8 mT ELF-MF [42]. The influence of ELF-MF on proliferation and apoptosis and the participation of $Ca^{2+}$ in these processes were also determined. In human neuroblastoma IMR32 and in rat pituitary GH3-cultured cells, the exposure to 1 mT 50 Hz ELF-MF caused the increased cell proliferation. At the same time, the increase of $Ca^{2+}$ current density and of voltage-gated $Ca^{2+}$ channel expression in the cell membrane was observed. In addition, blocking of $Ca^{2+}$ channels by 15 μM $Cd^{2+}$ alleviated the proliferative effect of ELF-MF. Apoptosis, induced by $H_2O_2$ or puromycin in IMR32 cells, was decreased after 72 h exposure to 1 mT 50 Hz ELF-MF. Blocking of L-type calcium channels by nifedipine also caused disappearance of the antiapoptotic effect of ELF-MF [106]. It has been also shown that ELF-MF influences calcium homeostasis in cultural entorhinal cortex neurons via calcium channel-independent mechanism. Twenty-four hour exposure to 1 or 3 mT ELF-MF does not affect voltage-gated calcium current and activity of calcium channels, but regulates intracellular calcium dynamics by decreasing the high-$K^+$-evoked intracellular calcium elevation [107]. In summary, this study suggests that the change in calcium currents through voltage-gated calcium channels is the mechanism responsible for the proliferation promotion and antiapoptotic effect of ELF-MF.

Measurements at a large glutamatergic synapse in the mammalian auditory brainstem—the calyx of Held—showed that vesicle endocytosis and synaptic transmission was enhanced in mice (8–10 postnatal days old) kept from birth under the influence of ELF-MF (50 Hz ELF-MF,1 mT). Moreover, in mice exposed to ELF-MF, the increase in expression of calcium channels at the presynaptic nerve terminal facilitating the influx of calcium was found. The observed mechanism is responsible for increasing endocytosis and synaptic plasticity [108].

Throughout the life course, new neurons are continuously formed in the hippocampus, which is therefore a major site of structural plasticity in the adult brain. The existence of a causal link between ELF-MF-enhanced synaptic plasticity and neurogenesis has been shown by a number of in vivo experimental studies. ELF-MF (60 Hz, 0.7 mT, applied over 21 days) improved neurological scores, enhanced neurotrophic factor levels, and reduced neuronal loss in a rat model of Huntington's disease [12]. In addition, prolonged exposure to ELF-MF (50 Hz, 100 μT; for 90 consecutive days; 2 h/day) increased LTP induction in rat's hippocampus [35]. In vivo exposure of adult mice to ELF-MF (50 Hz, 1 mT) produced a marked increase in the number of newly generated neurons in the granule cell layer of the dentate gyrus [34]. Although the ELF-MF of 1 mT (for 21 days) caused the decrease in the dendritic spine density of neurons in hippocampus after 7 and 10 days, the effects disappeared after 14 days [109]. Studies on the rat traumatic brain injury model [110] and on rat Alzheimer's disease model [49] have shown that ELF-MF reversed pathological brain damages and learning and memory abilities impairment. Similar effects were obtained after exposure of neurotoxin-injected mice to ELF-MF (50 Hz, 1 mT); the deficits such as neuronal maturation impairment, neurogenesis decrease, and memory disturbance decreased [111]. Studies on the beneficial effects of ELF-MF might yield fruitful insights related to clinical therapy of nervous-system-related diseases.

The cell differentiation at the expense of proliferation in different tissues and increased cell viability as an effect of exposure to low-frequency ELF-MFs is well evidenced in the literature. Collard et al. [112] reported an acceleration of proliferation and differentiation of human epidermis cells after exposure to low frequency (40 Hz) and demonstrated that the processes were related to a significant modification of gene expression. Falone et al. [85] found that ELF-MF (75 Hz, 2 mT) alone did not affect the viability of the human neuroblastoma SH-SY5Y cell line and that ELF-MF exposure prevented reduced cell viability after $H_2O_2$ application. Vannoni et al. [65] concluded that ELF-MF (100 Hz) stimulation is a useful tool to induce more divisions and thus to enhance cell proliferation of human osteoarthritic chondrocytes. The treatment of HeLa cells IMR-90 fibroblasts with ELF-MF (60 Hz, 6 mT) increased cell viability and activated cell cycle progression. In addition, ELF-MF mitigated the antiproliferative effect of GOx (agent stimulating $H_2O_2$ production) [83]. Di Loreto et al. [113] found that ELF-MF (50 Hz, 0.1–1 mT) had a positive effect on cell viability in primary cultures of maturing rat cortical neurons. The research of Ardeshirylajimi and Soleimani [105] on human pluripotent stem cells (iPSCs) suggested that ELF-MF (50 Hz, 1.5 mT) increases cell viability, division, proliferation, and mineralization of extracellular matrix. These results indicated that ELF-MF would improve the viability, proliferation, and differentiation of cells, and may be beneficial for the development of novel therapeutic approaches in regenerative medicine. The papers pointing to detrimental effect of ELF-MF on viability, differentiation, and proliferation should also be mentioned. It is, however, important that this effect is caused by high values of ELF-MF induction. Yin et al. [42] showed that the number of rat hippocampal neurons in G0/G1 phase was decreased and cells in S phase were accumulated as the effect of exposure to ELF-MF (50 Hz, 8 mT). The exposure of mesenchymal stem cells (bone marrow or adipose tissue derived) to ELF-MF of 20 mT (50 Hz) resulted in decreased cell proliferation [44,114]. This effect appeared to be related to the diminished expression of genes responsible for pluripotency and neuronal differentiation [44].

## 4. ELF-MF-Induced Changes in Levels of Neurotransmitters, Hormones, and Cytokines

ELF-MF-induced molecular changes modify to a certain extent some crucial neuronal processes. As it is commonly known, the communication between main groups of signaling substances: neurotransmitters, cytokines, and hormones, is of high importance for the maintenance of health status of an individual. We have a lot of data confirming the effect of ELF-MF on functioning of nervous, immune, and endocrinological systems. However, the mechanisms by which the magnetic stimulation modulates the activity of these systems and the interplay between them are open to be identified. Up-to-date results concluded that the exposure of rats to ELF-MF may be sufficient to induce significant changes in the content of neurotransmitters. The levels of major inhibitory and excitatory amino acids and neurotransmitters: glutamate (Glu), glutamine (Gln), glycine (Gly), tyrosine (Tyr), and $\gamma$-aminobutyric acid (GABA), were elevated in the thalamus after five days of exposure to ELF-MF (60 Hz, 2 mT). In the striatum, higher levels of Gln, Gly and GABA were found as well, whereas their concentrations were decreased in cortex, cerebellum, and hippocampus. Dopamine level was increased in the thalamus [115]. Extremely low-frequency magnetic field (10 Hz; 1.8–3.8 mT) exposure was found to alter turnover and receptor reactivity of serotoninergic and dopaminergic systems and some behavioral disturbances induced by these systems [116]. The rats receiving chronic (10 days) repetitive transcranial magnetic stimulation (rTMS) treatment showed the symptoms of anxiety, and it was shown that the rTMS-induced anxiety might involve the serotonergic system [117]. The continuous exposure of rats to ELF-MF (50 Hz, 0.5 mT) affected cortical serotoninergic neurotransmission, and intensity of these changes depended on ELF-MF exposure duration [118]. The data may indicate the ability of ELF-MF to modify the function of main neurotransmitter systems and thus to modulation of some physiological processes, such as memory, emotionality, mood changes, sleep, alertness, or stress response. The response of individual brain tissues to exposure was varied; the level of one neurotransmitter increased in a given tissue appeared to be decreased in another, suggesting that the radiation can induce varying responses in the nervous system [115].

As noted, the existing data indicate that the exposure to ELF-MF may count as a mild stress situation and could be a factor in the development of disturbances of brain stress systems: hypothalamo–pituitary–adrenal (HPA) axis and sympatho–adrenal–medullary (SAM) system [10,16,115,119,120]. Although some findings indicate the deteriorating effects of magnetic fields on hormonal stress response, others failed to exhibit any obvious effects. Continuous long-term (4–6 week) ELF-MF (50 Hz, 0.5 mT) treatment induced some signs of stress: HPA-axis activation (elevated blood glucose level, elevated POMC (the precursor protein for ACTH) mRNA level, and enhanced depression-like behavior in a forced swimming test), although other markers of stress (elevated basal ACTH and corticosterone secretion, adrenal gland hypertrophy, thymus involution, loss of weight gain, and anxiety-like behavior in elevated plus maze) were not observed. This confirms that ELF-MF of the abovementioned intensity creates a weak stress response [10]. In addition, 50 Hz ELF-MF (0.207 μT) significantly raised ACTH, cortisol, and glucose levels in guinea pigs [15]. The concentration of plasma corticosterone level was significantly higher and remained at a similar level in groups of rats exposed to restraint stress (RS) or ELF-MF [17]. Research by Mahdavi et al. [121] showed that exposure to both 1 and 5 Hz ELF-MF of 0.1 mT intensity caused an elevation of ACTH level in rats' plasma, whereas corticosterone level was reduced in both cases. In the animals exposed to 1 Hz ELF-MF, the concentration of adrenaline increased, but in rats exposed to 5 Hz, the level of adrenaline decreased. In rabbits exposed to ELF-MF (10 Hz ELF-MF), the level of blood corticosterone was increased in both the normal and high-cholesterol diet groups [122]. Chronic exposure (1 month) to 50 Hz 100/500 μT ELF-MF significantly raised corticosterone levels in rats' plasma [123]. In mice exposed to 10 μT ELF-MF (1, 4, or 24 h/day for 1 week—short-term exposure), no significant differences in CRH gene expression in hypothalamus were observed, whereas ACTH plasma level was lower regardless of the daily exposure time (1, 4, or 24 h/day for 1 week). Moreover, the expression of pituitary level of POMC was lower in an exposure time-dependent manner, and a statistically significant decrease appeared after 24 h/day exposure [124]. Mostafa et al. [125] showed that 2- and 4-week exposure of rats to ELF-MF (2 G, equivalent to 0.2 mT) significantly increased their plasma corticosterone level. Other studies on mouse model showed that even a relatively low level of EMF (12 nT) can cause corticosterone increase [126]. On the other hand, Kitaoka [16] revealed that levels of ACTH, the hormone that regulates corticosterone secretion, and hypothalamic CRH and pituitary POMC were not changed by ELF-MF (70 Hz, 3 mT). Significant changes were also found in the levels of noradrenaline in various parts of rats' brain: thalamus, hypothalamus, cerebellum and striatum, after 2 and 5 days exposure to 2 mT ELF-MF [115]. In the group of volunteers exposed to ELF-MF (50 Hz, 62 μT, for 2 h/day for 2 days with a 6-day interval) the cortisol level was increased at the beginning of ELF-MF exposure but later it diminished progressively [120]. The workers employed in the live-line procedures (132 kV high-voltage) for more than two years were found to be vulnerable for EM stress with altered adrenaline concentrations [40]. Moreover, exposure of turkey females to ELF-MF (50 Hz, 10 μT) caused NE-activated β-adrenoceptor function decrease, which is known to be involved in the formation of emotional disinterest and depression [126]. The data suggest that the exposure to ELF-MF can establish a new "set-point" for stress-system activity and the direction and dynamics of this process depend on the strength of the field and duration of exposure. The ELF-MF-induced changes in stress hormone levels will initiate cellular adaptation or damage by activation of intrinsic signaling pathways. Consequently, ELF-MF can change the vulnerability of the organism to subsequent stress factors and thus to diseases, mainly related to the nervous system.

Stress is known to strongly affect the immune system. It has been suggested that the potential contribution of ELF-MFs to anxiety or other stress associated disorders is also related to changes in the functioning of the immune system. Moreover, the chronic exposure to ELF-MF appears to also lead to immune system dysfunction, chronic allergic responses, inflammatory responses, and ill health [36]. However, as in other aspects of ELF-MF impact in organism, this factor can also be a double-edged sword and drive the survival-promoting processes. Importantly, the immune system and the stress systems—HPA axis and SAM—are closely linked to each other. Glucocorticoids and catecholamines are known to modify the secretion of cytokines: proteins that facilitate communication between the immune cells and the cells of the central nervous and endocrine systems [127]. Cytokines have the ability to modulate and activate the HPA axis. Proinflammatory cytokines: IL-1, IL-6, and TNF$\alpha$, induce corticotropin-releasing hormone (CRH) secretion and they are also involved at every stage of stress reaction [128]. Changes in plasma proinflammatory cytokines were observed after acute continuous exposure (24 h) to ELF-MF with magnetic intensity of 7 mT. The levels of IL-1$\beta$, IL-6, and IL-2 were elevated. The number of white and red blood cells and lymphocytes, and the hemoglobin concentration and hematocrit level were increased. However, the repetitive exposure to ELF-MF (1 h/day for 7 days) did not alter either cytokines levels or blood parameters [129]. The change in cytokine production was also noticed in stroke patients treated with ELF-MF (40 Hz, 5 mT ELF-MF) [130]. Following the exposure to ELF-MF, the plasma levels of IL-1$\beta$ and IL-2 cytokines and the level of IL-1$\beta$ mRNA expression were increased. In addition, ELF-MF exposure increases the levels of the anti-inflammatory transforming growth factor $\beta$ (TGF-$\beta$) and interleukin-18-binding protein [84]. Thus, the ELF-MF exposure can cause deregulation of the immune system, thereby increasing vulnerability to infectious and autoimmune diseases. Interesting results concerning the effect of ELF-MF on lymphocytes level were obtained by de Kleijn et al. [124]. In mice exposed to 10 $\mu$T ELF-MF (1, 4, or 24 h/day for 1 week—short-term exposure, or for 15-week long-term exposure) the increase in CD3+/CD4+ T-lymphocytes was observed only after short-term exposure to ELF-MF. The data suggest that the ELF-MF effect on immune and stress responses may be transient, because no changes in the number of immune cells were observed after long-term exposure. Several authors reported that the ELF-MF-evoked neuroplasticity can be mediated by the effect of magnetic radiation on cytokine level. Cytokines are found to influence the expression of neurotrophins and their receptors. This may indicate the role of inflammatory cytokines in the process of neuroplasticity [130]. The latest reports showed that 1 and 100 $\mu$T 50 Hz ELF-MF not only downregulates proinflammatory cytokines (IL-9 and TNF-$\alpha$), but also activates an inflammation-suppressing cytokine, IL-10. In this case, the most noticeable effect was obtained at the highest value of magnetic induction [131]. The ELF-MF (60 Hz, 10 mT) also mitigated the deficits in ischemic mice, among others in the context of immune function: the levels of inflammatory mediators MMP9 and IL-1$\beta$ were decreased [102]. The results demonstrate the recovery-stimulating potential of ELF-MF.

## 5. Association between ELF-MF Exposure and Emotional Behavior and Wellbeing

An association between ELF-MF exposure and emotional behavior has been indicated in many studies. The animal studies have shown that chronic exposure to ELF-MF may induce an anxiogenic and/or a depression-like effect. Dysfunction of stress systems can evoke negative emotional state and can potentiate fear- and anxiety-related behaviors [90,132]. Therefore, it is reasonable to speculate that elevation in ELF-MF-induced anxiety level may be attributed to the effect of ELF-MF on glucocorticoid release following activation of HPA axis and catecholaminergic sympathetic nervous system releasing adrenaline and noradrenaline. These pathways are key biological factors that modulate emotional behavior [90]. Liu et al. [133] reported that ELF-MF exposure (2 mT, 4 h/day for 25 days) had an anxiogenic effect in rats, such as anxiety-like behaviors in open field and elevated plus maze tests. Szemerszky et al. [10] demonstrated that ELF-MF exposure (0.5 mT, 4 weeks) in rats increased their immobility time in a forced swim test. The chronic exposure of mice to ELF-MF (3 mT, total exposure 200 h) induced the depression- and/or anxiety-like behavior (increase in total immobility time in a forced swim test and in the latency to enter the light box in a light–dark transition test). These behavioral disturbances were correlated with high corticosterone secretion [16]. Mice prenatally exposed to ELF-MF (50 Hz, 1 mT) lacked sociability and preference of social novelty, which can be a sign of autism-relevant social abnormalities; however, they did not show anxiety-like behavior [134]. The continuous (21 days) exposure to extremely low-frequency magnetic field (50 Hz, 10 mT) had no significant effect on activity and exploration activity but significantly increased stress and anxiety-related behavior in rats [135]. Quite similar observations in open field and elevated plus maze tests were described by Djordjevic et al. [73] after the exposure to ELF-MF (50 Hz, 10 mT) significantly reduced activity was observed. The noted effects of short-term ELF-MF exposure (50 Hz, 500 μT, 20 min) verified by behavioral tests in rats (elevated plus maze, novel object exploration) appear to suggest that these field parameters may cause some kind of discomfort, influence behavior, and increase passivity and situational anxiety [136]. Increased level of anxiety has also been found in rats exposed to ELF-MF of various flux density (50 Hz; 1, 100, 500, 2000 μT, and 2 mT) [56,137]. In accordance with this, Isogawa et al. [117] observed an anxiogenic effect of rTMS in rats tested in the elevated plus maze. It has also been shown that in rat pups after 6 weeks of exposure to ELF-MF (50 Hz, 3.5 mT, 1 h/day) behavior parameters, such as activity, motion, and response to sound and light, were decreased during exposure, but after exposure they settled back into normal control values [138]. However, the exposure of rats to ELF-MF of lower flux density (100 μT, 50 Hz, for 24 weeks) did not evoke any behavioral changes. The experimental group did not show any anxiety-like behaviors in open field or elevated plus maze. Similarly, depression-like behavior was not detected during tail suspension and forced swim tests [139]. Population studies paid attention to the role of ELF-MF in the development of sleep disorders, anxiety, and depression. It was shown that residential exposure to ELF-MF emitted by a radio–television broadcasting station could increase the anxiety in women [140]. Similarly, power plant workers (chronically exposed to ELF-MF) showed significantly poorer sleep quality than the unexposed group. Moreover, the level of depression symptoms in the exposed group was also significantly higher [14]. Interestingly, magnetic waves in the form of repetitive transcranial magnetic stimulation (rTMS) are used in therapy of depression. Three weeks of daily treatment caused a remission in a significant number of patients resistant to antidepressant treatment [13]. Similarly, 10-day treatment (20 × 2 s trains of 20 Hz stimulation with 58 s intervals) administered to patients with major depressive episodes significantly reduced scores in the Hamilton depression rating scale [141]. Exposure of post-stroke patients to 40 Hz, 7 mT ELF-MF 15 min/day for 4 weeks improved significantly cognitive functions and decreased up to 60% of depression syndromes [86]. As noted earlier, the possible explanation of beneficial effect of rTMS can be the ELF-MF-induced increase of mediators of corticosterone action in the hippocampus, i.e., neurotrophins, since these proteins appear to play a pivotal role in the structure and function of hippocampal neurons.

Behavioral effects of the ELF-MF depend on the length, frequency, and intensity of exposure, and on the initial balance of the brain transmitters [121,142]. Some authors noted reduced activity of animals after ELF-MF exposure, what is known as anxiety-like behavior. Others did not observe any changes following the exposure. Constant exposure to ELF-MF may also cause burdensome symptoms in humans, e.g., sleep disorders or depression. Notwithstanding, ELF-MF with high magnetic induction value (e.g., 7 mT—much higher than average exposure on a daily basis) appears to be effective in depression therapy. ELF-MF also improves cognitive functions in patients with a history of neurological injuries. However, again it should be noted that a considerable variety in the values of magnetic induction and exposure time were used in the research on ELF-MF effect.

## 6. Conclusions

Currently, people living in urbanized societies are exposed to the influence of various environmental stressors, including ELF-MF. The level of exposure can be different for individual groups and depends on the place of residence and occupation. The studies presented in this article indicate the possibility of changes at various levels of the organism organization as a result of exposure to ELF-MF. Effect of the field is observed in molecular and cellular responses, complex physiological processes such as activation of HPA axis and sympathetic system, as well as in behavioral and mood changes. As a result of exposure to

ELF-MF, the homeostasis is disturbed in a way that is similar to the effect of application of any other stressor. The effects of ELF-MF are associated with the occurrence of various ailments, such as anxiety and sleep or mood disorders, but this type of stimulation is also successfully used in the therapy of depressive disorders as an alternative for drug-resistant or post-stroke patients. There is considerable evidence that ELF-MF-induced processes include interplay between the monoaminergic system, glucocorticoids, and neurotrophins [143]. Therefore, it is conceivable that changes in any of these elements of stress response may ultimately lead to changes in brain function and can be reflected in behavior. The ELF-MF-evoked initial disruption in homeostasis triggers an overcompensation response to re-establish homeostasis, which results in a bidirectional effects at the subsequent stages of response. Recent findings have elucidated the cellular signaling pathways and molecular mechanisms that mediate the character of response, which typically involve free radicals, antioxidants, protein chaperones (e.g., Hsp70) and growth factors (e.g., BDNF), hormones (mainly corticosterone and noradrenaline), cytokines, and neurotransmitters. Despite many studies on bioeffects of ELF-MF exposure, the picture is still not clear and unambiguous. However, we can try to make some general conclusions.

Summarizing the observations cited above, when the organism is subjected to the influence of ELF-MF, typical reactions for stimulation with a stress factor can be observed; however, the changes can evolve in both directions: detrimental or beneficial. It is possible to hypothesize that, as in the case of other stressors, while the exposure to milder ELF-MF (lower intensity and shorter duration) could promote neural plasticity, the chronic stressful conditions (high intensity and long-term duration) could sensitize limbic circuits resulting in greater susceptibility to damage. Some existing studies suggest that ELF-MF of low density creates the weak stress response and improves the brain function [10,15,121], but the effects of high density ELF-MF are definitely stronger as far as stress systems activation and behavioral impairment are concerned [16,119,135]. According to Directive 2013/35/EU and ICNIRP, 2010, the ELF-MF of flux density below 1 mT does not cause any changes in the organism, the field in the range of 1–6 mT may induce some temporary changes in the functioning of the nervous system, but the consequence of higher values of ELF-MF flux density can be permanent. However, other factors such as temporal features of exposure or individual hypersensitivity [144] are also important and can determine the consequences of this kind of stress on the organism. The review of literature concerning the ELF-MF impact on the organism showed that biological effects are often discussed in relation to intensity (T), frequency (Hz) of the field, and duration of exposure. However, the quantification of the electromagnetic phenomena in the organism—dosimetry—is of high importance for proper determining ELF-MF effects. Establishing reliable and reproducible measurement procedures is required. Thus, experimental studies should include the detailed characterization of internal electromagnetic fields in addition to other parameters of ELF-MF exposure. Improvement in the process of validation of physical aspects related to ELF-MF exposure is necessary to achieve the reliable answers to questions concerning the effects of ELF-MF on organism.

Still, there is no answer to the question of where the threshold for ELF-MF exposure lies, above which the adaptive possibilities of the organism are exceeded and when the direction of ELF-MF-induced processes turn into pathology. A better understanding of effects of ELF-MF at the cellular, molecular, physiological, and behavioral levels fills the gap in our knowledge of ELF-MF effects on stress response of systems activity. It is important to recognize the risk concerning the effects of magnetic flux density of ELF-MF on development of stress-related and neurodegenerative disorders. Understanding the fundamental mechanisms of these differential responses in neurons will lead to a new approach in risk assessment of ELF-MF exposure. On the other hand, the property of ELF-MF supporting rehabilitation will be successfully used in the development of novel approaches for the prevention and treatment of many different diseases.

**Supplementary Materials:** The following are available online at https://www.mdpi.com/2076-3425/11/2/174/s1, Table S1: The influence of extremely low frequency magnetic field (ELF-MF) on different aspects of the organism function; http://doi.org/10.5281/zenodo.4457845.

**Author Contributions:** Visualization (schemes, tables, and graphical abstract) and writing—original draft preparation, A.K.; conceptualization, writing—original draft preparation, and supervision, J.R. Both authors have read and agreed to the published version of the manuscript

**Funding:** Project realized from NCS grant No. 2017/25/B/NZ7/00638 and POWER.03.05.00-00-Z302/17-00.

**Institutional Review Board Statement:** Not applicable.

**Informed Consent Statement:** Not applicable.

**Conflicts of Interest:** The authors declare no conflict of interest.

## References

1. SCENIHR (Scientific Committee on Emerging and Newly Identified Health Risks). *Potential Health Effects of Exposure to Electromagnetic Fields (EMF)*; European Commission: Luxembourg, 2015; pp. 1–288.
2. Touitou, Y.; Selmaoui, B. The effects of extremely low-frequency magnetic fields on melatonin and cortisol, two marker rhythms of the circadian system. *Dialogues Clin. Neurosci.* **2012**, *14*, 381–399. [CrossRef] [PubMed]
3. Giorgi, G.; Lecciso, M.; Capri, M.; Lukas Yani, S.; Virelli, A.; Bersani, F.; Del Re, B. An evaluation of genotoxicity in human neuronal-type cells subjected to oxidative stress under an extremely low frequency pulsed magnetic field. *Mutat. Res. Genet. Toxicol. Environ. Mutagen.* **2014**, *775*, 31–37. [CrossRef] [PubMed]
4. De Groot, M.W.; Kock, M.D.; Westerink, R.H. Assessment of the neurotoxic potential of exposure to 50Hz extremely low frequency electromagnetic fields (ELF-EMF) in naive and chemically stressed PC12 cells. *Neurotoxicology* **2014**, *44*, 358–364. [CrossRef] [PubMed]
5. Golbach, L.A.; Philippi, J.G.; Cuppen, J.J.; Savelkoul, H.F.; Verburg-van Kemenade, B.M. Calcium signalling in human neutrophil cell lines is not affected by low-frequency electromagnetic fields. *Bioelectromagnetics* **2015**, *36*, 430–443. [CrossRef] [PubMed]
6. Nakayama, M.; Nakamura, A.; Hondou, T.; Miyata, H. Evaluation of cell viability, DNA single-strand breaks, and nitric oxide production in LPS-stimulated macrophage RAW264 exposed to a 50-Hz magnetic field. *Int. J. Radiat. Biol.* **2016**, *92*, 583–589. [CrossRef]
7. Zhu, K.; Lv, Y.; Cheng, Q.; Hua, J.; Zeng, Q. Extremely Low Frequency Magnetic Fields Do Not Induce DNA Damage in Human Lens Epithelial Cells In Vitro. *Anat. Rec.* **2016**, *299*, 688–697. [CrossRef] [PubMed]
8. Burman, O.; Marsella, G.; Di Clemente, A.; Cervo, L. The effect of exposure to low frequency electromagnetic fields (EMF) as an integral part of the housing system on anxiety-related behaviour, cognition and welfare in two strains of laboratory mouse. *PLoS ONE* **2018**, *13*, e0197054. [CrossRef]
9. Baan, R.; Grosse, Y.; Lauby-Secretan, B.; El Ghissassi, F.; Bouvard, V.; Benbrahim-Tallaa, L.; Guha, N.; Islami, F.; Galichet, L.; Straif, K.; et al. Carcinogenicity of radiofrequency electromagnetic fields. *Lancet Oncol.* **2011**, *12*, 624–626. [CrossRef]
10. Szemerszky, R.; Zelena, D.; Barna, I.; Bárdos, G. Stress-related endocrinological and psychopathological effects of short- and long-term 50Hz electromagnetic field exposure in rats. *Brain Res. Bull.* **2010**, *81*, 92–99. [CrossRef]
11. Blank, M.; Goodman, R. Electromagnetic fields stress living cells. *Pathophysiology* **2009**, *16*, 71–78. [CrossRef]
12. Tasset, I.; Medina, F.J.; Jimena, I.; Agüera, E.; Gascón, F.; Feijóo, M.; Sánchez-López, F.; Luque, E.; Peña, J.; Drucker-Colín, R.; et al. Neuroprotective effects of extremely low-frequency electromagnetic fields on a Huntington's disease rat model: Effects on neurotrophic factors and neuronal density. *Neuroscience* **2012**, *209*, 54–63. [CrossRef] [PubMed]
13. George, M.S.; Lisanby, S.H.; Avery, D.; McDonald, W.M.; Durkalski, V.; Pavlicova, M.; Anderson, B.; Nahas, Z.; Bulow, P.; Zarkowski, P.; et al. Daily left prefrontal transcranial magnetic stimulation therapy for major depressive disorder: A sham-controlled randomized trial. *Arch. Gen. Psychiatry* **2010**, *67*, 507–516. [CrossRef] [PubMed]
14. Hosseinabadi, M.B.; Khanjani, N.; Ebrahimi, M.H.; Haji, B.; Abdolahfard, M. The effect of chronic exposure to extremely low-frequency electromagnetic fields on sleep quality, stress, depression and anxiety. *Electromagn. Biol. Med.* **2019**, *38*, 96–101. [CrossRef] [PubMed]
15. Sedghi, H.; Zare, S.; Hayatgeibi, H.; Alivandi, S.; Ebadi, A.G. Effects of 50 Hz Magnetic Field on Some Factors of Immune System in the Male Guinea Pigs. *Am. J. Immunol.* **2005**, *1*, 37–41. [CrossRef]
16. Kitaoka, K.; Kitamura, M.; Aoi, S.; Shimizu, N.; Yoshizaki, K. Chronic exposure to an extremely low-frequency magnetic field induces depression-like behavior and corticosterone secretion without enhancement of the hypothalamic-pituitary-adrenal axis in mice. *Bioelectromagnetics* **2013**, *34*, 43–51. [CrossRef]
17. Martínez-Sámano, J.; Flores-Poblano, A.; Verdugo-Díaz, L.; Juárez-Oropeza, M.A.; Torres-Durán, P.V. Extremely low frequency electromagnetic field exposure and restraint stress induce changes on the brain lipid profile of Wistar rats. *BMC Neurosci.* **2018**, *19*, 31. [CrossRef]

18. McEwen, B.S. Physiology and neurobiology of stress and adaptation: Central role of the brain. *Physiol. Rev.* **2007**, *87*, 873–904. [CrossRef] [PubMed]
19. Hajós-Korcsok, E.; Robinson, D.D.; Yu, J.H.; Fitch, C.S.; Walker, E.; Merchant, K.M. Rapid habituation of hippocampal serotonin and norepinephrine release and anxiety-related behaviors, but not plasma corticosterone levels, to repeated footshock stress in rats. *Pharmacol. Biochem. Behav.* **2003**, *74*, 609–616. [CrossRef]
20. Selye, H. *The Stress of Life*; McGraw-Hill: New York, NY, USA, 1956.
21. Selye, H. *Stress without Distress*; Springer: Boston, MA, USA, 1974.
22. Lazarus, R.S.; Folkman, S. *Stress, Appraisal, and Coping*; Springer: New York, NY, USA, 1984.
23. Calabrese, E.J.; Mattson, M.P. Hormesis provides a generalized quantitative estimate of biological plasticity. *J. Cell Commun. Signal.* **2011**, *5*, 25–38. [CrossRef]
24. Zimmermann, A.; Bauer, M.A.; Kroemer, G.; Madeo, F.; Carmona-Gutierrez, D. When less is more: Hormesis against stress and disease. *Microb. Cell* **2014**, *1*, 150–153. [CrossRef]
25. Krugers, H.J.; Karst, H.; Joels, M. Interactions between noradrenaline and corticosteroids in the brain: From electrical activity to cognitive performance. *Front. Cell Neurosci.* **2012**, *6*. [CrossRef] [PubMed]
26. Pierzchała-Koziec, K.; Zubel-Łojek, J.; Ocłoń, E.; Latacz, A.; Kępys, B. Emotional stress induces sex-specific sympatho-adrenomedullary responses in lambs. *Acta Biol. Crac. Série Zool.* **2015**, *57*, 39–45.
27. Tafet, G.E.; Bernardini, R. Psychoneuroendocrinological links between chronic stress and depression. *Prog. Neuropsychopharmacol. Biol. Psychiatry* **2003**, *27*, 893–903. [CrossRef]
28. Nesse, R.M.; Bhatnagar, S.; Young, E.A. Evolutionary Origins and Functions of the Stress Response. In *Stress: Concepts, Cognition, Emotion, and Behavior*; Fink, G., Ed.; Academic Press: London, UK, 2016; Volume 1, pp. 95–101.
29. Black, C.N.; Bot, M.; Révész, D.; Scheffer, P.G.; Penninx, B. The association between three major physiological stress systems and oxidative DNA and lipid damage. *Psychoneuroendocrinology* **2017**, *80*, 56–66. [CrossRef] [PubMed]
30. De Kloet, E.R.; Vreugdenhil, E.; Oitzl, M.S.; Joëls, M. Brain corticosteroid receptor balance in health and disease. *Endocr. Rev.* **1998**, *19*, 269–301. [CrossRef] [PubMed]
31. Rogalska, J. Mineralocorticoid and glucocorticoid receptors in hippocampus: Their impact on neurons survival and behavioral impairment after neonatal brain injury. *Vitam. Horm.* **2010**, *82*, 391–419. [CrossRef] [PubMed]
32. Arnett, M.G.; Muglia, L.M.; Laryea, G.; Muglia, L.J. Genetic Approaches to Hypothalamic-Pituitary-Adrenal Axis Regulation. *Neuropsychopharmacology* **2016**, *41*, 245–260. [CrossRef]
33. Mushak, P. Temporal stability of chemical hormesis (CH): Is CH just a temporary stop on the road to thresholds and toxic responses? *Sci. Total Environ.* **2016**, *569–570*, 1446–1456. [CrossRef]
34. Cuccurazzu, B.; Leone, L.; Podda, M.V.; Piacentini, R.; Riccardi, E.; Ripoli, C.; Azzena, G.B.; Grassi, C. Exposure to extremely low-frequency (50 Hz) electromagnetic fields enhances adult hippocampal neurogenesis in C57BL/6 mice. *Exp. Neurol.* **2010**, *226*, 173–182. [CrossRef]
35. Komaki, A.; Khalili, A.; Salehi, I.; Shahidi, S.; Sarihi, A. Effects of exposure to an extremely low frequency electromagnetic field on hippocampal long-term potentiation in rat. *Brain Res.* **2014**, *1564*, 1–8. [CrossRef]
36. Johansson, O. Disturbance of the immune system by electromagnetic fields-A potentially underlying cause for cellular damage and tissue repair reduction which could lead to disease and impairment. *Pathophysiology* **2009**, *16*, 157–177. [CrossRef] [PubMed]
37. Corallo, C.; Battisti, E.; Albanese, A.; Vannoni, D.; Leoncini, R.; Landi, G.; Gagliardi, A.; Landi, C.; Carta, S.; Nuti, R.; et al. Proteomics of human primary osteoarthritic chondrocytes exposed to extremely low-frequency electromagnetic fields (ELF EMFs) and to therapeutic application of musically modulated electromagnetic fields (TAMMEF). *Electromagn. Biol. Med.* **2014**, *33*, 3–10. [CrossRef] [PubMed]
38. Li, S.S.; Zhang, Z.Y.; Yang, C.J.; Lian, H.Y.; Cai, P. Gene expression and reproductive abilities of male Drosophila melanogaster subjected to ELF-EMF exposure. *Mutat. Res.* **2013**, *758*, 95–103. [CrossRef] [PubMed]
39. Merla, C.; Liberti, M.; Consales, C.; Denzi, A.; Apollonio, F.; Marino, C.; Benassi, B. Evidences of plasma membrane-mediated ROS generation upon ELF exposure in neuroblastoma cells supported by a computational multiscale approach. *Biochim. Biophys. Acta Biomembr.* **2019**, *1861*, 1446–1457. [CrossRef] [PubMed]
40. Tiwari, R.; Lakshmi, N.K.; Bhargava, S.C.; Ahuja, Y.R. Epinephrine, DNA integrity and oxidative stress in workers exposed to extremely low-frequency electromagnetic fields (ELF-EMFs) at 132 kV substations. *Electromagn. Biol. Med.* **2015**, *34*, 56–62. [CrossRef]
41. Duan, W.; Liu, C.; Zhang, L.; He, M.; Xu, S.; Chen, C.; Pi, H.; Gao, P.; Zhang, Y.; Zhong, M.; et al. Comparison of the genotoxic effects induced by 50 Hz extremely low-frequency electromagnetic fields and 1800 MHz radiofrequency electromagnetic fields in GC-2 cells. *Radiat. Res.* **2015**, *183*, 305–314. [CrossRef]
42. Yin, C.; Luo, X.; Duan, Y.; Duan, W.; Zhang, H.; He, Y.; Sun, G.; Sun, X. Neuroprotective effects of lotus seedpod procyanidins on extremely low frequency electromagnetic field-induced neurotoxicity in primary cultured hippocampal neurons. *Biomed. Pharmacother.* **2016**, *82*, 628–639. [CrossRef]
43. Ferroni, L.; Tocco, I.; De Pieri, A.; Menarin, M.; Fermi, E.; Piattelli, A.; Gardin, C.; Zavan, B. Pulsed magnetic therapy increases osteogenic differentiation of mesenchymal stem cells only if they are pre-committed. *Life Sci.* **2016**, *152*, 44–51. [CrossRef]
44. Haghighat, N.; Abdolmaleki, P.; Parnian, J.; Behmanesh, M. The expression of pluripotency and neuronal differentiation markers under the influence of electromagnetic field and nitric oxide. *Mol. Cell Neurosci.* **2017**, *85*, 19–28. [CrossRef]

45. Emre, M.; Cetiner, S.; Zencir, S.; Unlukurt, I.; Kahraman, I.; Topcu, Z. Oxidative stress and apoptosis in relation to exposure to magnetic field. *Cell Biochem. Biophys.* **2011**, *59*, 71–77. [CrossRef]
46. Juszczak, K.; Kaszuba-Zwoinska, J.; Thor, P.J. Pulsating electromagnetic field stimulation of urothelial cells induces apoptosis and diminishes necrosis: New insight to magnetic therapy in urology. *J. Physiol. Pharmacol.* **2012**, *63*, 397–401. [PubMed]
47. Rezaei-Tavirani, M.; Hasanzadeh, H.; Seyyedi, S.; Ghoujeghi, F.; Semnani, V.; Zali, H. Proteomic Analysis of Extremely Low-Frequency ElectroMagnetic Field (ELF-EMF) with Different Intensities in Rats Hippocampus. *Arch. Neurosci.* **2018**, *5*, e62954. [CrossRef]
48. Yang, M.L.; Ye, Z.M. Extremely low frequency electromagnetic field induces apoptosis of osteosarcoma cells via oxidative stress. *J. Zhejiang Univ.* **2015**, *44*, 323–328. (In Chinese)
49. Liu, X.; Zuo, H.; Wang, D.; Peng, R.; Song, T.; Wang, S.; Xu, X.; Gao, Y.; Li, Y.; Wang, S.; et al. Improvement of spatial memory disorder and hippocampal damage by exposure to electromagnetic fields in an Alzheimer's disease rat model. *PLoS ONE* **2015**, *10*, e0126963. [CrossRef] [PubMed]
50. Wei, J.; Tong, J.; Yu, L.; Zhang, J. EMF protects cardiomyocytes against hypoxia-induced injury via heat shock protein 70 activation. *Chem. Biol. Interact.* **2016**, *248*, 8–17. [CrossRef] [PubMed]
51. Huang, Y.; Liu, Z.; Liu, W.; Yin, C.; Ci, L.; Zhao, R.; Yang, X. Short communication: Salivary haptoglobin and chromogranin A as non-invasive markers during restraint stress in pigs. *Res. Vet. Sci.* **2017**, *114*, 27–30. [CrossRef]
52. D'amico, M.A.; Ghinassi, B.; Izzicupo, P.; Manzoli, L.; Di Baldassarre, A. Biological function and clinical relevance of chromogranin A and derived peptides. *Endocr. Connect.* **2014**, *3*, R45–R54. [CrossRef]
53. Touitou, Y.; Lambrozo, J.; Mauvieux, B.; Riedel, M. Evaluation in humans of ELF-EMF exposure on chromogranin A, a marker of neuroendocrine tumors and stress. *Chronobiol. Int.* **2020**, *37*, 60–67. [CrossRef]
54. Corraliza-Gomez, M.; Sanchez, D.; Ganfornina, M.D. Lipid-Binding Proteins in Brain Health and Disease. *Front. Neurol.* **2019**, *10*, 1152. [CrossRef]
55. Martínez-Sámano, J.; Torres-Durán, P.V.; Juárez-Oropeza, M.A.; Verdugo-Díaz, L. Effect of acute extremely low frequency electromagnetic field exposure on the antioxidant status and lipid levels in rat brain. *Arch. Med. Res.* **2012**, *43*, 183–189. [CrossRef]
56. Karimi, S.A.; Salehi, I.; Shykhi, T.; Zare, S.; Komaki, A. Effects of exposure to extremely low-frequency electromagnetic fields on spatial and passive avoidance learning and memory, anxiety-like behavior and oxidative stress in male rats. *Behav. Brain Res.* **2019**, *359*, 630–638. [CrossRef] [PubMed]
57. Sun, Y.; Huang, X.; Wang, Y.; Shi, Z.; Liao, Y.; Cai, P. Lipidomic alteration and stress-defense mechanism of soil nematode Caenorhabditis elegans in response to extremely low-frequency electromagnetic field exposure. *Ecotoxicol. Environ. Saf.* **2019**, *170*, 611–619. [CrossRef] [PubMed]
58. Che, Y.; Zhou, Z.; Shu, Y.; Zhai, C.; Zhu, Y.; Gong, S.; Cui, Y.; Wang, J.F. Chronic unpredictable stress impairs endogenous antioxidant defense in rat brain. *Neurosci. Lett.* **2015**, *584*, 208–213. [CrossRef] [PubMed]
59. Pejic, S.; Stojikjkovic, V.; Todorovic, A.; Gavrilovic, L.; Pavlovic, I.; Popovic, N.; Pajovic, S.B. Antioxidant Enzymes in Brain Cortex of Rats Exposed to Acute, Chronic and Combined Stress. *Folia Biol.* **2016**, *64*, 189–195. [CrossRef]
60. Pizzino, G.; Irrera, N.; Cucinotta, M.; Pallio, G.; Mannino, F.; Arcoraci, V.; Squadrito, F.; Altavilla, D.; Bitto, A. Oxidative Stress: Harms and Benefits for Human Health. *Oxid. Med. Cell Longev.* **2017**, *2017*, 8416763. [CrossRef]
61. Srivastava, K.K.; Kumar, R. Stress, oxidative injury and disease. *Indian J. Clin. Biochem.* **2015**, *30*, 3–10. [CrossRef]
62. Frahm, J.; Mattsson, M.O.; Simkó, M. Exposure to ELF magnetic fields modulate redox related protein expression in mouse macrophages. *Toxicol. Lett.* **2010**, *192*, 330–336. [CrossRef]
63. Garip, A.I.; Akan, Z. Effect of ELF-EMF on number of apoptotic cells; correlation with reactive oxygen species and HSP. *Acta Biol. Hung.* **2010**, *61*, 158–167. [CrossRef]
64. Mannerling, A.C.; Simkó, M.; Mild, K.H.; Mattsson, M.O. Effects of 50-Hz magnetic field exposure on superoxide radical anion formation and HSP70 induction in human K562 cells. *Radiat. Environ. Biophys.* **2010**, *49*, 731–741. [CrossRef]
65. Vannoni, D.; Albanese, A.; Battisti, E.; Aceto, E.; Giglioni, S.; Corallo, C.; Carta, S.; Ferrata, P.; Fioravanti, A.; Giordano, N. In vitro exposure of human osteoarthritic chondrocytes to ELF fields and new therapeutic application of musically modulated electromagnetic fields: Biological evidence. *J. Biol. Regul. Homeost. Agents* **2012**, *26*, 39–49.
66. Calcabrini, C.; Mancini, U.; De Bellis, R.; Diaz, A.R.; Martinelli, M.; Cucchiarini, L.; Sestili, P.; Stocchi, V.; Potenza, L. Effect of extremely low-frequency electromagnetic fields on antioxidant activity in the human keratinocyte cell line NCTC 2544. *Biotechnol. Appl. Biochem.* **2017**, *64*, 415–422. [CrossRef] [PubMed]
67. Sun, Y.; Shi, Z.; Wang, Y.; Tang, C.; Liao, Y.; Yang, C.; Cai, P. Coupling of oxidative stress responses to tricarboxylic acid cycle and prostaglandin E2 alterations in Caenorhabditis elegans under extremely low-frequency electromagnetic field. *Int. J. Radiat. Biol.* **2018**, *94*, 1159–1166. [CrossRef] [PubMed]
68. Akdag, M.Z.; Dasdag, S.; Ulukaya, E.; Uzunlar, A.K.; Kurt, M.A.; Taşkin, A. Effects of extremely low-frequency magnetic field on caspase activities and oxidative stress values in rat brain. *Biol. Trace Elem. Res.* **2010**, *138*, 238–249. [CrossRef] [PubMed]
69. Goraca, A.; Ciejka, E.; Piechota, A. Effects of extremely low frequency magnetic field on the parameters of oxidative stress in heart. *J. Physiol. Pharmacol.* **2010**, *61*, 333–338. [PubMed]
70. Budziosz, J.; Stanek, A.; Sieroń, A.; Witkoś, J.; Cholewka, A.; Sieroń, K. Effects of Low-Frequency Electromagnetic Field on Oxidative Stress in Selected Structures of the Central Nervous System. *Oxid. Med. Cell Longev.* **2018**, *2018*, 1427412. [CrossRef] [PubMed]

71. Duan, Y.; Wang, Z.; Zhang, H.; He, Y.; Lu, R.; Zhang, R.; Sun, G.; Sun, X. The preventive effect of lotus seedpod procyanidins on cognitive impairment and oxidative damage induced by extremely low frequency electromagnetic field exposure. *Food Funct.* **2013**, *4*, 1252–1262. [CrossRef]
72. Luo, X.; Chen, M.; Duan, Y.; Duan, W.; Zhang, H.; He, Y.; Yin, C.; Sun, G.; Sun, X. Chemoprotective action of lotus seedpod procyanidins on oxidative stress in mice induced by extremely low-frequency electromagnetic field exposure. *Biomed. Pharmacother.* **2016**, *82*, 640–648. [CrossRef]
73. Djordjevic, N.Z.; Paunović, M.G.; Peulić, A.S. Anxiety-like behavioural effects of extremely low-frequency electromagnetic field in rats. *Environ. Sci. Pollut. Res. Int.* **2017**, *24*, 21693–21699. [CrossRef]
74. Martínez-Sámano, J.; Torres-Durán, P.V.; Juárez-Oropeza, M.A.; Elías-Viñas, D.; Verdugo-Díaz, L. Effects of acute electromagnetic field exposure and movement restraint on antioxidant system in liver, heart, kidney and plasma of Wistar rats: A preliminary report. *Int. J. Radiat. Biol.* **2010**, *86*, 1088–1094. [CrossRef]
75. Kuzay, D.; Ozer, C.; Sirav, B.; Canseven, A.G.; Seyhan, N. Oxidative effects of extremely low frequency magnetic field and radio frequency radiation on testes tissues of diabetic and healthy rats. *Bratisl. Lek. Listy* **2017**, *118*, 278–282. [CrossRef]
76. El-Helaly, M.; Abu-Hashem, E. Oxidative stress, melatonin level, and sleep insufficiency among electronic equipment repairers. *Indian J. Occup. Environ. Med.* **2010**, *14*, 66–70. [CrossRef] [PubMed]
77. Zhang, Y.; Zhang, D.; Zhu, B.; Zhang, H.; Sun, Y.; Sun, C. Effects of dietary green tea polyphenol supplementation on the health of workers exposed to high-voltage power lines. *Environ. Toxicol. Pharmacol.* **2016**, *46*, 183–187. [CrossRef] [PubMed]
78. Hosseinabadi, M.B.; Khanjani, N. The Effect of Extremely Low-Frequency Electromagnetic Fields on the Prevalence of Musculoskeletal Disorders and the Role of Oxidative Stress. *Bioelectromagnetics* **2019**, *40*, 354–360. [CrossRef] [PubMed]
79. Morabito, C.; Rovetta, F.; Bizzarri, M.; Mazzoleni, G.; Fanò, G.; Mariggiò, M.A. Modulation of redox status and calcium handling by extremely low frequency electromagnetic fields in C2C12 muscle cells: A real-time, single-cell approach. *Free Radic. Biol. Med.* **2010**, *48*, 579–589. [CrossRef] [PubMed]
80. Ehnert, S.; Fentz, A.K.; Schreiner, A.; Birk, J.; Wilbrand, B.; Ziegler, P.; Reumann, M.K.; Wang, H.; Falldorf, K.; Nussler, A.K. Extremely low frequency pulsed electromagnetic fields cause antioxidative defense mechanisms in human osteoblasts via induction of •$O^{2-}$ and $H_2O_2$. *Sci. Rep.* **2017**, *7*, 14544. [CrossRef] [PubMed]
81. Patruno, A.; Tabrez, S.; Pesce, M.; Shakil, S.; Kamal, M.A.; Reale, M. Effects of extremely low frequency electromagnetic field (ELF-EMF) on catalase, cytochrome P450 and nitric oxide synthase in erythro-leukemic cells. *Life Sci.* **2015**, *121*, 117–123. [CrossRef]
82. Lewicka, M.; Henrykowska, G.A.; Pacholski, K.; Szczęsny, A.; Dziedziczak-Buczyńska, M.; Buczyński, A. The impact of electromagnetic radiation of different parameters on platelet oxygen metabolism—In vitro studies. *Adv. Clin. Exp. Med.* **2015**, *24*, 31–35. [CrossRef]
83. Song, K.; Im, S.H.; Yoon, Y.J.; Kim, H.M.; Lee, H.J.; Park, G.S. A 60 Hz uniform electromagnetic field promotes human cell proliferation by decreasing intracellular reactive oxygen species levels. *PLoS ONE* **2018**, *13*, e0199753. [CrossRef]
84. Reale, M.; Kamal, M.A.; Patruno, A.; Costantini, E.; D'Angelo, C.; Pesce, M.; Greig, N.H. Neuronal cellular responses to extremely low frequency electromagnetic field exposure: Implications regarding oxidative stress and neurodegeneration. *PLoS ONE* **2014**, *9*, e104973. [CrossRef]
85. Falone, S.; Marchesi, N.; Osera, C.; Fassina, L.; Comincini, S.; Amadio, M.; Pascale, A. Pulsed electromagnetic field (PEMF) prevents pro-oxidant effects of H2O2 in SK-N-BE(2) human neuroblastoma cells. *Int. J. Radiat. Biol.* **2016**, *92*, 281–286. [CrossRef]
86. Cichoń, N.; Bijak, M.; Miller, E.; Saluk, J. Extremely low frequency electromagnetic field (ELF-EMF) reduces oxidative stress and improves functional and psychological status in ischemic stroke patients. *Bioelectromagnetics* **2017**, *38*, 386–396. [CrossRef] [PubMed]
87. Cichoń, N.; Rzeźnicka, P.; Bijak, M.; Miller, E.; Miller, S.; Saluk, J. Extremely low frequency electromagnetic field reduces oxidative stress during the rehabilitation of post-acute stroke patients. *Adv. Clin. Exp. Med.* **2018**, *27*, 1285–1293. [CrossRef] [PubMed]
88. Fulda, S.; Gorman, A.M.; Hori, O.; Samali, A. Cellular Stress Responses: Cell Survival and Cell Death. *Int. J. Cell Biol.* **2010**. [CrossRef] [PubMed]
89. Soleimani, A.F.; Zulkifli, I.; Omar, A.R.; Raha, A.R. The relationship between adrenocortical function and Hsp70 expression in socially isolated Japanese quail. *Comp. Biochem. Physiol. A Mol. Integr. Physiol.* **2012**, *161*, 140–144. [CrossRef] [PubMed]
90. Cirulli, F.; Alleva, E. The NGF saga: From animal models of psychosocial stress to stress-related psychopathology. *Front. Neuroendocrinol.* **2009**, *30*, 379–395. [CrossRef] [PubMed]
91. Hagena, H.; Hansen, N.; Manahan-Vaughan, D. β-Adrenergic Control of Hippocampal Function: Subserving the Choreography of Synaptic Information Storage and Memory. *Cereb. Cortex* **2016**, *26*, 1349–1364. [CrossRef]
92. Amaroli, A.; Chessa, M.G.; Bavestrello, G.; Bianco, B. Effects of an extremely low-frequency electromagnetic field on stress factors: A study in Dictyostelium discoideum cells. *Eur. J. Protistol.* **2013**, *49*, 400–405. [CrossRef]
93. Wyszkowska, J.; Shepherd, S.; Sharkh, S.; Jackson, C.W.; Newland, P.L. Exposure to extremely low frequency electromagnetic fields alters the behaviour, physiology and stress protein levels of desert locusts. *Sci. Rep.* **2016**, *6*, 36413. [CrossRef]
94. Zeni, O.; Simkó, M.; Scarfi, M.R.; Mattsson, M.O. Cellular Response to ELF-MF and Heat: Evidence for a Common Involvement of Heat Shock Proteins? *Front. Public Health* **2017**, *5*, 280. [CrossRef]
95. Sharma, D.; Masison, D.C. Hsp70 structure, function, regulation and influence on yeast prions. *Protein Pept. Lett.* **2009**, *16*, 571–581. [CrossRef]

96. Perez, F.P.; Zhou, X.; Morisaki, J.; Jurivich, D. Electromagnetic field therapy delays cellular senescence and death by enhancement of the heat shock response. *Exp. Gerontol.* **2008**, *43*, 307–316. [CrossRef] [PubMed]
97. Gutzeit, H.O. Biological Effects of ELF-EMF Enhanced Stress Response: New Insights and New Questions. *Electro Magn.* **2001**, *20*, 15–26. [CrossRef]
98. Goodman, R.; Lin-Ye, A.; Geddis, M.S.; Wickramaratne, P.J.; Hodge, S.E.; Pantazatos, S.P.; Blank, M.; Ambron, R.T. Extremely low frequency electromagnetic fields activate the ERK cascade, increase hsp70 protein levels and promote regeneration in Planaria. *Int. J. Radiat. Biol.* **2009**, *85*, 851–859. [CrossRef] [PubMed]
99. Miranda, M.; Morici, J.F.; Zanoni, M.B.; Bekinschtein, P. Brain-Derived Neurotrophic Factor: A Key Molecule for Memory in the Healthy and the Pathological Brain. *Front. Cell Neurosci.* **2019**, *13*, 363. [CrossRef]
100. Bathina, S.; Das, U.N. Brain-derived neurotrophic factor and its clinical implications. *Arch. Med. Sci.* **2015**, *11*, 1164–1178. [CrossRef] [PubMed]
101. Li, Y.; Yan, X.; Liu, J.; Li, L.; Hu, X.; Sun, H.; Tian, J. Pulsed electromagnetic field enhances brain-derived neurotrophic factor expression through L-type voltage-gated calcium channel- and Erk-dependent signaling pathways in neonatal rat dorsal root ganglion neurons. *Neurochem. Int.* **2014**, *75*, 96–104. [CrossRef]
102. Cichoń, N.; Bijak, M.; Czarny, P.; Miller, E.; Synowiec, E.; Sliwinski, T.; Saluk-Bijak, J. Increase in Blood Levels of Growth Factors Involved in the Neuroplasticity Process by Using an Extremely Low Frequency Electromagnetic Field in Post-stroke Patients. *Front. Aging Neurosci.* **2018**, *10*, 294. [CrossRef]
103. Urnukhsaikhan, E.; Mishig-Ochir, T.; Kim, S.C.; Park, J.K.; Seo, Y.K. Neuroprotective Effect of Low Frequency-Pulsed Electromagnetic Fields in Ischemic Stroke. *Appl. Biochem. Biotechnol.* **2017**, *181*, 1360–1371. [CrossRef]
104. Catterall, W.A.; Few, A.P. Calcium channel regulation and presynaptic plasticity. *Neuron* **2008**, *59*, 882–901. [CrossRef]
105. Ardeshirylajimi, A.; Soleimani, M. Enhanced growth and osteogenic differentiation of Induced Pluripotent Stem cells by Extremely Low-Frequency Electromagnetic Field. *Cell Mol. Biol.* **2015**, *61*, 36–41.
106. Grassi, C.; D'Ascenzo, M.; Torsello, A.; Martinotti, G.; Wolf, F.; Cittadini, A.; Azzena, G.B. Effects of 50 Hz electromagnetic fields on voltage-gated Ca2+ channels and their role in modulation of neuroendocrine cell proliferation and death. *Cell Calcium* **2004**, *35*, 307–315. [CrossRef] [PubMed]
107. Luo, F.L.; Yang, N.; He, C.; Li, H.L.; Li, C.; Chen, F.; Xiong, J.X.; Hu, Z.A.; Zhang, J. Exposure to extremely low frequency electromagnetic fields alters the calcium dynamics of cultured entorhinal cortex neurons. *Environ. Res.* **2014**, *135*, 236–246. [CrossRef] [PubMed]
108. Sun, Z.C.; Ge, J.L.; Guo, B.; Guo, J.; Hao, M.; Wu, Y.C.; Lin, Y.A.; La, T.; Yao, P.T.; Mei, Y.A.; et al. Extremely Low Frequency Electromagnetic Fields Facilitate Vesicle Endocytosis by Increasing Presynaptic Calcium Channel Expression at a Central Synapse. *Sci. Rep.* **2016**, *6*, 21774. [CrossRef] [PubMed]
109. Zhao, Q.R.; Lu, J.M.; Yao, J.J.; Zhang, Z.Y.; Ling, C.; Mei, Y.A. Neuritin reverses deficits in murine novel object associative recognition memory caused by exposure to extremely low-frequency (50 Hz) electromagnetic fields. *Sci. Rep.* **2015**, *5*, 11768. [CrossRef]
110. Yang, Y.; Li, L.; Wang, Y.G.; Fei, Z.; Zhong, J.; Wei, L.Z.; Long, Q.F.; Liu, W.P. Acute neuroprotective effects of extremely low-frequency electromagnetic fields after traumatic brain injury in rats. *Neurosci. Lett.* **2012**, *516*, 15–20. [CrossRef]
111. Sakhaie, M.H.; Soleimani, M.; Pourheydar, B.; Majd, Z.; Atefimanesh, P.; Asl, S.S.; Mehdizadeh, M. Effects of Extremely Low-Frequency Electromagnetic Fields on Neurogenesis and Cognitive Behavior in an Experimental Model of Hippocampal Injury. *Behav. Neurol.* **2017**, *2017*, 9194261. [CrossRef]
112. Collard, J.F.; Lazar, C.; Nowé, A.; Hinsenkamp, M. Statistical validation of the acceleration of the differentiation at the expense of the proliferation in human epidermal cells exposed to extremely low frequency electric fields. *Prog. Biophys. Mol. Biol.* **2013**, *111*, 37–45. [CrossRef]
113. Di Loreto, S.; Falone, S.; Caracciolo, V.; Sebastiani, P.; D'Alessandro, A.; Mirabilio, A.; Zimmitti, V.; Amicarelli, F. Fifty hertz extremely low-frequency magnetic field exposure elicits redox and trophic response in rat-cortical neurons. *J. Cell Physiol.* **2009**, *219*, 334–343. [CrossRef]
114. Fathi, E.; Farahzadi, R. Zinc Sulphate Mediates the Stimulation of Cell Proliferation of Rat Adipose Tissue-Derived Mesenchymal Stem Cells Under High Intensity of EMF Exposure. *Biol. Trace Elem. Res.* **2018**, *184*, 529–535. [CrossRef]
115. Chung, Y.H.; Lee, Y.J.; Lee, H.S.; Chung, S.J.; Lim, C.H.; Oh, K.W.; Sohn, U.D.; Park, E.S.; Jeong, J.H. Extremely low frequency magnetic field modulates the level of neurotransmitters. *Korean J. Physiol. Pharmacol.* **2015**, *19*, 15–20. [CrossRef]
116. Sieroń, A.; Labus, Ł.; Nowak, P.; Cieślar, G.; Brus, H.; Durczok, A.; Zagził, T.; Kostrzewa, R.M.; Brus, R. Alternating extremely low frequency magnetic field increases turnover of dopamine and serotonin in rat frontal cortex. *Bioelectromagnetics* **2004**, *25*, 426–430. [CrossRef] [PubMed]
117. Isogawa, K.; Fujiki, M.; Akiyoshi, J.; Tsutsumi, T.; Horinouchi, Y.; Kodama, K.; Nagayama, H. Anxiety induced by repetitive transcranial magnetic stimulation is suppressed by chronic treatment of paroxetine in rats. *Pharmacopsychiatry* **2003**, *36*, 7–11. [CrossRef] [PubMed]
118. Janać, B.; Tovilović, G.; Tomić, M.; Prolić, Z.; Radenović, L. Effect of continuous exposure to alternating magnetic field (50 Hz, 0.5 mT) on serotonin and dopamine receptors activity in rat brain. *Gen. Physiol. Biophys.* **2009**, *28*, 41–46. [PubMed]
119. Jadidi, M.; Firoozabadi, S.M.; Rashidy-Pour, A.; Sajadi, A.A.; Sadeghi, H.; Taherian, A.A. Acute exposure to a 50 Hz magnetic field impairs consolidation of spatial memory in rats. *Neurobiol. Learn. Mem.* **2007**, *88*, 387–392. [CrossRef]

120. Kirschenlohr, H.; Ellis, P.; Hesketh, R.; Metcalfe, J. Gene expression profiles in white blood cells of volunteers exposed to a 50 Hz electromagnetic field. *Radiat. Res.* **2012**, *178*, 138–149. [CrossRef]
121. Mahdavi, S.M.; Sahraei, H.; Yaghmaei, P.; Tavakoli, H. Effects of electromagnetic radiation exposure on stress-related behaviors and stress hormones in male wistar rats. *Biomol. Ther.* **2014**, *22*, 570–576. [CrossRef]
122. Hosseini, E.; Nafisi, S.; Zare, S. The effects of electromagnetic fields on plasma levels of corticosterone, free-T3, free-T4 malonyl-dialdehyde in white male rabbit with normal diet and hyperchlostrol diet. *Vet. Res. Forum* **2011**, *2*, 222–225.
123. Afhami, M.; Bahaoddini, A.; Saadat, M. P20: Investigation the Effect of EMF on Plasma Levels of Corticosterone, Testosterone and Testicular Gene Expression of Gstt1 of Male Rats. *Shefaye Khatam* **2016**, *4*, 43.
124. de Kleijn, S.; Ferwerda, G.; Wiese, M.; Trentelman, J.; Cuppen, J.; Kozicz, T.; de Jager, L.; Hermans, P.W.; Verburg-van Kemenade, B.M. A short-term extremely low frequency electromagnetic field exposure increases circulating leukocyte numbers and affects HPA-axis signaling in mice. *Bioelectromagnetics* **2016**, *37*, 433–443. [CrossRef]
125. Mostafa, R.M.; Mostafa, Y.M.; Ennaceur, A. Effects of exposure to extremely low-frequency magnetic field of 2 G intensity on memory and corticosterone level in rats. *Physiol. Behav.* **2002**, *76*, 589–595. [CrossRef]
126. Laszlo, A.M.; Ladanyi, M.; Boda, K.; Csicsman, J.; Bari, F.; Serester, A.; Molnar, Z.; Sepp, K.; Galfi, M.; Radacs, M. Effects of extremely low frequency electromagnetic fields on turkeys. *Poult. Sci.* **2018**, *97*, 634–642. [CrossRef] [PubMed]
127. Tian, R.; Hou, G.; Li, D.; Yuan, T.F. A possible change process of inflammatory cytokines in the prolonged chronic stress and its ultimate implications for health. *Sci. World J.* **2014**, *2014*, 780616. [CrossRef] [PubMed]
128. Silverman, M.N.; Pearce, B.D.; Biron, C.A.; Miller, A.H. Immune modulation of the hypothalamic-pituitary-adrenal (HPA) axis during viral infection. *Viral. Immunol.* **2005**, *18*, 41–78. [CrossRef] [PubMed]
129. Wyszkowska, J.; Jędrzejewski, T.; Piotrowski, J.; Wojciechowska, A.; Stankiewicz, M.; Kozak, W. Evaluation of the influence of in vivo exposure to extremely low-frequency magnetic fields on the plasma levels of pro-inflammatory cytokines in rats. *Int. J. Radiat. Biol.* **2018**, *94*, 909–917. [CrossRef] [PubMed]
130. Cichoń, N.; Saluk-Bijak, J.; Miller, E.; Sliwinski, T.; Synowiec, E.; Wigner, P.; Bijak, M. Evaluation of the effects of extremely low frequency electromagnetic field on the levels of some inflammatory cytokines in post-stroke patients. *J. Rehabil. Med.* **2019**, *51*, 854–860. [CrossRef] [PubMed]
131. Mahaki, H.; Jabarivasal, N.; Sardanian, K.; Zamani, A. Effects of Various Densities of 50 Hz Electromagnetic Field on Serum IL-9, IL-10, and TNF-α Levels. *Int. J. Occup. Environ. Med.* **2020**, *11*, 24–32. [CrossRef]
132. Sterner, E.Y.; Kalynchuk, L.E. Behavioral and neurobiological consequences of prolonged glucocorticoid exposure in rats: Relevance to depression. *Prog. Neuropsychopharmacol. Biol. Psychiatry* **2010**, *34*, 777–790. [CrossRef]
133. Liu, T.; Wang, S.; He, L.; Ye, K. Anxiogenic effect of chronic exposure to extremely low frequency magnetic field in adult rats. *Neurosci. Lett.* **2008**, *434*, 12–17. [CrossRef]
134. Alsaeed, I.; Al-Somali, F.; Sakhnini, L.; Aljarallah, O.S.; Hamdan, R.M.; Bubishate, S.A.; Sarfaraz, Z.K.; Kamal, A. Autism-relevant social abnormalities in mice exposed perinatally to extremely low frequency electromagnetic fields. *Int. J. Dev. Neurosci.* **2014**, *37*, 58–64. [CrossRef]
135. Korpinar, M.A.; Kalkan, M.T.; Tuncel, H. The 50 Hz (10 mT) sinusoidal magnetic field: Effects on stress-related behavior of rats. *Bratisl. Lek. Listy* **2012**, *113*, 521–524. [CrossRef]
136. Balassa, T.; Szemerszky, R.; Bárdos, G. Effect of short-term 50 Hz electromagnetic field exposure on the behavior of rats. *Acta Physiol. Hung.* **2009**, *96*, 437–448. [CrossRef] [PubMed]
137. He, L.H.; Shi, H.M.; Liu, T.T.; Xu, Y.C.; Ye, K.P.; Wang, S. Effects of extremely low frequency magnetic field on anxiety level and spatial memory of adult rats. *Chin. Med. J.* **2011**, *124*, 3362–3366. [PubMed]
138. Mattar, F.A.; Bareedy, M.H.; El-Dosouky, M.E.M.; Zaghloul, M.S.; Mahmod, M.S.M. Effects of 6 Weeks Exposure of 3.5 mT (ELF EMF) on Some Animal behaviors in White Albino Rat (Sprague Dawley) Pups. *Zagazig Vet. J.* **2014**, *42*, 74–80. [CrossRef]
139. Lai, J.; Zhang, Y.; Liu, X.; Zhang, J.; Ruan, G.; Chaugai, S.; Chen, C.; Wang, D.W. Effects of extremely low frequency electromagnetic fields (100μT) on behaviors in rats. *Neurotoxicology* **2016**, *52*, 104–113. [CrossRef] [PubMed]
140. Boscolo, P.; Di Giampaolo, L.; Di Donato, A.; Antonucci, A.; Paiardini, G.; Morelli, S.; Vasile, R.; Spagnoli, G.; Reale, M.; Dadorante, V.; et al. The immune response of women with prolonged exposure to electromagnetic fields produced by radiotelevision broadcasting stations. *Int. J. Immunopathol. Pharmacol.* **2006**, *19* (Suppl. 4), 43–48.
141. Berman, R.M.; Narasimhan, M.; Sanacora, G.; Miano, A.P.; Hoffman, R.E.; Hu, X.S.; Charney, D.S.; Boutros, N.N. A randomized clinical trial of repetitive transcranial magnetic stimulation in the treatment of major depression. *Biol. Psychiatry* **2000**, *47*, 332–337. [CrossRef]
142. Janać, B.; Pesić, V.; Jelenković, A.; Vorobyov, V.; Prolić, Z. Different effects of chronic exposure to ELF magnetic field on spontaneous and amphetamine-induced locomotor and stereotypic activities in rats. *Brain Res. Bull.* **2005**, *67*, 498–503. [CrossRef]
143. Roszkowski, M.; Manuella, F.; von Ziegler, L.; Durán-Pacheco, G.; Moreau, J.L.; Mansuy, I.M.; Bohacek, J. Rapid stress-induced transcriptomic changes in the brain depend on beta-adrenergic signaling. *Neuropharmacology* **2016**, *107*, 329–338. [CrossRef]
144. Genuis, S.J.; Lipp, C.T. Electromagnetic hypersensitivity: Fact or fiction? *Sci. Total Environ.* **2012**, *414*, 103–112. [CrossRef]

*Article*

# Cognitive Improvement Effects of Electroacupuncture Combined with Computer-Based Cognitive Rehabilitation in Patients with Mild Cognitive Impairment: A Randomized Controlled Trial

**Jae-Hong Kim [1,2], Jae-Young Han [3,*], Gwang-Cheon Park [2] and Jeong-Soon Lee [4]**

1 Department of Acupuncture and Moxibustion Medicine, College of Korean Medicine, DongShin University, Naju City 58245, Korea; nahonga@hanmail.net

2 Clinical Research Center, DongShin University Gwangju Korean Medicine Hospital, 141, Wolsan-ro, Nam-gu, Gwangju City 61619, Korea; smailcc@nate.com

3 Department of Physical and Rehabilitation Medicine, Chonnam National University Medical School and Hospital, Gwangju City 61469, Korea

4 Department of Nursing, Christian College of Nursing, Gwangju City 61662, Korea; mishilee@ccn.ac.kr

* Correspondence: rmhanjy@daum.net; Tel.: +82-62-220-5186

Received: 14 November 2020; Accepted: 11 December 2020; Published: 14 December 2020

**Abstract:** This outcome assessor-blinded, randomized controlled clinical trial investigated the effects of electroacupuncture combined with computer-based cognitive rehabilitation (EA-CCR) on mild cognitive impairment (MCI). A per-protocol analysis was employed to compare the efficacy of EA-CCR to that of computer-based cognitive rehabilitation (CCR). Thirty-two patients with MCI completed the trial (EA-CCR group, 16; CCR group, 16). Patients received EA-CCR or CCR treatment once daily three days per week for eight weeks. Outcome (primary, ADAS-K-cog; secondary, MoCA-K, CES-D, K-ADL, K-IADL, and EQ-5D-5L) measurements were performed at baseline (week 0), at the end of the intervention (week 8), and at 12 weeks after completion of the intervention (week 20). Both groups showed significant changes in ADAS-K-cog score (EA-CCR, $p < 0.001$; CCR, $p < 0.001$) and MoCA-K (EA-CCR, $p < 0.001$; CCR, $p < 0.001$). Only the EA-CCR group had a significant change in CES-D ($p = 0.024$). No significant differences in outcomes and in the results of a subanalysis based on age were noted between the groups. These results indicate that EA-CCR and CCR have beneficial effects on improving cognitive function in patients with MCI. However, electroacupuncture in EA-CCR showed no positive add-on effects on improving cognitive function, depression, activities of daily living, and quality of life in patients with MCI.

**Keywords:** mild cognitive impairment; electroacupuncture; computer-based cognitive rehabilitation; randomized controlled trial

## 1. Introduction

Mild cognitive impairment (MCI) is a condition in which individuals demonstrate a slight objective impairment in cognition (typically memory) that does not require help with the performance of activities of daily living [1–3]. MCI is considered an intermediate stage between the expected cognitive decline of normal aging and Alzheimer's disease (AD), with a conversion rate of 5–10% per year [4–7]. Thus, MCI is a target for the prevention of AD development [8].

No high-quality evidence exists to support pharmacologic treatments for MCI [9]. Systematic reviews and meta-analyses evaluating the efficacy of cholinesterase inhibitors for MCI treatment have concluded that there is no convincing evidence that cholinesterase inhibitors have an effect on cognitive test scores or the progression of MCI to AD [10,11]. Some non-pharmacologic interventions, such as computerized cognitive training [12,13], exercise training [14], aerobic dance routine [15], and acupuncture [16,17], may be beneficial for patients with MCI. However, currently, no treatment method for MCI has been established [18].

Acupuncture is a common traditional Chinese medicine technique that is used for the treatment of various kinds of neurological disorders, including MCI [16]. Electroacupuncture (EA) treatment refers to the insertion of more than two needles into the skin and applying weak electricity through the needle [17]. EA has been reported to produce greater effect on neuroblast plasticity in the dentate gyrus [19], more widespread signal increases in the human brain as measured by functional magnetic resonance imaging [20] than acupuncture alone. A systematic review and clinical trials suggest that EA may have a beneficial effect on MCI [17,21–24]. Moreover, cognitive training could enhance brain activation in the areas related to memory [25]. Computer-based cognitive rehabilitation (CCR) has generated considerable attention as a safe, relatively inexpensive, and scalable intervention that aims to reduce cognitive decline in older adults [13]. Systematic reviews have demonstrated that CCR may generate some positive effects on patients with MCI or dementia [12,13].

Although evidence suggests that both EA and CCR have benefits on cognitive functions, evidence regarding the efficacy and safety of EA combined with CCR (EA-CCR) for treating MCI is insufficient. Hence, this study was performed to investigate the efficacy and safety of EA-CCR for the treatment of MCI and to determine whether EA has add-on effects by comparing EA-CCR with CCR alone in patients with MCI.

## 2. Materials and Methods

This study followed the Standard Protocol Items: Recommendations for Interventional Trials (SPIRIT) and Consolidated Standards of Reporting Trials (CONSORT) statements (Table S1). Details of the methods in this study have been reported previously [26].

### *2.1. Study Design*

This study was a prospective, outcome assessor-blinded, single-center (DongShin University Gwangju Korean Medicine Hospital, Republic of Korea), randomized controlled trial with a 1:1 allocation ratio. A total of 36 participants who met the inclusion and exclusion criteria were randomly allocated to either the EA-CCR or the CCR group ($n$ = 18 each group). Participants in the CCR group received RehaCom cognitive rehabilitation only, while those in the EA-CCR group had EA at the following acupoints: Baihui (GV20), Sishencong (EX-HN1), Fengchi (GB20), and Shenting (GV24), and RehaCom cognitive rehabilitation. The treatment duration was 8 weeks in both groups. Outcome measures were determined at baseline (week 0), 8 weeks after the first intervention (week 8; i.e., at the end of the intervention), and 12 weeks after completion of the intervention (week 20). The study design is summarized in Table 1.

### *2.2. Ethical Considerations*

This study was conducted in accordance with the Declaration of Helsinki, and the protocol of this study (ver. 1.1) was approved by the Ministry of Food and Drug Safety (Medical Device Clinical Trial Plan approval number: 859; approval date: 24 July 2018) and the Institutional Review Board (IRB) of DongShin University Gwangju Korean Medicine Hospital (approval number: DSGOH-050; approval date: 17 September 2018) before the trial began. This trial was registered at the Clinical Research Information Service (cris.nih.go.kr; registration number: KCT0003415; registration date: 4 January 2019). The purpose and potential risks of this study were fully explained to the participants. All participants provided written informed consent before participating in this study.

**Table 1.** Standard protocol items: Recommendations for Interventional Trials (SPIRIT) statement.

| | STUDY PERIOD | | | | | | | | | | |
|---|---|---|---|---|---|---|---|---|---|---|---|
| | **Enrolment** | **Allocation** | **Post-Allocation** | | | | | | | | **Close-Out** |
| **TIMEPOINT** | **Screening** | | Visit1-3 | Visit4-6 | Visit7-9 | Visit10-12 | Visit13-15 | Visit16-18 | Visit19-21 | Visit22-24 | Visit25 |
| | Week | | 1 | 2 | 3 | 4 | 5 | 6 | 7 | 8 | 20 |
| ENROLMENT | | | | | | | | | | | |
| Informed consent | X | | | | | | | | | | |
| Sociodemographic profile | X | | | | | | | | | | |
| Medical history | X | | | | | | | | | | |
| Vital signs | X | X | X | X | X | X | X | X | X | X | X |
| Inclusion/exclusion criteria | X | | | | | | | | | | |
| Allocation | | X | | | | | | | | | |
| K-MMSE, MoCA-K | X | | | | | | | | | | |
| INTERVENTIONS | | | | | | | | | | | |
| CCR | | | X | X | X | X | X | X | X | X | |
| EA-CCR | | | X | X | X | X | X | X | X | X | |
| ASSESSMENTS | | | | | | | | | | | |
| Change of medical history | | | X | X | X | X | X | X | X | X | X |
| Safety assessment | | | X | X | X | X | X | X | X | X | X |
| ADAS-K-cog | | | X | | | | | | | X | X |
| MoCA-K | | | X | | | | | | | X | X |
| CES-D | | | X | | | | | | | X | X |
| K-ADL, K-IADL | | | X | | | | | | | X | X |
| EQ-5D-5L | | | X | | | | | | | X | X |

The table shows the enrollment, interventions, and data collection protocols. K-MMSE, Korean version of Mini-Mental State Examination; MoCA-K, Korean version of the Montreal Cognitive Assessment; CCR, computer-based cognitive rehabilitation; EA-CCR, electroacupuncture combined with computer-based cognitive rehabilitation; ADAS-K-cog, Korean version of Alzheimer's Disease Assessment Scale—cognitive subscale; CES-D, Center for Epidemiological Studies—Depression Scale; K-ADL, Korean Activities of Daily Living; K-IADL Korean Instrumental Activities of Daily Living; EQ-5D-5L, European Quality of Life Five Dimension-Five Level Scale.

### 2.3. Participant Recruitment

Participants were recruited at DongShin University Gwangju Korean Medicine Hospital. We submitted our study protocol to the Clinical Research Information Service on 12 November 2018. Considering the possibility of recruitment within the study period, we began the recruitment on 29 November 2018, which was before the trial registration. This study was advertised via local newspapers, the Internet, and posters in communities and hospitals. Participants received an explanation about the study from the clinical research coordinator (CRC) and were requested to voluntarily sign an informed consent form before participation. All recruited individuals were screened by the Korean version of the Mini-Mental State Examination (K-MMSE) and of the Montreal Cognitive Assessment (MoCA-K) to ensure that all inclusion criteria are met. The CRC monitored the medical conditions of the enrolled participants to maximize adherence to intervention protocols.

### 2.4. Participation

Participants who met all of the following criteria were included in the study: (1) Age 55–85 years; (2) fulfillment of the Peterson diagnostic criteria for MCI [1,2], with memory impairment for at least 3 months; (3) K-MMSE score of 20–23; (4) MoCA-K scale score of 0–22; (5) adequate Korean language fluency, for reliable completion of all study assessments; and (6) voluntary provision of informed consent.

The exclusion criteria were as follows: (1) Diagnosis of dementia according to the Diagnostic and Statistical Manual of Mental Disorders-IV; (2) history of structural brain lesions that could cause cognitive impairment, such as traumatic brain injury, stroke, intracranial space-occupying lesions, and congenital mental retardation; (3) presence of cancer and/or serious cardiovascular, cerebrovascular, liver, or kidney diseases; (4) history of treatment for alcohol or drug dependency or mental diseases, such as schizophrenia, serious anxiety, and depression in the past 6 months; (5) ongoing treatment for MCI, such as medication, acupuncture, and cognitive training); (6) difficulties in assessment due to visual and hearing impairments; (7) presence of contraindications for EA, such as blood clotting abnormalities (e.g., hemophilia), infection of the skin over the head, and presence of a pacemaker); and (8) concurrent participation in other clinical trials.

### 2.5. Randomization and Blinding

Following the acquisition of written informed consent, the practitioners who would perform the intervention conducted a screening interview. Thereafter, the assessor performed baseline measurements for the participants who met the inclusion criteria. The 36 enrolled participants were immediately assigned serial numbers generated using SPSS version 21 software (IBM Corp., Armonk, NY, USA) and were randomly allocated to one of the two study groups (n = 18 each group). The serial number codes were inserted into opaque envelopes that were sealed and kept in a double-locked cabinet; the envelopes were opened by the principal investigator or practitioner who would perform the intervention in the presence of the patient and a guardian.

We could only adopt a single outcome assessor-blinding approach because sham treatment was impossible because of the characteristics of EA application, which included insertion and electric stimulation. During the study, there was no contact between the assessor and any participant at any time point other than the time of assessment. Data analysts without conflicts of interest were involved in this study.

### 2.6. Implementation

The CRC generated the allocation sequence, enrolled the participants, and assigned participants to interventions.

## 2.7. Intervention

Participants in the CCR group received RehaCom cognitive rehabilitation (30 min) once a day, 3 days per week for 8 weeks. Participants in the EA-CCR group received EA (30 min) at Baihui (GV20), Sishencong (EX-HN1), Fengchi (GB20), and Shenting (GV24) in addition to RehaCom cognitive rehabilitation (30 min) once a day, 3 days per week for 8 weeks. EA was performed first, followed by CCR. The treatments were administered by Korean medicine doctors with 6 years of formal university training in Korean medicine and a license to administer treatment. To ensure strict adherence to the study protocol, the doctors received training together and used the same techniques.

### 2.7.1. Electroacupuncture Treatment

EA was performed at the following acupoints: Baihui (GV20), Sishencong (EX-HN1), Fengchi (GB20), and Shenting (GV24) [24]. Only sterile, stainless-steel, disposable acupuncture needles (size, 0.25 × 30 mm; Dong Bang Acupuncture, Inc., Boryeong, Republic of Korea; product no.: A84010.02) with guide tubes and an EA stimulator [CELLMAC PLUS (STN-330); Stratek, Co., Ltd., Anyang, Republic of Korea; product no.: A16010.04] were used. With the patients in the sitting position, the needles were inserted at an angle of 15–30° along the scalp. GB20 was punctured 17–30 mm in the direction of the tip of the nose. GV24, the anterior EX-HN1, and GV20 were punctured in the forward direction, while the left, right, and posterior EX-HN1 points were punctured in the direction of GV20. The depths of insertion were 9–24 mm, depending on the location of the needle [27]. After insertion, the needles were left in position for 30 min. Manual stimulation was not used. GV24 and GV20, the left and right EX-HN1, the anterior and posterior EX-HN1, and the left and right GB20 were subjected to EA under the following parameters: AC; continuous waves; frequency, 3 Hz; and intensity, between 2–4 mA such that the patient could feel it. Each participant received a total of 24 30-min sessions (three times per week for 8 weeks) [24] (Table S2).

### 2.7.2. RehaCom Cognitive Rehabilitation

All participants received RehaCom cognitive rehabilitation in the sitting position. Six different therapeutic programs to restore attention, memory, and executive functions were employed. Each program has one to four different tasks from which participants could choose during each therapy session. We mainly used topological memory, physiognomic memory, memory of words, and figural memory tasks of memory program and shopping, logical reasoning, and calculation tasks of executive function program. Each participant received a total of 24 30-min sessions (three times per week for 8 weeks).

During the clinical trial period, all participants were allowed to use routine management regimens, existing medications (e.g., those for hypertension, diabetes, or hyperlipidemia), and medications for maintaining and improving their health status. However, they were not permitted to engage in other treatments for ameliorating MCI symptoms. All medical devices, including the acupuncture needles, EA stimulator, and RehaCom software (HASOMED GmbH., Magdeburg, Germany), were inspected by the investigators, who recorded check-up results in the management register.

## 2.8. Outcome Measurements

Scores for the Korean version of Alzheimer's Disease Assessment Scale—cognitive subscale (ADAS-K-cog), MoCA-K, Center for Epidemiological Studies—Depression Scale (CES-D), Korean Activities of Daily Living (K-ADL) scale, Korean Instrumental Activities of Daily Living (K-IADL) scale, and European Quality of Life Five Dimension Five Level Scale (EQ-5D-5L) were recorded before treatment, at the end of treatment, and at 12 weeks after treatment completion.

The primary outcome was improvement in cognitive function as assessed using the ADAS-K-cog, which is a tool that is considered the gold standard for assessing the efficacy of various anti-dementia treatments [28]. Particularly, it is known to be sensitive to the treatment responses of patients with MCI or early dementia [29].

The secondary outcomes included changes in the MoCA-K scale, CES-D, K-ADL scale, K-IADL scale, and EQ-5D-5L scores over time. The MoCA-K scale is a clinician-friendly, validated, brief instrument with high sensitivity and specificity for detecting MCI [30]. The CES-D is a short self-report scale designed to measure the current level of depressive symptomatology, with emphasis on the affective component (i.e., depressed mood) [31]. The K-ADL and K-IADL scales are used to assess physical function. The K-ADL scale is used to assess basic activities of elderly individuals; the K-IADL scale, to estimate complex activities representing instrumental self-maintenance and social behavior [32]. The EQ-5D-5L is a generic instrument for assessing health-related quality of life, which comprises five dimensions [33].

### *2.9. Sample Size Calculation*

Because of the lack of adequate preliminary studies and limited research funds, study period, and recruitment opportunities, we have adopted a pilot study design with 18 participants in each group. As our study was a pilot study, the sample size was not sufficient to provide information on the efficacy of EA-CCR on MCI. Nevertheless, our study could provide an indication on the feasibility of a randomized trial of EA-CCR treatment for MCI and could determine whether EA-CCR is an acceptable treatment for patients with MCI.

### *2.10. Statistical Analyses*

With the approval of the IRB, the statistical analysis in the study protocol was revised. We performed per-protocol (PP) analyses for the assessment of efficacy and a supplementary full analysis (FA) set. Missing values were imputed by the last observation carried forward method. We compared the results of the PP analyses and those of the supplementary FA set. If there was a significant difference between the PP and FA groups, the cause was reviewed and reflected during the efficacy assessment. Analysis was performed by blinded biostatisticians using SPSS version 20.0 software (SPSS Inc., Chicago, IL, USA); two-sided significance tests with a 5% significance level were employed. Continuous variables were presented as means and standard deviations and categorical variables as count frequencies and percentages.

Baseline data were obtained and compared using independent $t$-test, $\chi^2$ test, and Fisher's exact test. Differences in all outcome value changes in the two groups were compared using Wilcoxon signed-rank test and repeated-measures analysis of variance (ANOVA; Friedman tests). ADAS-K-cog, MoCA-K, CES-D, K-ADL scale, K-IADL scale, and EQ-5D-5L values were compared by repeated-measures ANOVA across two to three testing time points (i.e., week 0, week 8, and week 20). Differences in outcome value changes between the two groups (week 0 vs. week 8, week 0 vs. week 20, and week 8 vs. week 20) were compared using the Mann–Whitney U-test (nonparametric test). Moreover, the participants were divided into two groups according to age: <70 and >70 groups, and a subanalysis was conducted to investigate the differences in ADAS-K-cog, MoCA-K, CES-D, K-ADL scale, K-IADL scale, and EQ-5D-5L changes (week 0 vs. week 8, week 0 vs. week 20, and week 8 vs. week 20) between the two groups.

## 3. Results

### *3.1. Particpants*

We recruited participants between 29 November 2018 and 23 October 2019. Of the 476 patients assessed for eligibility, 440 were excluded. Thirty-six patients were included in this study and were randomly assigned to either the EA-CCR group ($n$ = 18) or CCR group ($n$ = 18). Two participants in both groups did not complete the treatment. Results of the PP analysis for the assessment of efficacy were not different from those of the supplementary FA set. Thus, data of 32 patients with MCI (EA-CCR group, $n = 16$; CCR group, $n = 16$) were used in the final analysis (Figure 1).

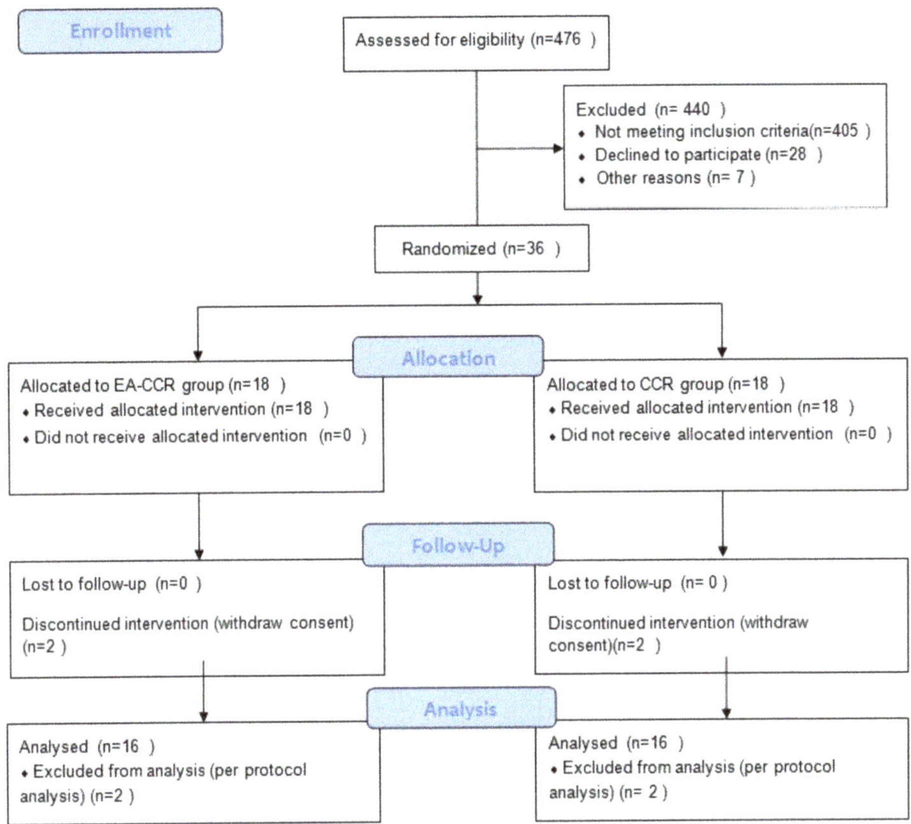


**Figure 1.** CONSORT 2010 flow diagram.

### 3.2. Baseline Charactersitics

The baseline demographic characteristics and study variables of the 32 patients in the two groups are presented in Table 2. No significant differences in the baseline demographic characteristics and study variables were detected between the two groups ($p > 0.05$; Table 2).

### 3.3. Primary and Secondary Outcomes

After eight weeks of intervention, we observed significant improvements in both groups (changes in ADAS-K-cog and MoCA-K) and in the EA-CCR group (changes in CES-D) (Table 3).

Repeated-measures ANOVA showed no significant interaction between time and group with respect to all study variables (Table 4).

No significant differences in the changes in ADAS-K-cog, MoCA-K, CES-D, K-ADL, K-IADL, and EQ-5D-5L (week 0 vs. week 8, week 0 vs. week 20, and week 8 vs. week20) were found between the two groups (Table 5).

A subanalysis based on patient age (i.e., <70 years and >70 years) showed no significant differences in all variables between the two groups (Tables 6 and 7).

**Table 2.** Homogeneity tests for baseline demographic characteristics and study variables for 32 patients with Mild Cognitive Impairment.

| Dependent Variables | EA-CCR ($n$ = 16) | CCR ($n$ = 16) | $p$ or $x^2$ ($P$) |
|---|---|---|---|
| | Mean (SD) or n (%) | Mean (SD) or n (%) | |
| Age (y) | 69.94 (5.94) | 74.25 (5.39) | 21.33 (0.166) ǂ |
| Gender (Female) | 14 (87.5) | 14 (87.5) | 0.00 (1.000) ǂ |
| Education | 9.13 (4.83) | 8.69 (5.08) | 4.84 (0.679) ǂ |
| ADAS-K-cog | 11.13 ± 4.10 | 11.19 ± 6.22 | −0.03 (0.973) * |
| MoCA-K | 18.75 ± 2.54 | 19.31 ± 2.92 | −0.58 (0.565) * |
| CES-D | 14.25 ± 5.91 | 11.50 ± 6.94 | 1.21 (0.237) * |
| K-ADL | 7.19 ± 0.40 | 7.13 ± 0.34 | 0.47 (0.640) * |
| K-IADL | 11.00 ± 3.03 | 11.50 ± 3.10 | −0.46 (0.648) * |
| EQ-5D-5L | 6.38 ± 1.36 | 7.31 ± 2.52 | −1.31 (0.201) * |

* $t$-test; ǂ $x^2$-test.

**Table 3.** Changes in outcome measures (week 0 vs. week 8, week 0 vs. week 20) after treatment completion in patients who received electroacupuncture combined with computer-based cognitive rehabilitation (EA-CCR) and computer-based cognitive rehabilitation (CCR) (n = 16 each) for Mild Cognitive Impairment.

| Groups | Dependent Variables | Week 0 (M ± SD) | Week 8 (M ± SD) | Week 20 (M ± SD) | Difference (w8-w0) | Z ($p$) * | Difference (w20-w0) | Z ($p$) * | $x^2$ ($p$) ǂ |
|---|---|---|---|---|---|---|---|---|---|
| EA-CCR group ($n$ = 16) | ADAS-K -cog | 11.13 ± 4.10 | 7.19 ± 4.75 | 6.19 ± 3.95 | −3.94 ± 2.57 | −3.42 (0.001) | −4.94 ± 3.45 | −3.42 (0.001) | 21.08 (<0.001) |
| | MoCA-K | 18.75 ± 2.54 | 24.25 ± 3.26 | 24.56 ± 4.21 | 5.50 ± 2.48 | −3.53 (<0.001) | 5.81 ± 3.69 | −3.41 (0.001) | 24.10 (<0.001) |
| | CES-D | 14.25 ± 5.91 | 10.94 ± 6.18 | 10.75 ± 6.02 | −3.31 ± 4.77 | −2.39 (0.017) | −3.50 ± 6.20 | −2.01 (0.038) | 7.48 (0.024) |
| | K-ADL | 7.19 ± 0.40 | 7.06 ± 0.25 | 7.13 ± 0.34 | −0.13 ± 0.34 | −1.41 (0.157) | −0.06 ± 0.44 | −0.58 (0.564) | 2.00 (0.368) |
| | K-IADL | 11.00 ± 3.03 | 10.69 ± 1.85 | 10.44 ± 0.96 | −0.31 ± 3.03 | −0.54 (0.593) | −0.56 ± 2.45 | −0.73 (0.465) | 0.15 (0.926) |
| | EQ-5D-5L | 6.38 ± 1.36 | 6.06 ± 1.24 | 5.75 ± 1.00 | 0.31 ± 1.49 | −0.93 (0.351) | 0.63 ± 1.54 | −1.40 (0.161) | 3.56 (0.169) |
| CCR group ($n$ = 16) | ADAS-K -cog | 11.19 ± 6.22 | 7.38 ± 3.86 | 5.81 ± 2.76 | -3.81 ± 3.47 | −3.33 (0.001) | −5.38 ± 5.19 | −3.42 (0.001) | 19.22 (<0.001) |
| | MoCA-K | 19.31 ± 2.92 | 24.19 ± 2.48 | 25.13 ± 1.89 | 4.88 ± 2.45 | −3.53 (<0.001) | 5.81 ± 2.37 | −3.53 (<0.001) | 26.00 (<0.001) |
| | CES-D | 11.50 ± 6.94 | 11.00 ± 6.69 | 9.75 ± 6.77 | −0.50 ± 7.43 | −0.00 (1.00) | −1.75 ± 8.70 | −1.02 (0.306) | 1.86 (0.395) |
| | K-ADL | 7.13 ± 0.34 | 7.06 ± 0.25 | 7.06 ± 0.25 | −0.06 ± 0.25 | −1.00 (0.317) | −0.06 ± 0.25 | −1.00 (0.317) | 2.00 (0.368) |
| | K-IADL | 11.50 ± 3.10 | 11.44 ± 3.08 | 11.62 ± 3.12 | −0.06 ± 0.93 | −0.45 (0.655) | −0.13 ± 1.31 | −0.54 (0.593) | 0.20 (0.905) |
| | EQ-5D-5L | 7.31 ± 2.52 | 6.12 ± 1.24 | 6.31 ± 1.25 | 1.00 ± 2.92 | −2.21 (0.027) | −1.00 ± 2.92 | −1.29 (0.196) | 5.28 (0.071) |

* Wilcoxon signed-rank test; ǂ Repeated measures ANOVA (Friedman test).

**Table 4.** Results of repeated-measures ANOVA for the outcomes of treatment between patients who received EA-CCR and CCR for Mild Cognitive Impairment ($n$ = 16 each).

| Dependent Variables | Group (n) | Week 0 (M ± SD) | Week 8 (M ± SD) | Week 20 (M ± SD) | Source | SS | df | Mean Square | F | $p$ |
|---|---|---|---|---|---|---|---|---|---|---|
| ADAS-K-cog | EA-CCR ($n$ = 16) | 11.13 ± 4.10 | 7.19 ± 4.75 | 6.19 ± 3.95 | Time | 461.27 | 2 | 230.64 | 38.16 | <0.001 |
| | CCR ($n$ = 16) | 11.19 ± 6.22 | 7.38 ± 3.86 | 5.81 ± 2.76 | Group Time | 1.396 | 2 | 0.70 | 0.12 | 0.891 |
| MoCA-K | EA-CCR ($n$ = 16) | 18.75 ± 2.54 | 24.25 ± 3.26 | 24.56 ± 4.21 | Time | 651.58 | 2 | 325.79 | 86.11 | <0.001 |
| | CCR ($n$ − 16) | 19.31 ± 2.92 | 24.19 ± 2.48 | 25.13 ± 1.89 | Group Time | 2.08 | 2 | 1.04 | 0.28 | 0.760 |
| CES-D | EA-CCR ($n$ = 16) | 14.25 ± 5.91 | 10.94 ± 6.18 | 10.75 ± 6.02 | Time | 117.77 | 2 | 58.89 | 2.52 | 0.089 |
| | CCR ($n$ = 16) | 11.50 ± 6.94 | 11.00 ± 6.69 | 9.75 ± 6.77 | Group Time | 32.27 | 2 | 16.14 | 0.69 | 0.506 |
| K-ADL | EA-CCR ($n$ = 16) | 7.19 ± 0.40 | 7.06 ± 0.25 | 7.13 ± 0.34 | Time | 0.15 | 2 | 0.07 | 1.75 | 0.183 |
| | CCR ($n$ = 16) | 7.13 ± 0.34 | 7.06 ± 0.25 | 7.06 ± 0.25 | Group Time | 0.02 | 2 | 0.01 | 0.25 | 0.780 |
| K-IADL | EA-CCR ($n$ = 16) | 11.00 ± 3.03 | 10.69 ± 1.85 | 10.44 ± 0.96 | Time | 0.90 | 2 | 0.45 | 0.26 | 0.770 |
| | CCR ($n$ = 16) | 11.50 ± 3.10 | 11.44 ± 3.08 | 11.62 ± 3.12 | Group Time | 1.94 | 2 | 0.97 | 0.57 | 0.570 |
| EQ-5D-5L | EA-CCR ($n$ = 16) | 6.38 ± 1.36 | 6.06 ± 1.24 | 5.75 ± 1.00 | Time | 13.08 | 2 | 6.54 | 3.83 | 0.027 |
| | CCR ($n$ = 16) | 7.31 ± 2.52 | 6.12 ± 1.24 | 6.31 ± 1.25 | Group Time | 3.08 | 2 | 1.54 | 0.90 | 0.411 |

**Table 5.** Comparison of changes in outcome measurements between patients who received EA-CCR and CCR for Mild Cognitive Impairment ($n$ = 16 each).

| Dependent Variables | Group (n) | Week 0 (M ± SD) | Difference (w8-w0) | Z (p) * | Difference (w20-w0) | Z (p) * | Difference (w20-w8) | Z (p) * |
|---|---|---|---|---|---|---|---|---|
| ADAS-K -cog | EA-CCR ($n$ = 16) | 11.13 ± 4.10 | −3.94 ± 2.57 | −0.38 (0.703) | −4.94 ± 3.45 | −0.21 (0.835) | −1.00 ± 2.66 | −0.25 (0.804) |
| | CCR (n = 16) | 11.19 ± 6.22 | −3.81 ± 3.47 | | −5.38 ± 5.19 | | 1.56 ± 2.83 | |
| MoCA-K | EA-CCR ($n$ = 16) | 18.75 ± 2.54 | 5.50 ± 2.48 | −0.72 (0.470) | 5.81 ± 3.69 | −0.23 (0.819) | 0.31 ± 2.98 | −0.74 (0.459) |
| | CCR (n = 16) | 19.31 ± 2.92 | 4.88 ± 2.45 | | 5.81 ± 2.37 | | 0.94 ± 2.26 | |
| CES-D | EA-CCR ($n$ = 16) | 14.25 ± 5.91 | −3.31 ± 4.77 | −1.32 (0.186) | −3.50 ± 6.20 | −0.63 (0.533) | −0.19 ± 4.05 | −1.19 (0.234) |
| | CCR (n = 16) | 11.50 ± 6.94 | −0.50 ± 7.43 | | −1.75 ± 8.70 | | −1.25 ± 8.50 | |
| K-ADL | EA-CCR ($n$ = 16) | 7.19 ± 0.40 | −0.13 ± 0.34 | −0.60 (0.551) | −0.06 ± 0.44 | −0.03 (0.974) | 0.06 ± 0.25 | −1.00 (0.317) |
| | CCR (n = 16) | 7.13 ± 0.34 | −0.06 ± 0.25 | | −0.06 ± 0.25 | | 0.00 ± 0.00 | |
| K-IADL | EA-CCR ($n$ = 16) | 11.00 ± 3.03 | −0.31 ± 3.03 | −0.45 (0.655) | −0.56 ± 2.45 | −0.42 (0.677) | −0.25 ± 1.29 | −0.07 (0.948) |
| | CCR (n = 16) | 11.50 ± 3.10 | −0.06 ± 0.93 | | −0.13 ± 1.31 | | 0.19 ± 1.05 | |
| EQ−5D−5L | EA-CCR ($n$ = 16) | 6.38 ± 1.36 | 0.31 ± 1.49 | −1.31 (0.189) | 0.63 ± 1.54 | −0.21 (0.832) | −0.31 ± 0.95 | −1.17 (0.243) |
| | CCR ($n$ = 16) | 7.31 ± 2.52 | −1.19 ± 1.91 | | −1.00 ± 2.92 | | 0.19 ± 1.68 | |

* Mann-Whitney U-test.

**Table 6.** Comparison of changes in outcome measurement between patients who received EA-CCR ($n$ = 9) and CCR ($n$ = 4) for under 70 years old group.

| Dependent Variables | Group ($n$) | Week 0 (M ± SD) | Week 8 (M ± SD) | Week 20 (M ± SD) | Difference (w8-w0) | $Z$ ($p$) * | Difference (w20-w0) | $Z$ ($p$) * | Difference (w20-w8) | $Z$ ($p$) * |
|---|---|---|---|---|---|---|---|---|---|---|
| ADAS-cog | EA-CCR ($n$ = 9) | 11.56 ± 3.94 | 7.56 ± 5.66 | 7.00 ± 4.64 | −4.00 ± 3.24 | −0.70 (0.483) | −4.56 ± 4.03 | −0.94 (0.349) | −0.56 ± 2.74 | −0.80 (0.424) |
| | CCR ($n$ = 4) | 7.00 ± 2.94 | 4.75 ± 3.40 | 4.75 ± 3.40 | −2.25 ± 1.71 | | −2.25 ± 0.96 | | 0.00 ± 2.00 | |
| MoCA-K | EA-CCR ($n$ = 9) | 18.33 ± 2.50 | 23.44 ± 3.47 | 23.00 ± 4.33 | 5.11 ± 2.57 | −0.47 (0.639) | 4.67 ± 3.24 | −1.02 (0.308) | −0.44 ± 2.88 | −0.78 (0.438) |
| | CCR ($n$ = 4) | 19.75 ± 2.22 | 25.50 ± 2.52 | 26.25 ± 1.71 | 5.75 ± 2.06 | | 6.50 ± 1.00 | | 0.75 ± 1.26 | |
| CES-D | EA-CCR ($n$ = 9) | 13.78 ± 4.47 | 11.22 ± 6.24 | 12.22 ± 4.32 | −2.56 ± 3.64 | −0.16 (0.876) | −1.56 ± 5.27 | −0.70 (0.486) | 1.00 ± 4.72 | −0.46 (0.643) |
| | CCR ($n$ = 4) | 10.50 ± 7.59 | 11.50 ± 9.29 | 13.00 ± 7.75 | 1.00 ± 8.68 | | 2.50 ± 14.57 | | 1.50 ± 16.11 | |
| K-ADL | EA-CCR ($n$ = 9) | 7.00 ± 0.00 | 7.00 ± 0.00 | 7.11 ± 0.33 | 0.00 ± 0.00 | 0.00 (1.000) | 0.11 ± 0.33 | −0.67 (0.505) | 0.11 ± 0.33 | −0.68 (0.505) |
| | CCR ($n$ = 4) | 7.00 ± 0.00 | 7.00 ± 0.00 | 7.00 ± 0.00 | 0.00 ± 0.00 | | 0.00 ± 0.00 | | 0.00 ± 0.00 | |
| K-IADL | EA-CCR ($n$ = 9) | 11.78 ± 3.96 | 11.22 ± 2.39 | 10.78 ± 1.20 | −0.56 ± 4.13 | −0.42 (0.676) | −1.00 ± 3.28 | −0.79 (0.429) | −0.44 ± 1.74 | −0.94 (0.348) |
| | CCR ($n$ = 4) | 10.75 ± 1.50 | 10.75 ± 1.50 | 11.75 ± 2.06 | 0.00 ± 0.00 | | 1.00 ± 2.00 | | 1.00 ± 2.00 | |
| EQ−5D−5L | EA-CCR ($n$ = 9) | 6.44 ± 1.59 | 5.78 ± 1.09 | 5.33 ± 0.71 | −0.67 ± 1.41 | −0.24 (0.810) | −1.11 ± 1.69 | −0.24 (0.813) | −0.44 ± 1.13 | −0.50 (0.614) |
| | CCR ($n$ = 4) | 6.75 ± 2.06 | 6.00 ± 1.41 | 6.50 ± 1.73 | −0.75 ± 0.96 | | −0.25 ± 3.10 | | 0.50 ± 2.52 | |

* Mann-Whitney U-test.

**Table 7.** Comparison of changes in outcome measurements between patients who received EA-CCR ($n$ = 7) and CCR ($n$ = 12) for over 70 years old group.

| Dependent Variables | Group ($n$) | Week 0 (M ± SD) | Week 8 (M ± SD) | Week 20 (M ± SD) | Difference (w8-w0) | Z ($p$) * | Difference (w20-w0) | Z ($p$) * | Difference (w20-w8) | Z ($p$) * |
|---|---|---|---|---|---|---|---|---|---|---|
| ADAS-cog | EA-CCR ($n$ = 7) | 10.57 ± 4.54 | 6.71 ± 3.64 | 5.14 ± 2.85 | −3.86 ± 1.57 | −0.86 (0.931) | −5.43 ± 2.76 | −0.85 (0.932) | −1.57 ± 2.64 | −0.43 (0.668) |
| | CCR ($n$ = 12) | 12.58 ± 6.47 | 8.25 ± 3.72 | 6.17 ± 2.55 | −4.33 ± 3.80 | | −6.42 ± 5.63 | | −2.08 ± 2.94 | |
| MoCA-K | EA-CCR ($n$ = 7) | 19.29 ± 2.69 | 25.29 ± 2.87 | 26.57 ± 3.31 | 6.00 ± 2.45 | −1.41 (0.158) | 7.29 ± 3.95 | −1.11 (0.265) | 1.29 ± 3.04 | −0.86 (0.932) |
| | CCR ($n$ = 12) | 19.17 ± 3.19 | 23.75 ± 2.42 | 24.75 ± 1.86 | 4.58 ± 2.57 | | 5.58 ± 2.68 | | 1.00 ± 2.56 | |
| CES-D | EA-CCR ($n$ = 7) | 14.86 ± 7.73 | 10.57 ± 6.58 | 8.86 ± 7.63 | −4.29 ± 6.10 | −1.32 (0.188) | −6.00 ± 6.78 | −0.93 (0.352) | −1.71 ± 2.56 | −0.55 (0.579) |
| | CCR ($n$ = 12) | 11.83 ± 7.03 | 10.83 ± 6.12 | 8.67 ± 6.40 | −1.00 ± 7.32 | | −3.17 ± 6.04 | | −2.17 ± 4.91 | |
| K-ADL | EA-CCR ($n$ = 7) | 7.43 ± 0.53 | 7.14 ± 0.38 | 7.14 ± 0.38 | −0.29 ± 0.49 | 1.14 (0.256) | −0.29 ± 0.49 | −1.14 (0.258) | 0.00 ± 0.00 | −0.00 (1.000) |
| | CCR ($n$ = 12) | 7.17 ± 0.39 | 7.08 ± 0.29 | 7.08 ± 0.29 | −0.08 ± 0.29 | | −0.08 ± 0.29 | | 0.00 ± 0.00 | |
| K-IADL | EA-CCR ($n$ = 7) | 10.00 ± 0.00 | 10.00 ± 0.00 | 10.00 ± 0.00 | 0.00 ± 0.00 | −0.00 (1.000) | 0.00 ± 0.00 | −0.00 (1.000) | −0.00 ± 0.00 | −0.76 (0.445) |
| | CCR ($n$ = 12) | 11.75 ± 3.49 | 11.67 ± 3.47 | 11.58 ± 3.48 | −0.08 ± 1.08 | | −0.17 ± 0.94 | | −0.08 ± 0.29 | |
| EQ–5D–5L | EA-CCR ($n$ = 7) | 6.29 ± 1.11 | 6.43 ± 1.40 | 6.29 ± 1.11 | 0.14 ± 1.57 | −1.45 (0.146) | 0.00 ± 1.15 | −0.78 (0.433) | −0.14 ± 0.69 | −0.83 (0.409) |
| | CCR ($n$ = 12) | 7.50 ± 2.71 | 6.17 ± 1.19 | 6.25 ± 1.14 | −1.33 ± 2.15 | | −1.25 ± 2.96 | | 0.08 ± 1.44 | |

* Mann-Whitney U-test.

### 3.4. Safety Evaluation

Adverse events in this study were recorded on a case report form, and their relationship with the intervention was evaluated. No adverse events related to the intervention occurred in this study.

## 4. Discussion

To the best of our knowledge, this is the first randomized controlled study to investigate the effects of EA-CCR on cognitive function, depression, activities of daily living, and quality of life in patients with MCI by comparing the effects of EA-CCR with those of CCR alone. Our study design, which includes eight weeks of treatment [16,24], specific acupoints for acupuncture [16,17,24], and EA treatment method [24], was based on a previous study.

We observed significant improvements in both groups (i.e., changes in ADAS-cog and MoCA-K) and in EA-CCR group (i.e., changes in CES-D). However, EA in EA-CCR showed no positive add-on effects on cognitive function, depression, activities of daily living, and quality of life in patients with MCI. A subanalysis according to age also demonstrated no positive add-on effects of EA.

Systematic reviews reported that both EA and CCR could improve cognitive function and thus are effective treatments for patients with MCI [12,17]. However, in the EA-CCR group, no positive add-on cognitive improvement effects were observed. We postulate several reasons for our results. First, the small sample size, inclusion criteria, and treatment frequency possibly influenced the results. MCI is a neurodegenerative disease that is slowly progressive; thus, eight weeks of EA may not be enough to improve cognitive function. The EA-CCR group (n = 18) in our study received a total of 24 30-min sessions (once daily, three times per week for eight weeks). In a previous study that showed positive synergistic effects of acupuncture and CCR on cognitive function improvement, the combination treatment group (poststroke patients, n = 60) received a total of 60 30-min sessions (once daily, 5 days per week for 12 weeks) [34]. Second, the intervention that is combined with EA may affect the results. In previous studies that showed positive add-on effects of acupuncture on MCI [35], EA was combined with pharmacological treatment. In our study, we used CCR with EA to investigate the effects of combinational treatment using non-pharmacologic interventions on MCI. The interaction between EA and CCR may affect the results. Third, acupoint specificity may have affected the results. The selection and compatibility of acupoints have a direct effect on therapeutic effects. According to the concept of "holism" in traditional Chinese medicine, acupoints in limbs, especially those located below the elbow and knee joints, are extremely important for managing organ and meridian diseases. These points could be therapeutic for local and systemic problems [36]. In another systematic review that reported the cognitive improvement effects of EA in patients with MCI, four of the five studies used both scalp and body acupuncture and one study used scalp acupuncture only [17]. In our study, we used scalp acupuncture only based on a previous study [24].

Our study has some limitations. First, we adopted a single outcome assessor-blinded approach because sham treatment was impossible given the characteristics of EA application. This limitation may have led to a bias in the results of the study. Second, because of limited research funds, study period, and recruitment opportunities, our study did not have enough sample size and long follow-up period to investigate the cognitive improvement effects and long-term effects of EA on MCI. This limitation also may have affected the results of this study. Thus, it is necessary to conduct further studies with enough sample size and long follow-up period to investigate the cognitive improvement effects and long-term effects of EA on MCI. Third, we did not investigate the add-on effect of EA through various acupuncture methods. Apart from needle insertion, issues such as needling sensation, psychological factors, acupoint specificity, acupuncture manipulation, and needle duration also have relevant influences on the therapeutic effects of acupuncture [37]. While several different acupuncture methods for treating MCI exist, we only performed EA at Baihui (GV20), Sishencong (EX-HN1), Fengchi (GB20), and Shenting (GV24) for 30 min. Thus, further studies on effective acupuncture methods are warranted. Fourth, this study was a pilot study to investigate the cognitive improvement effects of EA-CCR, so we did not evaluate the cost-effectiveness of EA-CCR. Thus, further studies on cost-effectiveness of EA-CCR are needed.

## 5. Conclusions

Results in our study indicate that EA-CCR and CCR have beneficial effects on cognitive function improvement in patients with MCI. However, EA-CCR did not show positive add-on effects of EA on the improvement of cognitive function, depression, activities of daily living, and quality of life in patients with MCI. Moreover, no significant differences in outcomes between the two treatments were noted.

Nevertheless, we believe that the results of our study could have varied greatly depending on sample size, frequency and total number of sessions, intervention that is combined with EA, and acupoint specificity. We hope that well-designed RCTs with enough sample size aimed at investigating possible effects or add-on effects of EA on MCI will be conducted in the future.

**Supplementary Materials:** The following are available online at http://www.mdpi.com/2076-3425/10/12/984/s1, Table S1: CONSORT 2010 checklist of information to include when reporting a randomized trial and Table S2: STRICTA checklist.

**Author Contributions:** J.-H.K. and J.-Y.H. designed and conceptualized the trial, wrote the initial draft, and analyzed the data. J.-H.K. and G.-C.P. designed the trial and conducted the trial. J.-S.L. is responsible for planning the data analysis and interpreting the data from the trial. All authors have read and agreed to the published version of the manuscript.

**Funding:** This research was supported by a grant from the Korea Health Technology R&D Project through the Korea Health Industry Development Institute (KHIDI) (https://www.khidi.or.kr/kps) funded by the Ministry of Health & Welfare, Republic of Korea (grant number: HI18C0546). The funder had no role in the design of the study; data collection, analysis, interpretation; decision to publish; and preparation of the manuscript.

**Acknowledgments:** The authors express their sincere gratitude to their colleagues and the staff at DongShin University Gwangju Korean Medicine Hospital for their support.

**Conflicts of Interest:** The authors declare no conflict of interest.

## References

1. Petersen, R.C. Mild cognitive impairment as a diagnostic entity. *J. Intern. Med.* **2004**, *256*, 183–194. [CrossRef] [PubMed]
2. Petersen, R.C.; Smith, G.E.; Waring, S.C.; Ivnik, R.J.; Tangalos, E.G.; Kokmen, E. Mild cognitive impairment: Clinical characterization and outcome. *Arch. Neurol.* **1999**, *56*, 303–308. [CrossRef] [PubMed]
3. Langa, K.M.; Levine, D.A. The diagnosis and management of mild cognitive impairment: A clinical review. *JAMA* **2014**, *312*, 2551–2561. [CrossRef] [PubMed]
4. Manly, J.J.; Tang, M.X.; Schupf, N.; Stern, Y.; Vonsattel, J.P.; Mayeux, R. Frequency and course of mild cognitive impairment in a multiethnic community. *Ann. Neurol.* **2008**, *63*, 494–506. [CrossRef]
5. Panza, F.; D'Introno, A.; Colacicco, A.M.; Capurso, C.; Del Parigi, A.; Caselli, R.J.; Pilotto, A.; Argentieri, G.; Scapicchio, P.L.; Scafato, E.; et al. Current epidemiology of mild cognitive impairment and other predementia syndromes. *Am. J. Geriatr. Psychiatry* **2005**, *13*, 633–644. [CrossRef]
6. Roberts, R.O.; Knopman, D.S.; Mielke, M.M.; Cha, R.H.; Pankratz, V.S.; Christianson, T.J.; Geda, Y.E.; Boeve, B.F.; Ivnik, R.J.; Tangalos, E.G.; et al. Higher risk of progression to dementia in mild cognitive impairment cases who revert to normal. *Neurology* **2014**, *82*, 317–325. [CrossRef]
7. Eshkoor, S.A.; Hamid, T.A.; Mun, C.Y.; Ng, C.K. Mild cognitive impairment and its management in older people. *Clin. Interv. Aging* **2015**, *10*, 687–693. [CrossRef]
8. Bahar-Fuchs, A.; Clare, L.; Woods, B. Cognitive training and cognitive rehabilitation for mild to moderate Alzheimer's disease and vascular dementia. *Cochrane Database Syst. Rev.* **2013**, *2013*, CD003260. [CrossRef]
9. Petersen, R.C.; Lopez, O.; Armstrong, M.J.; Getchius, T.S.D.; Ganguli, M.; Gloss, D.; Gronseth, G.S.; Marson, D.; Pringsheim, T.; Day, G.S.; et al. Practice guideline update summary: Mild cognitive impairment. *Neurology* **2018**, *90*, 126–135. [CrossRef]
10. Russ, T.C.; Morling, J.R. Cholinesterase inhibitors for mild cognitive impairment. *Cochrane Database Syst. Rev.* **2012**, *2012*, CD009132. [CrossRef]

11. Tricco, A.C.; Soobiah, C.; Berliner, S.; Ho, J.M.; Ng, C.H.; Ashoor, H.M.; Chen, M.H.; Hemmelgarn, B.; Straus, S.E. Efficacy and safety of cognitive enhancers for patients with mild cognitive impairment: A systematic review and meta-analysis. *CMAJ* **2013**, *185*, 1393–1401. [CrossRef]
12. Klimova, B.; Maresova, P. Computer-based training programs for older people with mild cognitive impairment and/or dementia. *Front. Hum. Neurosci.* **2017**, *11*, 262. [CrossRef] [PubMed]
13. Hill, N.T.; Mowszowski, L.; Naismith, S.L.; Chadwick, V.L.; Lampit, A. Computerized cognitive training in older adults with mild cognitive impairment or dementia: A systematic review and meta-analysis. *Am. J. Psychiatry* **2017**, *174*, 329–340. [CrossRef] [PubMed]
14. Gates, N.; FiataroneSingh, M.A.; Sachdev, P.S.; Valenzuela, M. The effects of exercise training on cognitive function in older adults with mild cognitive impairment: A meta-analysis of randomized controlled trials. *Am. J. Geriatr. Psychiatry* **2013**, *21*, 1086–1097. [CrossRef] [PubMed]
15. Zhu, Y.; Wu, H.; Qu, M.; Wang, S.; Zhang, Q.; Zhou, L.; Wang, S.; Wang, W.; Wu, T.; Xiao, M.; et al. Effects of a specially designed aerobic dance routine on mild cognitive impairment. *Clin. Interv. Aging* **2018**, *13*, 1691–1700. [CrossRef] [PubMed]
16. Deng, M.; Wang, X.F. Acupuncture for amnestic mild cognitive impairment: A meta-analysis of randomized controlled trials. *Acupunct. Med.* **2016**, *34*, 342–348. [CrossRef] [PubMed]
17. Kim, H.; Kim, H.K.; Kim, S.Y.; Kim, Y.I.; Yoo, H.R.; Jung, I.C. Cognitive improvement effects of electroacupuncture for the treatment of MCI compared with western medications: A systematic review and meta-analysis. *BMC Complement. Altern. Med.* **2019**, *19*, 13. [CrossRef]
18. Kwon, C.Y.; Lee, B.; Suh, H.W.; Chung, S.Y.; Kim, J.W. Efficacy and safety of auricular acupuncture for cognitive impairment and dementia: A systematic review. *Evid. Based Complement. Alternat. Med.* **2018**, *2018*, 3426079. [CrossRef]
19. Hwang, I.K.; Chung, J.Y.; Yoo, D.Y.; Yi, S.S.; Youn, H.Y.; Seong, J.K.; Yoon, Y.S. Comparing the effects of acupuncture and electroacupuncture at Zusanli and Baihui on cell proliferation and neuroblast differentiation in the rat hippocampus. *J. Vet. Med. Sci.* **2010**, *72*, 279–284. [CrossRef]
20. Napadow, V.; Makris, N.; Liu, J.; Kettner, N.W.; Kwong, K.K.; Hui, K.K. Effects of electroacupuncture versus manual acupuncture on the human brain as measured by fMRI. *Hum. Brain Mapp.* **2005**, *24*, 193–205. [CrossRef]
21. Zhao, Y.; Zhang, H.; Zhao, L. Observations on the efficacy of electric scalp acupuncture in treating mild cognitive impairment. *J. Sichuan Tradit. Chin. Med.* **2012**, *30*, 112–114.
22. Zhao, L.; Zhang, F.W.; Zhang, H.; Zhao, Y.; Zhaou, B.; Chen, W.Y.; Zhu, M.J. Mild cognitive impairment disease treated with electroacupuncture: A multi-center randomized controlled trial. *Zhongguo Zhen Jiu* **2012**, *32*, 779–784. [PubMed]
23. Luo, Z.; Zheng, L.; Li, H.; Peng, S.; Mao, R.; Xiong, D.; Tan, T. Efficacy of combined treatment of drugs and electric acupuncture on cognitive function of patients with mild cognitive impairment. *China Med. Her.* **2013**, *10*, 118–120.
24. Zhang, H.; Zhao, L.; Yang, S.; Chen, Z.; Li, Y.; Peng, X.; Yang, Y.; Zhu, M. Clinical observation on effect of scalp electroacupuncture for mild cognitive impairment. *J. Tradit. Chin. Med.* **2013**, *33*, 46–50. [CrossRef]
25. Simon, S.S.; Yokomizo, J.E.; Bottino, C.M. Cognitive intervention in amnestic mild cognitive impairment: A systematic review. *Neurosci. Biobehav. Rev.* **2012**, *36*, 1163–1178. [CrossRef] [PubMed]
26. Kim, J.H.; Han, J.Y.; Park, G.C.; Lee, J.S. Effects of electroacupuncture combined with computer-based cognitive rehabilitation on mild cognitive impairment: Study protocol for a pilot randomized controlled trial. *Trials* **2019**, *20*, 478. [CrossRef]
27. WHO Western Pacific Region. *WHO Standard Acupuncture Point Locations in the Western Pacific Region*; Elsevier Korea L.C.C.: Seoul, Korea, 2009; Volume 185, pp. 217–219.
28. Kueper, J.K.; Speechley, M.; Montero-Odasso, M. The Alzheimer's disease assessment scale-cognitive subscale (ADAS-cog): Modifications and responsiveness in pre-dementia populations. A narrative review. *J. Alzheimer's Dis.* **2018**, *63*, 423–444. [CrossRef] [PubMed]
29. Jung, E.S.; Lee, J.H.; Kim, H.T.; Park, S.S.; Kim, J.E.; Cha, J.Y.; Seol, I.C.; Choi, Y.E.; Yoo, H.R. Effects of acupuncture on patients with mild cognitive impairment assessed using functional near-infrared spectroscopy on week 12(close-out): A pilot study protocol. *Integr. Med. Res.* **2018**, *7*, 287–295.

30. Nasreddine, Z.S.; Phillips, N.A.; Bédirian, V.; Charbonneau, S.; Whitehead, V.; Collin, I.; Cummings, J.L.; Chertkow, H. The Montreal Cognitive Assessment, MoCA: A brief screening tool for mild cognitive impairment. *J. Am. Geriatr. Soc.* **2005**, *53*, 695–699. [CrossRef]
31. Radloff, L.S. The CES-D scale: A self-report depression scale for research in the general population. *Appl. Psychol. Meas.* **1977**, *1*, 385–401. [CrossRef]
32. Won, C.W.; Yang, K.Y.; Rho, Y.G.; Kim, S.Y.; Lee, E.J.; Yoon, J.L.; Cho, K.H.; Shin, H.C.; Cho, B.R.; Oh, J.R.; et al. The development of Korean activities of daily living (K-ADL) and Korean instrumental activities of daily living (K-IADL) scale. *J. Korean Geriatr. Soc.* **2002**, *6*, 107–120.
33. Herdman, M.; Gudex, C.; Lloyd, A.; Janssen, M.F.; Kind, P.; Parkin, D.; Bonsel, G.; Badia, X. Development and preliminary testing of the new five-level version of EQ-5D (EQ-5D-5L). *Qual. Life Res.* **2011**, *20*, 1727–1736. [CrossRef] [PubMed]
34. Jiang, C.; Yang, S.; Tao, J.; Huang, J.; Li, Y.; Ye, H.; Chen, S.; Hong, W.; Chen, L. Clinical efficacy of acupuncture treatment in combination with RehaCom cognitive training for improving cognitive function in stroke: A 2×2 factorial design randomized controlled trial. *J. Am. Med. Assoc.* **2016**, *17*, 1114–1122. [CrossRef] [PubMed]
35. Wang, S.; Yang, H.; Zhang, J.; Zhang, B.; Liu, T.; Gan, L.; Zheng, J. Efficacy and safety assessment of acupuncture and nimodipine to treatment mild cognitive impairment after cerebral infarction: A randomized controlled trial. *BMC Complement. Altern. Med.* **2016**, *16*, 361. [CrossRef]
36. World Health Organization Regional Office for the Western Pacific. *WHO International Standard Terminologies on Traditional Medicine in the Western Pacific Region*; World Health Organization Regional Office for the Western Pacific: Manila, Philippines, 2007.
37. Shi, G.X.; Yang, X.M.; Liu, C.Z.; Wang, L.P. Factors contributing to therapeutic effects evaluated in acupuncture clinical trials. *Trials* **2012**, *13*, 42. [CrossRef]

*Article*

# Emotional Components of Pain Perception in Borderline Personality Disorder and Major Depression—A Repetitive Peripheral Magnetic Stimulation (rPMS) Study

**Kathrin Malejko [1,*], André Huss [2], Carlos Schönfeldt-Lecuona [1], Maren Braun [1] and Heiko Graf [1]**

1 Department of Psychiatry and Psychotherapy III, University of Ulm, 89075 Ulm, Germany; carlos.schoenfeldt@uni-ulm.de (C.S.-L.); marenbraun@gmx.de (M.B.); heiko.graf@uni-ulm.de (H.G.)

2 Department of Neurology, University of Ulm, 89081 Ulm, Germany; andre.huss@uni-ulm.de

* Correspondence: kathrin.malejko@uni-ulm.de; Tel.: +49-0731-5006-1401; Fax: +49-0731-5006-1402

Received: 2 October 2020; Accepted: 23 November 2020; Published: 24 November 2020

**Abstract:** Various studies suggested alterations in pain perception in psychiatric disorders, such as borderline personality disorder (BPD) and major depression (MD). We previously investigated affective components of pain perception in BPD compared to healthy controls (HC) by increasing aversive stimulus intensities using repetitive peripheral magnetic stimulation (rPMS) and observed alterations in emotional rather than somatosensory components in BPD. However, conclusions on disorder specific alterations in these components of pain perception are often limited due to comorbid depression and medication in BPD. Here, we compared 10 patients with BPD and comorbid MD, 12 patients with MD without BPD, and 12 HC. We applied unpleasant somatosensory stimuli with increasing intensities by rPMS and assessed pain threshold (PT), cutaneous sensation, emotional valence, and arousal by a Self-Assessments Manikins scale. PTs in BPD were significantly higher compared to HC. The somatosensory discrimination of stimulus intensities did not differ between groups. Though elevated rPMS intensities led to increased subjective aversion and arousal in MD and HC, these emotional responses among intensity levels remained unchanged in BPD. Our data give further evidence for disorder-specific alterations in emotional components of pain perception in BPD with an absent emotional modulation among varying aversive intensity levels.

**Keywords:** repetitive peripheral magnetic stimulation; rPMS; pain; borderline personality disorder; depression

---

## 1. Introduction

Pain is considered to be an unpleasant emotional and sensory experience accompanied by potential or actual tissue damage. Apart from the somatosensory pathway, affective and cognitive components crucially modulate individual pain experience and perception [1]. Various studies have assumed clinically relevant alterations in pain perception in several psychiatric disorders and, in particular, in borderline personality disorder (BPD) [2–6] and major depressive disorder (MD) [7–12]. However, evidence on disorder specific alterations in pain perception in these two disorders is ambiguous depending on the different components of pain perception that are investigated and due to comorbidities, medication, and varying stimulus modalities.

Thus, previous studies investigating the experience of pain in MD revealed inconsistent findings. Elevated pain thresholds (PT) and attenuated pain sensitivity in MD were observed relative to healthy controls (HC) [7,9,10,12], particularly when pressure, thermal, or electrical stimuli were applied to the skin [8,11]. In contrast, one study demonstrated hyperalgesia to ischemic muscle pain in MD compared to HC [8]. Of note, most of these studies focused on somatosensory components of pain perception, whereas reliable evidence regarding alterations in emotional and cognitive components of pain perception in MD remains scarce.

In BPD, one core symptom is non-suicidal self-injury (NSSI) [13], which often manifests as cutting or burning, which is thought to regulate and relieve aversive inner tension [14,15]. Here, BPD patients frequently report hypo- or analgesia during NSSI [16]. Accordingly, various studies on pain perception observed reduced pain sensitivity in BPD compared to HC [2–6]. Whereas basic somatosensory stimulus perception and processing in BPD is thought to be unaffected, alterations in pain experience due to differences in affective or cognitive components of pain seem plausible [3,5,17]. We previously investigated affective components of pain perception in BPD compared to HC by parametrically increasing aversive stimulus intensities using repetitive peripheral magnetic stimulation (rPMS) [3]. The capability to discriminate different stimulus intensities did not differ between BPD and HC, but an elevation in levels of subjective aversion and arousal corresponding to intensity levels was solely observed in HC but not in BPD. These observations suggest preponderant alterations in emotional rather than somatosensory processes of pain perception in BPD, however, the specificity of these results was limited due to comorbid depression and/or medication in the investigated sample of BPD.

Based on those remaining and unanswered issues, we investigated a cohort of patients with BPD with comorbid MD, a sample of patients with MD without BPD, and HC to account for effects of depression and medication on emotional experience of pain in BPD. To warrant comparability with our previous findings, we applied unpleasant electrical stimuli with increasing stimulus intensities by rPMS and assessed participants' (i) cutaneous sensation, (ii) emotional valence, and (iii) arousal using a Self-Assessments Manikins (SAM) scale. We assumed alterations regarding pain thresholds as well as emotional valence and arousal level during aversive stimulation specifically in BPD and intended to disentangle effects by comparisons with a clinical group diagnosed with MD without BPD with similar depressive symptoms and under medication.

## 2. Materials and Methods

### 2.1. Subjects

To account for gender differences and to minimize sample heterogeneity, we analyzed 34 females aged 18–55 years. Of those, 10 patients were diagnosed with BPD and comorbid MD and 12 patients with MD without BPD. A total of 12 healthy controls (HC) were investigated by a physician and served as a control group with no current or lifetime psychiatric diagnoses. Participants in the clinical groups and HC were matched for the highest degree of education and age. Participants were recruited from the inpatient units of the Department of Psychiatry and Psychotherapy III of the University Hospital Ulm. All patients in the MD- and 8 patients of the BPD-group took antidepressant medication, mainly selective serotonin reuptake inhibitors (SSRIs) or serotonin and norepinephrine reuptake inhibitors (SNRIs) (see Supplementary Material Section Table S1). Antidepressant medication was not interrupted but held stable for four weeks prior to the measurements. All participants were right-handed according to the Edinburgh Handedness Inventory. Participants with any severe medical disorder, epilepsy, current substance use disorder, and psychotic disorders were excluded from the study. All participants gave written informed consent prior to the study that was approved by the local ethical committee of Ulm University (Ethical approval code 52/09) and conducted in accordance with the Declaration of Helsinki.

### 2.2. Psychometric Measurements

All participants were screened by using the Structured Clinical Interview for DSM-IV (SCID-I and -II [18]), and clinical diagnoses of patients with MD and BPD were verified by one of the study psychologists or physicians. Current depressive symptoms were assessed by using the Beck Depression

Inventory (second edition, BDI-II [19]) in its German version [20]. NSSI was assessed by the German version of the Modified Ottawa/Ulm Self-injury Inventory (MOUSI) [21]. All patients in the BPD group committed NSSI at least once per week during the preceding 6 months. Psychological and somatoform dissociative features were assessed before and after rPMS procedures by the Dissociation Tension Scale-acute (DSS-acute) [22,23]. The DSS-acute is a self-rating instrument on a 10-level Likert scale (0–9). Ten items refer to psychological phenomena of dissociation, nine items include physical characteristics of dissociation, and two items describe borderline-specific symptoms. The total value is calculated from the 21 items and divided by the number of items. The presence of dissociative symptoms is assumed for patients who achieved higher values than 1.57 (unpublished cut-off value) [23].

### 2.3. Study Design

We examined pain thresholds (PT) on two consecutive days ($T_1$&$T_2$) while there were no significant differences between $T_1$&$T_2$ for all groups (PT: $p = 0.09$). We computed mean values to account for intraindividual variations in pain perception [24,25]. For rPMS procedures, all participants were seated in a comfortable chair and wore earplugs as well as headphones to reduce acoustic artefacts from magnetic impulses. The non-dominant arm rested extended on a table that was placed in front of the chair. With a pillow underneath, the palm pointed upwards in the direction of the ceiling. A circular parabolic coil (MMC-140 MacVenture, 140 mm, 33 kT/s$^{-1}$) was placed in the palm with the handle pointing towards the opposite direction of the proband's arm. The stimulator was a MagPro-X100 (2 Tesla) that was used to evoke an aversive sensation. An individual baseline of PT was defined anew before rPMS was applied in different intensities with a frequency of 25 Hz for 1 s. Intertrial interval (ITI) was 15 s. The starting point was at 10% of the stimulator's maximum output intensity and with each step, stimulation intensity was increased by 10%. Immediately after each burst, participants were asked to evaluate its unpleasantness. Once the applied burst was described as unpleasant and almost painful, it was defined as one's individual PT. When 100% of the maximum output was reached but not described as painful, this level was used as the reference baseline. After establishing one's individual PT prior to each session, 50 bursts (25 Hz, 1 s) of rPMS with an ITI of 15 s were randomly delivered at five different intensities. Those five different intensities were selected as follows: the intensity of the burst that was described as painful (PT level) was set as a value of 5; all other levels of intensities (subthreshold values: 4–1) were applied in decreasing steps of 10%. The local aversive rPMS stimuli were evaluated immediately after each burst through Self-Assessments Manikins (SAM) [26] with three visual analogue scales (range: 1–9) representing the dimensions "cutaneous sensation" (scale 1; from "no pain at all" = 1 to "very painful" = 9), "emotional valence" (scale 2; from "pleasant" = 1 to "very unpleasant" = 9), and "level of arousal" (scale 3; from "I feel very calm" = 1 to "I feel an unbearable tension" = 9).

### 2.4. Data Analysis

Datasets were analyzed for normal distribution by a Shapiro–Wilk test, and respective statistical tests were chosen based on its outcome. Accordingly, descriptive statistics provided median values with 25% and 75% percentile or mean values with standard error of mean (SEM), respectively. Inter- and intra-group comparisons were performed by two-way analysis of variance (ANOVA) with Tukey's multiple comparison test. All statistical tests were carried out by using GraphPad Prism 8 software (GraphPad Software Inc., La Jolla, CA, USA). A $p$-value $\leq 0.05$ was considered as statistically significant.

## 3. Results

### 3.1. Demographic and Behavioral Data

A total of 12 patients with MD ($M_{age} = 31.8$ ($SD = 10.0$)), 10 patients with BPD and comorbid MD ($M_{age} = 31.2$ ($SD = 8.1$)) and 12 HCs ($M_{age} = 30.0$ ($SD = 4.4$)) completed the study protocol and served for final data analysis. In line with the clinical diagnosis and the high comorbidity of MD in

BPD [27], BDI scores indicated a moderate or severe major depression in both the MD and the BPD groups. Of note, BDI scores did not significantly differ between the BPD and the MD groups ($t = 1.92$; $p = 0.069$).

Regarding dissociative symptoms, we observed significantly higher DSS scores before than immediately after rPMS in the BPD group ($p = 0.032$, $t = 2.65$), while DSS scores between pre- and post-rPMS did not differ in the MD group ($p = 0.703$, $t = 0.39$) or between MD and BPD (pre-rPMS: $p = 0.278$, $t = 1.11$; post-rPMS: $p = 0.576$, $t = 0.56$) (see Supplementary Material Table S1).

### 3.2. rPMS Data

#### Pain Thresholds (PT)

An ANOVA revealed significant group-by-PTs interaction ($F = 6.85$, $p = 0.004$). PTs (in %) were significantly higher in BPD than in HC ($p = 0.002$), while there were no significant differences in PTs between BPD and MD ($p = 0.215$) or MD compared to HC ($p = 0.130$; see Figure 1).

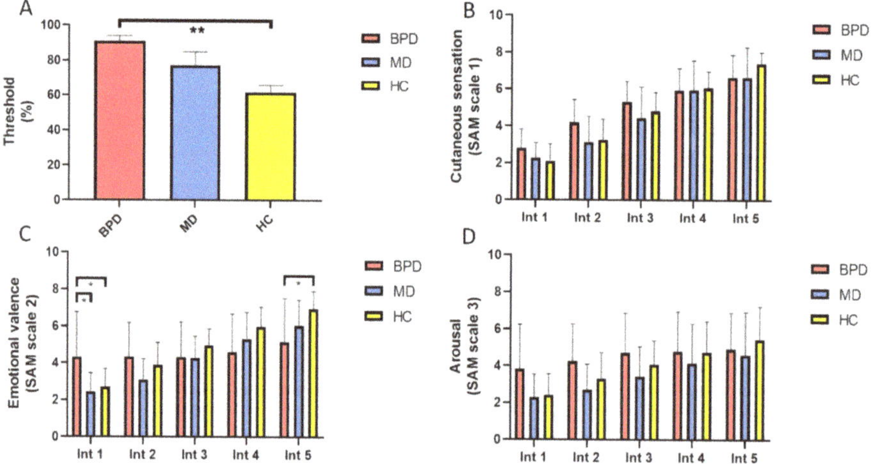


**Figure 1.** (**A**) Pain threshold (PT) of patients with borderline personality disorder (BPD) and comorbid major depression, patients with major depressive disorder (MD) without BPD, and healthy controls (HC). (**B–D**) Differences between groups regarding their subjective rating of Self-Assessments Manikins (SAM) scale 1, 2, and 3 for all levels of increasing stimulus intensities. * = $p < 0.05$; ** = $p < 0.01$; Int = intensity level.

### 3.3. Pain Perception

#### 3.3.1. Cutaneous Sensation (SAM Scale 1)

*Between groups:* No significant differences were observed between groups regarding the rating of cutaneous sensation for all intensity levels (see Figure 1).

*Within group:* We observed significant differences between cutaneous sensation for almost all intensity levels within all groups (see Table 1), indicating a similar capability to discriminate different and increasing levels of stimulus intensities.

**Table 1.** Differences between intensity levels within group of patients with BPD and comorbid major depression, patients with MD without BPD, and HC.

| | BPD ($n$ = 10) | | | MD ($n$ = 12) | | | HC ($n$ = 12) | | |
|---|---|---|---|---|---|---|---|---|---|
| **SAM Scale** | **1** | **2** | **3** | **1** | **2** | **3** | **1** | **2** | **3** |
| Int. 1 vs. Int. 2 | ns | ns | ns | ns | ns | ns | ns | ns | ns |
| Int. 1 vs. Int. 3 | **** | ns | ns | *** | * | ns | **** | ** | ns |
| Int. 1 vs. Int. 4 | **** | ns | ns | **** | **** | ns | **** | **** | * |
| Int. 1 vs. Int. 5 | **** | ns | ns | **** | **** | * | **** | **** | *** |
| Int. 2 vs. Int. 3 | ns | ns | ns | ns | ns | ns | * | ns | ns |
| Int. 2 vs. Int. 4 | * | ns | ns | **** | ** | ns | **** | ** | ns |
| Int. 2 vs. Int. 5 | **** | ns | ns | **** | **** | ns | **** | **** | * |
| Int. 3 vs. Int. 4 | ns | ns | ns | * | ns | ns | ns | ns | ns |
| Int. 3 vs. Int. 5 | ns | ns | ns | *** | * | ns | **** | * | ns |
| Int. 4 vs. Int. 5 | ns | ns | ns | ns | ns | ns | ns | ns | ns |

Cutaneous sensation = SAM scale 1; emotional valence = SAM scale 2; arousal = SAM scale 3; Int. = intensity level; ns = not significant; * = $p < 0.05$; ** = $p < 0.01$; *** = $p < 0.001$; **** = $p < 0.0001$.

### 3.3.2. Emotional Valence (SAM Scale 2)

*Between groups:* Regarding the emotional valence rating, significantly higher values were found in BPD compared to MD ($p = 0.010$) as well as compared to HC ($p = 0.031$) at intensity level 1, while at intensity level 5, corresponding emotional valence was significantly lower in BPD compared to HC ($p = 0.016$) (see Figure 1).

*Within group:* No significant differences regarding emotional valence with increasing stimulus intensities were observed for BPD, while an increase in stimulus intensity was accompanied by elevated levels of aversion in patients with MD and HC (see Table 1).

### 3.3.3. Arousal (SAM Scale 3)

*Between groups:* Regarding levels of subjective arousal, no significant differences were observed between groups for all intensity levels (see Figure 1).

*Within group:* No significant differences were observed within the BPD-group, while HC showed significant and patients with MD trendwise differences in their level of arousal at several intensity levels. Thus, HC and patients with MD revealed higher levels of arousal with higher levels of stimulation intensity (see Table 1).

## 4. Discussion

Here, we show patients with BPD and comorbid MD, patients with MD without BPD, and HC under different levels of unpleasant somatosensory rPMS-stimuli to elucidate previous reports regarding alterations of affective components of pain perception in BPD. By including two clinical samples in one study, we aimed to correct for potential confounding factors arising from MD and to disentangle disorder-specific alterations in pain processing in BPD. Psychometric measures revealed comparable depressive symptoms in both BPD with comorbid MD and MD without BPD. During rPMS, increasing aversive stimulus levels were similarly discriminated by all participants (SAM scale 1). While increasing rPMS stimulation intensity led to elevated subjective aversion and arousal levels in MD and HC, subjective emotional reactions were not modulated by unpleasant stimulus intensities in BPD (SAM scale 2 and 3).

In line with previous observations [4,6], we observed significantly higher PTs in BPD compared to HC, referring to a reduced pain sensitivity. Albeit not statistically significant, it is of note that we found a trend to higher PTs in BPD compared to MD and a similar trend towards increased PTs in MD compared to HC. However, particularly the lack of significant differences in PTs between BPD and MD may imply that PTs per se may not sufficiently differentiate between these disorders. In addition, the observation of significantly elevated PTs in BPD is not enough to conclude a generalized somatosensory deficit in BPD [28]. This is supported by the results of an unimpaired sensory discrimination of increasing somatosensory aversive stimuli intensities compared to MD and HC (SAM scale 1). In line with this observation, a recent neuroimaging study observed similar sensory stimulus intensity encoding neural activation within brain regions related to neural pain processing in BPD compared to HC [17].

We observed significant group differences regarding the assessment of emotional valence and arousal according to increasing stimulus intensities applied by rPMS. Whereas subjective emotional valence and arousal increased with elevating aversive stimulus intensities in both MD and HC, this intensity related modulation of emotional valence and arousal was not evident in BPD. In particular, we observed an increased subjective emotional valence and arousal in BPD compared to HC at low aversive rPMS stimulus levels and lower emotional valence and arousal in BPD compared to HC at high rPMS stimulus levels. This pattern led to the assumption of a disorder specific alteration in affective appraisal of pain in BPD that also differentiates patients with BPD and comorbid MD from MD without BPD compared to HC. Thus, our data strengthen recent findings from neuroimaging studies which demonstrated an altered neural pain processing in BPD with increased neural activation within the prefrontal cortex but attenuated activation of the anterior cingulate cortex and limbic regions [4,29,30]. Accordingly, this pattern was interpreted as a neuroanatomical proxy of an anti-nociceptive mechanism through downregulation of the emotional aspects of pain processing by increased top-down regulation. Alternatively, the lack of variation in emotional valence with changing intensity levels in BPD may be due to altered emotional perception or the ability to differentiate emotions, which is clinically often observed.

However, it has to be mentioned that we were only able to investigate a comparably small sample size that compromises the generalizability of our results. The small sample size may also account for the lack of statistically significant differences in PTs between MD and HC. Nevertheless, the trend of higher PTs in MD compared to HC found in our study is in line with previous findings [7,9,10,12]. Moreover, considering the antinociceptive effects of antidepressants, especially SSRIs [31,32], it is of note that we investigated patients with BPD and MD under medication that may potentially confound our results. However, alterations in affective-motivational components of pain in BPD have also been observed in absence of medication [4,29,30,33] which support our findings. Another shortcoming that needs be mentioned is the fact that we cannot exclude trial-by-trial time effects on subjective evaluation. To correct for individual fluctuations regarding the subjective evaluation of intensity, unpleasantness, and arousal over time, we performed these measures on two different (consecutive) days and could not detect significant differences in these inter-day comparisons.

## 5. Conclusions

We investigated patients with BPD and comorbid MD and patients with MD without BPD and compared them to HC to elucidate BPD-specific alterations in pain perception by controlling for potential confounds owing to comorbidity and medication. Increasing levels of unpleasant stimuli were applied via rPMS, and we assessed participants' PT, subjective cutaneous sensation, emotional valence, and arousal level by SAM scales. Our study supports previous results of elevated pain thresholds in BPD compared to HC. In addition, we did not find any significant differences compared to MD, which could indicate that this finding may not be a disorder specific alteration. During rPMS, we found no significant differences between groups regarding the somatosensory discrimination of increasing stimulus intensities levels (SAM scale 1). Increasing levels of stimuli intensities led to elevated emotional valence and arousal level only in the MD and the HC groups, whereas, in BPD patients, responses remained unchanged among different intensity levels. BPD patients did not show a modulation in their emotional reaction to increasing intensity levels of unpleasant somatosensory stimulation. Thus, we provide further evidence regarding disorder-specific alterations in emotional components of pain perception in BPD with an absent emotional modulation among varying aversive intensity levels.

**Supplementary Materials:** The following are available online at http://www.mdpi.com/2076-3425/10/12/905/s1, Table S1: Characteristics of patients with borderline personality disorder (BPD) and comorbid major depression, patients with major depressive disorder (MD) without BPD, and healthy controls (HC).

**Author Contributions:** H.G. and K.M. contributed substantially to the work. They analyzed and interpreted the data and drafted the manuscript. C.S.-L. and A.H. interpreted the data and revised the manuscript critically for important intellectual content. C.S.-L. and M.B. obtained the data. All the authors approved the final version to be published and agreed to be accountable for all aspects of the work in ensuring that questions related to the accuracy or integrity of any part of the work are appropriately investigated and resolved. All authors have read and agreed to the published version of the manuscript.

**Funding:** This research received no external funding.

**Acknowledgments:** We kindly thank Paula Klassen for the linguistic and grammatical revision of the manuscript.

**Conflicts of Interest:** The authors declare no conflict of interest.

## References

1. Peters, M.L. Emotional and Cognitive Influences on Pain Experience. *Inflamm. Psychiatry* **2015**, *30*, 138–152. [CrossRef]
2. Bohus, M.; Limberger, M.; Ebner, U.; Glocker, F.X.; Schwarz, B.; Wernz, M.; Lieb, K. Pain perception during self-reported distress and calmness in patients with borderline personality disorder and self-mutilating behavior. *Psychiatry Res.* **2000**, *95*, 251–260. [CrossRef]
3. Cárdenas-Morales, L.; Fladung, A.-K.; Kammer, T.; Schmahl, C.; Plener, P.L.; Connemann, B.J.; Schönfeldt-Lecuona, C. Exploring the affective component of pain perception during aversive stimulation in borderline personality disorder. *Psychiatry Res.* **2011**, *186*, 458–460. [CrossRef] [PubMed]
4. Schmahl, C.; Bohus, M.; Esposito, F.; Treede, R.-D.; Di Salle, F.; Greffrath, W.; Ludaescher, P.; Jochims, A.; Lieb, K.; Scheffler, K.; et al. Neural Correlates of Antinociception in Borderline Personality Disorder. *Arch. Gen. Psychiatry* **2006**, *63*, 659–666. [CrossRef] [PubMed]
5. Schmahl, C.; Greffrath, W.; Baumgärtner, U.; Schlereth, T.; Magerl, W.; Philipsen, A.; Lieb, K.; Bohus, M.; Treede, R.-D. Differential nociceptive deficits in patients with borderline personality disorder and self-injurious behavior: Laser-evoked potentials, spatial discrimination of noxious stimuli, and pain ratings. *Pain* **2004**, *110*, 470–479. [CrossRef] [PubMed]
6. Ludäscher, P.; Bohus, M.; Lieb, K.; Philipsen, A.; Jochims, A.; Schmahl, C. Elevated pain thresholds correlate with dissociation and aversive arousal in patients with borderline personality disorder. *Psychiatry Res.* **2007**, *149*, 291–296. [CrossRef] [PubMed]
7. Adler, G.; Gattaz, W.F. Pain perception threshold in major depression. *Biol. Psychiatry* **1993**, *34*, 687–689. [CrossRef]
8. Bär, K.-J.; Brehm, S.; Boettger, M.K.; Boettger, S.; Wagner, G.; Sauer, H. Pain perception in major depression depends on pain modality. *Pain* **2005**, *117*, 97–103. [CrossRef]
9. Bär, K.-J.; Greiner, W.; Letsch, A.; Köbele, R.; Sauer, H. Influence of gender and hemispheric lateralization on heat pain perception in major depression. *J. Psychiatr. Res.* **2003**, *37*, 345–353. [CrossRef]
10. Dickens, C.; McGowan, L.; Dale, S. Impact of Depression on Experimental Pain Perception: A Systematic Review of the Literature with Meta-Analysis. *Psychosom. Med.* **2003**, *65*, 369–375. [CrossRef]
11. Lautenbacher, S.; Spernal, J.; Schreiber, W.; Krieg, J.-C. Relationship between Clinical Pain Complaints and Pain Sensitivity in Patients with Depression and Panic Disorder. *Psychosom. Med.* **1999**, *61*, 822–827. [CrossRef] [PubMed]
12. Schwier, C.; Kliem, A.; Boettger, M.K.; Bär, K.-J. Increased Cold-Pain Thresholds in Major Depression. *J. Pain* **2010**, *11*, 287–290. [CrossRef] [PubMed]
13. Zanarini, M.C.; Frankenburg, F.R.; Reich, D.B.; Fitzmaurice, G.; Weinberg, I.; Gunderson, J.G. The 10-year course of physically self-destructive acts reported by borderline patients and axis II comparison subjects. *Acta Psychiatr. Scand.* **2008**, *117*, 177–184. [CrossRef] [PubMed]
14. Kleindienst, N.; Bohus, M.; Ludäscher, P.; Limberger, M.F.; Kuenkele, K.; Ebner-Priemer, U.W.; Chapman, A.L.; Reicherzer, M.; Stieglitz, R.-D.; Schmahl, C. Motives for Nonsuicidal Self-Injury Among Women With Borderline Personality Disorder. *J. Nerv. Ment. Dis.* **2008**, *196*, 230–236. [CrossRef] [PubMed]
15. Shearer, S.L. Phenomenology of self-injury among inpatient women with borderline personality disorder. *J. Nerv. Ment. Dis.* **1994**, *182*, 524–526. [PubMed]

16. Leibenluft, E.; Gardner, D.L.; Cowdry, R.W. Special Feature the Inner Experience of the Borderline Self-Mutilator. *J. Pers. Disord.* **1987**, *1*, 317–324. [CrossRef]
17. Malejko, K.; Neff, D.; Brown, R.C.; Plener, P.L.; Bonenberger, M.; Abler, B.; Grön, G.; Graf, H. Somatosensory Stimulus Intensity Encoding in Borderline Personality Disorder. *Front. Psychol.* **2018**, *9*, 9. [CrossRef]
18. First, M.B.; Spitzer, R.L.; Gibbon, M.; Williams, J.B. *User's Guide for the Structured Clinical Interview for DSM-IV Axis I Disorders SCID-I: Clinician Version*; American Psychiatric Pubulisher: Washington, DC, USA, 1997.
19. Beck, A.T.; Steer, R.A.; Brown, G.K. *Manual for the Beck Depression Inventory-II*; Psychological Corporation: San Antonio, TX, USA, 1996; Volume 1, p. 82.
20. Hautzinger, M.; Keller, F.; Kühner, C. *Beck Depressions-Inventar (BDI-II)*; Harcourt Test Services: Frankfurt, Germany, 2006.
21. Nixon, M.K.; Cloutier, P.; Aggarwal, S. Affect Regulation and Addictive Aspects of Repetitive Self-Injury in Hospitalized Adolescents. *J. Am. Acad. Child Adolesc. Psychiatry* **2002**, *41*, 1333–1341. [CrossRef]
22. Stiglmayr, C.; Schimke, P.; Wagner, T.; Braakmann, D.; Schweiger, U.; Sipos, V.; Fydrich, T.; Schmahl, C.; Ebner-Priemer, U.; Kleindienst, N.; et al. Development and Psychometric Characteristics of the Dissociation Tension Scale. *J. Pers. Assess.* **2010**, *92*, 269–277. [CrossRef]
23. Stiglmayr, C.E.; Braakmann, D.; Haaf, B.; Stieglitz, R.D.; Bohus, M. [Development and characteristics of Dissociation-Tension-Scale acute (DSS-Akute)]. *Psychother. Psychosom. Med. Psychol.* **2003**, *53*, 287–294.
24. Schulz, E.; Tiemann, L.; Schuster, T.; Gross, J.; Ploner, M. Neurophysiological Coding of Traits and States in the Perception of Pain. *Cereb. Cortex* **2011**, *21*, 2408–2414. [CrossRef] [PubMed]
25. Tiemann, L.; May, E.S.; Postorino, M.; Schulz, E.; Nickel, M.M.; Bingel, U.; Ploner, M. Differential neurophysiological correlates of bottom-up and top-down modulations of pain. *Pain* **2015**, *156*, 289–296. [CrossRef]
26. Bradley, M.M.; Lang, P.J. Measuring emotion: The self-assessment manikin and the semantic differential. *J. Behav. Ther. Exp. Psychiatry* **1994**, *25*, 49–59. [CrossRef]
27. Distel, M.A.; Smit, J.H.; Spinhoven, P.; Penninx, B.W. Borderline personality features in depressed or anxious patients. *Psychiatry Res.* **2016**, *241*, 224–231. [CrossRef] [PubMed]
28. Pavony, M.T.; Lenzenweger, M.F. Somatosensory processing and borderline personality disorder features: A signal detection analysis of proprioception and exteroceptive sensitivity. *J. Pers. Disord.* **2013**, *27*, 208–221. [CrossRef]
29. Niedtfeld, I.; Schmitt, R.; Winter, D.; Bohus, M.; Schmahl, C.; Herpertz, S.C. Pain-mediated affect regulation is reduced after dialectical behavior therapy in borderline personality disorder: A longitudinal fMRI study. *Soc. Cogn. Affect. Neurosci.* **2017**, *12*, 739–747. [CrossRef]
30. Niedtfeld, I.; Schulze, L.; Kirsch, P.; Herpertz, S.C.; Bohus, M.; Schmahl, C. Affect Regulation and Pain in Borderline Personality Disorder: A Possible Link to the Understanding of Self-Injury. *Biol. Psychiatry* **2010**, *68*, 383–391. [CrossRef]
31. Singh, V.P.; Jain, N.K.; Kulkarni, S.K. On the antinociceptive effect of fluoxetine, a selective serotonin reuptake inhibitor. *Brain Res.* **2001**, *915*, 218–226. [CrossRef]
32. Sikka, P.; Kumar, G.; Bindra, V.K.; Kaushik, S.; Kapoor, S.; Saxena, K.K. Study of antinociceptive activity of SSRI (fluoxetine and escitalopram) and atypical antidepressants (venlafaxine and mirtazepine) and their interaction with morphine and naloxone in mice. *J. Pharm. Bioallied Sci.* **2011**, *3*, 412–416. [CrossRef]
33. Ludäscher, P.; Valerius, G.; Stiglmayr, C.; Mauchnik, J.; Lanius, R.A.; Bohus, M.; Schmahl, C. Pain sensitivity and neural processing during dissociative states in patients with borderline personality disorder with and without comorbid posttraumatic stress disorder: A pilot study. *J. Psychiatry Neurosci.* **2010**, *35*, 177–184. [CrossRef]


**Publisher's Note:** MDPI stays neutral with regard to jurisdictional claims in published maps and institutional affiliations.

*Article*

# Low Intensity, Transcranial, Alternating Current Stimulation Reduces Migraine Attack Burden in a Home Application Set-Up: A Double-Blinded, Randomized Feasibility Study

**Andrea Antal *, Rebecca Bischoff, Caspar Stephani, Dirk Czesnik, Florian Klinker, Charles Timäus, Leila Chaieb and Walter Paulus**

Department of Clinical Neurophysiology, University Medical Center, Georg-August University, 37075 Göttingen, Germany; rebecca.bischoff@stud.uni-goettingen.de (R.B.); Cstephani@med.uni-goettingen.de (C.S.); dczesnik@gwgd.de (D.C.); fklinker@med.uni-goettingen.de (F.K.); ctimeus@med.uni-goettingen.de (C.T.); Leila.chaieb@med.uni-goettingen.de (L.C.); wpaulus@med.uni-goettingen.de (W.P.)

* Correspondence: AAntal@gwdg.de; Tel.: +49-551-398461; Fax: +49-551-398126

Received: 6 October 2020; Accepted: 19 November 2020; Published: 21 November 2020

**Abstract: Background:** Low intensity, high-frequency transcranial alternating current stimulation (tACS) applied over the motor cortex decreases the amplitude of motor evoked potentials. This double-blind, placebo-controlled parallel group study aimed to test the efficacy of this method for acute management of migraines. **Methods:** The patients received either active (0.4 mA, 140 Hz) or sham stimulation for 15 min over the visual cortex with the number of terminated attacks two hours post-stimulation as the primary endpoint, as a home therapy option. They were advised to treat a maximum of five migraine attacks over the course of six weeks. **Results:** From forty patients, twenty-five completed the study, sixteen in the active and nine in the sham group with a total of 102 treated migraine attacks. The percentage of terminated migraine attacks not requiring acute rescue medication was significantly higher in the active (21.5%) than in the sham group (0%), and the perceived pain after active stimulation was significantly less for 2–4 h post-stimulation than after sham stimulation. **Conclusion:** tACS over the visual cortex has the potential to terminate migraine attacks. Nevertheless, the high drop-out rate due to compliance problems suggests that this method is impeded by its complexity and time-consuming setup.

**Keywords:** tACS; migraine; acute treatment; visual cortex; transcranial stimulation

## 1. Introduction

Transcranial magnetic (TMS) and direct current stimulation (tDCS) applied over the visual or motor areas have shown efficacy in the acute and prophylactic treatment of migraines in placebo-controlled studies [1–12] (for a recent meta-analyses see [13]). The application of two-pulses of TMS over the visual cortex or over the painful area has been claimed to ameliorate or terminate migraine pain [3,5]. This effect is assumed to be based on influencing neuronal activity and, in the case of an aura, interfering with the occurrence of cortical spreading depression in the early phase of the migraine attack [14].

In healthy subjects, transcranial alternating current stimulation (tACS) with 0.4 mA at 140 Hz applied over the primary motor cortex (M1) can significantly decrease the amplitude of motor evoked potentials (MEPs) at rest [15]. In the present study, we aimed to target the visual cortex of migraine patients at the onset of the migraine attack by having the patient apply tACS at home. We applied this kind of "inhibitory" stimulation based on the results of previous studies, suggesting that the migraine is associated with higher visual neuronal excitability or responsiveness (e.g., [16–23]). Although there are no studies in which this kind of stimulation was applied over the visual cortex, we hypothesized that modifying cortical activity through the application of high-frequency transcranial oscillations might adjust behaviorally "maladaptive" brain states and induce a new balance, forcing the network to restore adequate synchronization and excitation/inhibition balance.

Transcranial stimulation, including tACS, is normally administered by medical professionals in a clinical setting to ensure correct administration of the treatment. The necessity to visit the hospital immediately to treat a migraine attack makes this type of treatment unpractical. The necessity of repeated visits may also increase drop-out rates in long-term studies, e.g., in depression [24], or even interfere with patient-recruitment. Self-administration of tACS by the patients or with the help of their relatives would counteract this disadvantage. The feasibility of this approach, using tDCS, has been demonstrated for several disorders, including depression, Parkinson's disease, Alzheimer's disease, trigeminal neuralgia, and menstrual migraine [25–28]. Furthermore, home stimulation can significantly reduce personal costs and more importantly, the therapy of the patients can be continued and remotely supervised even during a pandemic [29].

The present study is aimed at promoting a safe and feasible protocol for self-administered tACS in the home therapy of migraine attacks. This protocol has not been used in patients before. Special attention was paid to optimal user training for a maximally standardized and reproducible transcranial stimulation setup.

## 2. Methods

All aspects of this study conformed to the Declaration of Helsinki; written informed consent was given by all study participants. The experimental protocol was approved by the ethics committee of the Medical Faculty of the University of Göttingen (code: 1/5/03, amendment: 19.04.12).

### 2.1. Patients

Forty migraine patients were recruited from outpatient clinics and private practices for the study. This estimation was based on the sample size of previous feasibility studies, treating acute migraine attacks with brain stimulation methods (for a review see [13]). At this stage, we aimed to prove the feasibility of the methodology. Inclusion criteria were migraine with or without aura and disease duration ≥6 months [30]. Exclusion criteria were significant chronic health disorders, diagnosed neuropsychiatric disorders, pregnancy or breast feeding, history of substance abuse or dependence, a history of neurological disorders other than migraine, an implanted pacemaker and cranial metallic hardware. All patients were naïve to transcranial stimulation and none took prophylactic migraine medication during the study period. If applicable, female patients were advised to continue contraception (that was started at least 6 months prior to enrollment into the study) during the whole study period. None of the patients had a history of acute migraine medication overuse.

### 2.2. Experimental Design

The primary endpoint of this double-blind, placebo-controlled study was the termination of the migraine attacks within two hours post-stimulation (numerical analogue scale (NAS) values <1). If the pain after this period was still present and cannot be tolerated by the patients, the patients were allowed to take their regular acute migraine medications.

Patients were asked to maintain a headache diary throughout the study duration. During the study, the frequency of the migraine attacks was recorded, including onset and duration of the pain, number of migraine-related days and the type of analgesics taken. Patients were advised to document the degree of pain on a NAS with severity ratings ranging from 0 to 10 at onset of a migraine attack as well as 1 h and 2, 4, 8, 24, and 48 h thereafter. A NAS is frequently used as a valid and reliable measurement of migraine pain [31].

## 2.3. Transcranial Alternating Current Stimulation

The patients were assigned to receive either treatment "A", referring to real or "B", referring to sham stimulation, according to a computer randomization list. The battery-driven stimulators (NeuroConn, Ilmenau, Germany) were preprogrammed and coded by the coordinating investigator, who had no contact with the patients. During programming, the type of the stimulation can be saved; however, during stimulation, no differences between real and sham stimulations on the screen can be detected. Unknown to the patients, the parameters used during the home stimulation sessions, including the time and duration of the sessions, were stored in the stimulator.

The stimulation was then applied by the patient at home. Since electrode preparation and positioning are essential factors in reproducible remotely-supervised treatment [32], the patients were given detailed instructions and a training session in the department before being allowed to use the stimulator. Saline-soaked sponge electrodes were used. The stimulating electrode (4 × 4 cm) was placed over the Oz and the return electrode (5 × 7 cm) over the Cz electrode positions and fixed with the elastic band. This was done by the patients, without help.

According to a modeling study, these electrode positions present current densities in the range of 0.05–0.15 $A/m^2$, the higher intensities being allocated to the medial, as compared to the lateral occipital cortex [33]. tACS with 0.4 mA was applied for 15 min, including 20 s ramp-up and ramp-down phases. For sham stimulation, the electrodes were placed in the same positions as for active stimulation, but the stimulator was turned off automatically after 30 s of stimulation. Both the patients and the training investigator were blinded with regard to the type of tACS applied. The patients were instructed to start the stimulation session at the beginning of the migraine attack (e.g., after the appearance of aura or pain). The patients were aware of the fact that they would receive either sham or real stimulation.

Since any potential adverse effects (AEs) of this technique in a patient population are not yet known, the patients were asked to report AEs during the whole study period and they were instructed on what do to do in case of the occurrence of severe AEs. Furthermore, they completed a questionnaire after the whole stimulation session. The questionnaire contained rating scales for the presence of discomforting sensations such as pain, tingling, itching or burning under the electrodes due to tACS [34] (1 = very mild and 5 = extremely strong intensity).

## 2.4. Statistical Analysis

Repeated measures ANOVA was used to test for differences in pain perception with the factors "type of stimulation" (active and sham) and "time" (before and after treatment, hours). Mann–Whitney U test was used to compare the number of terminated attacks (with and without medication) in the active and sham groups. With regard to the primary endpoint, a $p$-value of $\leq 0.05$ was considered significant. All other analyses are considered exploratory and confidence intervals as well as $p$-values are reported without correction for multiple testing.

The incidences of AEs were coded in a binary system (no = 0, yes = 1) and the severities of the AEs were rated using a NAS from one to five, one being very mild and five being of an extremely strong intensity of any given AE.

# 3. Results

Forty patients were randomized using a computer algorithm to get real (25 patients) or sham (15 patients) stimulation. Fifteen patients, nine from the active and six from the sham group, had to be excluded from evaluation during the course of the study. Eight were excluded because they failed to perform any stimulation at home. Four of these were only identified by analyzing the stimulator memory. Four patients had no migraine attacks during the study period, two patients decided to withdraw without giving any reason and one patient experienced a panic attack before the stimulation. Therefore, only twenty-five patients returned a valid migraine diary, the demographical characteristics and medical history of which are summarized in Table 1. The demographical characteristics of the patients related to the disease (duration of the disease and number of attacks/year) did not differ significantly between the active and sham stimulation groups (t-test, $p > 0.1$).

**Table 1.** Demographics and medical history of the patients.

| | tACS ($n = 16$) | Sham ($n = 9$) |
|---|---|---|
| **With aura** | 9 | 5 |
| **Without aura** | 7 | 4 |
| **Mean age (SD)** | 31.1 (8.9) | 28.1 (10.5) |
| **Mean duration in years (SD)** | 13.7 (7.8) | 14.8 (10.3) |
| **Mean number of attacks/year (SD)** | 28.7 (18.5) | 42.8 (42.2) |
| **Pain localization** | | |
| unilateral | 11 | 5 |
| bilateral | 5 | 4 |
| **with Family history** | 9 | 7 |
| **Medication** | | |
| Acetylsalicylic acid (Aspirin) | 2 | 1 |
| Triptans | 4 | 3 |
| Ibuprofen | 2 | 1 |
| Paracetamol | 4 | 3 |
| Others<br>-Antidepressants<br>-Metamizole<br>-Thyroid Hormone | <br>2<br>1<br>1 | <br>0<br>0<br>0 |
| **Oral contraception** | 9 | 5 |
| **Smokers** | 4 | 1 |

These 25 patients suffered a total of 102 documented migraine attacks during the study: 65 migraine attacks were treated in the active group (mean: 4.06 attacks/patient, range: 1–5) and 37 in the sham group (mean = 4.11 attacks/patient, range 1–5). In the active group, 27 attacks were treated with drugs within two hours after the stimulation compared to 14 in the sham group (41.5% vs. 37.8%) (Figure 1).

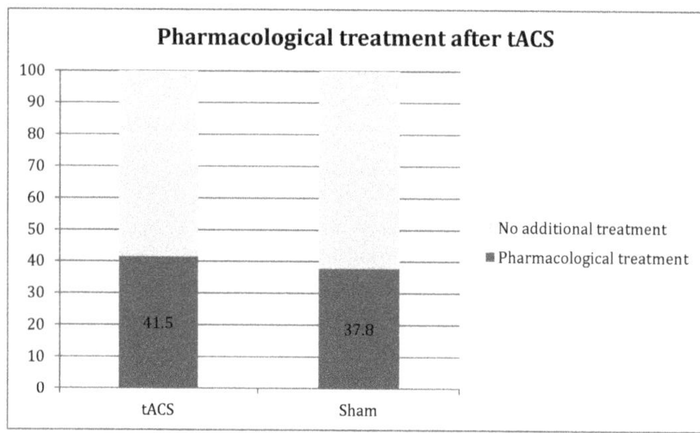


**Figure 1.** Medication use after transcranial alternating current stimulation (tACS) treatment. The Y axis represents 100% of the migraine attacks.

During the attacks without pharmacological interventions, the pain abated within two hours post-stimulation in 14 of the 38 attacks in the active group, but in none of the 23 attacks in the sham group, showing a statistical significant difference between the two groups ($p < 0.001$). If we consider the pain severity in both groups, it was significantly lower after tACS than after sham stimulation in the first two–four hours (main effect: $F(1,35) = 9.173$, $p < 0.0045$; interaction: $F(7,245) = 6.62$, $p < 0.00001$) (Figure 2). According to the pain diaries, none of the documented attacks reoccurred after 24 and 48 h.

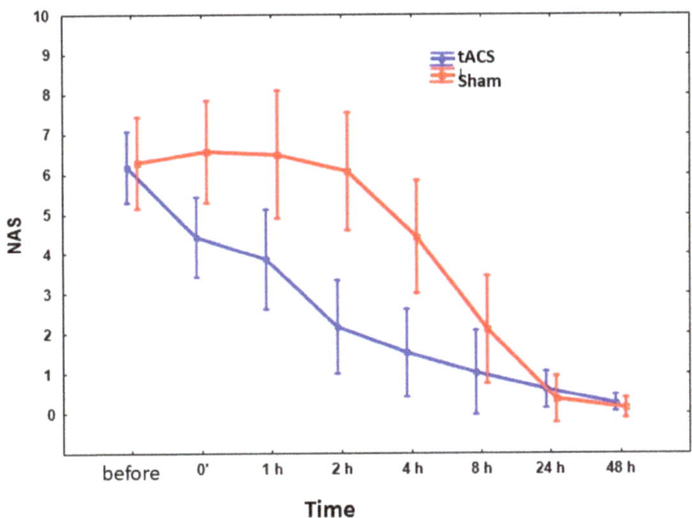


**Figure 2.** Effect of 140 Hz 0.4 mA tACS applied over the visual cortex on pain severity during migraine attack. X axis represents numerical analogue scale (NAS) values before and after stimulations. Bars represent 95% confidence intervals.

With regard to the presence of aura, it was not possible to make reliable statistical analysis because of the low number of patients in each group, however, no different effects were observed in patients with or without aura.

AEs of tACS: No medical interventions other than acute migraine medications were required; 23 patients completed the questionnaire, 15 in the active and 8 in the sham group. Table 2 summarizes the AEs due to stimulation.

**Table 2.** Adverse effects of tACS reported after stimulation. N: number of patients; MI = mean intensity (1 = very mild and 5 = extremely strong intensity).

| | Pain under the Electrodes | | Tingling | | Itching | |
|---|---|---|---|---|---|---|
| | N | MI | N | MI | N | MI |
| **tACS ($n = 16$)** | 1 | 2 | 5 | 1.8 | 5 | 1.4 |
| **Sham ($n = 9$)** | 1 | 3 | 4 | 1.8 | 2 | 1.3 |
| | **Nervousness** | | **Fatigue** | | **Unpleasantness** | |
| | N | MI | N | MI | N | MI |
| **tACS ($n = 16$)** | 1 | 4.0 | 6 | 2.2 | 2 | 2.0 |
| **Sham ($n = 9$)** | 3 | 2.5 | 6 | 2.2 | 5 | 3.3 |

## 4. Discussion

Our hypothesis that "inhibitory" tACS over the visual cortex could be an effective acute treatment option was based on data suggesting that migraine is associated with higher neuronal excitability or responsiveness (e.g., [16–23]) and the observation that 0.4 mA 140 Hz tACS over the motor cortex probably decreases cortical excitability [15]. Accordingly, we found that a significantly higher percentage of migraine attacks were terminated within two hours post-stimulation in the tACS group. Nevertheless, only less than one in four of the attacks could be completely terminated by this intervention; in almost half of the attacks additional acute medication was required. In the sham group, 38% of the attacks were treated with drugs and none of them responded to the sham stimulation.

Despite the variety of pharmaceutical options available for the prophylaxis or acute treatment of migraines, a substantial proportion of patients remains resistant to drug therapy. The efficacy response rates for these therapies range around 40–50% in most studies, suggesting that responsive patients generally represent less than half of the population [35–37]. Several non-pharmaceutical alternatives, such as exercise and acupuncture, have been compared with common prophylactic medications [38–40] and seem to offer some benefit for migraine patients [41]. Non-invasive neuromodulation, including magnetic and low intensity electric stimulation, is an emerging treatment strategy for migraine headache disorders. These methods provide the distinctive opportunity of avoiding disparate medication AEs and interactions. Two pulses of TMS at low or high intensity were applied in an open-label study during acute migraine attacks. Stimulation was over the region of pain in patients without aura, or over the visual cortex in patients with aura [5]. Pain intensity was reduced by 75% up to 20 min post-TMS. Furthermore, 32% of patients reported no further headache for up to 24 h after one treatment, 29% after two treatments and 40% after three treatments. In another study, 164 patients with aura were stimulated over the visual cortex within one hour of aura onset using a randomized, double-blind, parallel-group, sham-controlled design [3]. Up to three attacks were treated over a three-month period. Real TMS was more effective than sham stimulation in alleviating pain at two hours (39% vs. 22%), and for sustained pain relief at 24 h (29% vs. 16%) and 48 h (27% vs. 13%). Based on telephone interviews, single pulse TMS in 190 episodic or chronic migraine patients reduced the number of headache days after 12 weeks of treatment in nearly 60% of patients in whom acute medications were contraindicated or ineffective [42]. Nevertheless, the discontinuation rate was 55% in this study. Repetitive TMS (rTMS) as a preventative treatment both for episodic and chronic migraine has resulted in mixed outcomes [4,43,44]. Generally, rTMS is promising with moderate evidence in acute and prophylactic treatment that it contributes to reductions in headache frequency, duration, intensity, abortive medication use, depression, and functional impairment compared to baseline, when the M1 or the frontal cortex were stimulated, using "excitatory" frequencies [45]. Nevertheless, many of the studies reported non-significant changes compared to sham treatment.

So far, the efficacy of prophylactic anodal and cathodal tDCS has been primarily tested with diverse results, mainly due to the different setting procedure and location of electrodes [1,6,10,46,47]. Nevertheless, in different studies, it has been observed that the anodal stimulation of the primary motor cortex and the cathodal stimulation of the occipital cortex are associated with a significant reduction in the number of headache days, consumption of tablets, and pain intensity, and a significant increase in the number of headache-free days [48].

To our knowledge, tACS has never been employed before in patients with migraine. Previous data suggest that stimulating the motor cortex of healthy young subjects with 140 Hz tACS at 0.4 mA can decrease the amplitude of MEPs [15] for more than one hour after stimulation. However, it is not clear what is the exact neuronal underlying mechanism. It is hypothesized that 140 Hz (at this lower intensity) only facilitated intracortical inhibitory networks of corticospinal neurons and may have inhibited intracortical facilitatory influences on corticospinal neurons. Using lower frequencies in the alpha range and higher intensities, tACS induced increased alpha power [49,50]. We assume that tACS over the visual cortex may not only reduce local excitability but possibly modify the activity of the brainstem through nociceptive pathways [30]. It is suggested that there is a functional connection between the visual cortex and brainstem second-order nociceptors in the spinal trigeminal nucleus. Therefore, inhibiting the projection from the visual cortex to the brainstem might result in less pain during attacks. With regard to the stimulation montage, the placement of return electrode over the motor cortex that was achieved in previous tDCS studies [46], and based on the fact that 140 Hz stimulation effectively modified the size of MEP amplitudes [15], might result in better clinical efficacy.

Home therapy was well tolerated by the patients who used the stimulator. The majority of user feedbacks after stimulation concerning the efficacy was either positive or neutral. Nevertheless, the main reason for the substantial fraction of non-compliance might be the time-consuming task of positioning the electrodes before stimulation as compared to taking a pill. This is a general indicator for a problematic feasibility of this kind of intervention. In future studies, family members should be involved in the training sessions, when stimulation is to be performed at home as an acute intervention. In addition to this, although the patients were instructed to start the stimulation immediately after the first signs of the migraine attack appeared, many of them probably did not do that, reflected by their relatively high baseline NAS values. Indeed, if the attack starts at work or in other situations other than at home, the stimulation cannot be started immediately. Furthermore, due to the high drop-out rate, the current study is limited by the small remaining sample size.

## 5. Conclusions

In summary, acute application of tACS over the visual cortex (0.4 mA, 140 Hz) for 15 min was able to terminate migraine attacks. Despite home treatment, the logistic effort was high with strict training and supervision by healthcare professionals. Improved strategies to further simplify the procedure will certainly reduce the drop-out rate. Fine tuning of dose titration may also increase efficacy. Furthermore, strategies to increase efficacy in combination of neuroplasticity modification with migraine prophylactic drugs warrant further investigations.

**Author Contributions:** Each author made substantial contributions to the manuscript, all of them accepted the final version. Conceptualization: A.A., R.B., W.P.; methodology: A.A., R.B., L.C., D.C., W.P.; investigation: A.A., R.B., C.S., C.T., D.C., F.K., L.C., W.P.; analysis: A.A., R.B., L.C., W.P.; writing and interpretation of data: A.A., R.B., C.S., C.T., D.C., F.K., L.C., W.P. All authors have read and agreed to the published version of the manuscript.

**Funding:** This study (A.A., W.P.) was supported by the Migraine Research Foundation (N.Y.). We acknowledge support by the German Research Foundation and the Open Access Publication Funds of the University of Göttingen, Germany.

**Conflicts of Interest:** The authors declare no conflict of interest.

**Disclosures:** A.A. has received honoraria as a speaker for NeuroCare (Munich, Germany) and as consultant from Savir GmbH (Magdeburg, Germany). W.P. is a member of the scientific advisory board of Precisis AG.

## References

1. Antal, A.; Kriener, N.; Lang, N.; Boros, K.; Paulus, W. Cathodal transcranial direct current stimulation of the visual cortex in the prophylactic treatment of migraine. *Cephalalgia* **2011**, *31*, 820–828. [CrossRef] [PubMed]
2. Lipton, R.B.; Pearlman, S.H. Transcranial magnetic simulation in the treatment of migraine. *Neurotherapeutics* **2010**, *7*, 204–212. [CrossRef] [PubMed]
3. Lipton, R.B.; Dodick, D.W.; Silberstein, S.D.; Saper, J.R.; Aurora, S.K.; Pearlman, S.H.; Fischell, R.E.; Ruppel, P.L.; Goadsby, P.J. Single-pulse transcranial magnetic stimulation for acute treatment of migraine with aura: A randomised, double-blind, parallel-group, sham-controlled trial. *Lancet Neurol.* **2010**, *9*, 373–380. [CrossRef]
4. Teepker, M.; Hötzel, J.; Timmesfeld, N.; Reis, J.; Mylius, V.; Haag, A.; Oertel, W.H.; Rosenow, F.; Schepelmann, K. Low-frequency rTMS of the vertex in the prophylactic treatment of migraine. *Cephalalgia* **2010**, *30*, 137–144. [CrossRef] [PubMed]
5. Clarke, B.M.; Upton, A.R.M.; Kamath, M.V.; Al-Harbi, T.M.; Castellanos, C.M. Transcranial magnetic stimulation for migraine: Clinical effects. *J. Headache Pain* **2006**, *7*, 341–346. [CrossRef]

6. Viganò, A.; D'Elia, T.S.; Sava, S.; Auvé, M.; De Pasqua, V.; Colosimo, A.; Di Piero, V.; Schoenen, J.; Magis, D. Transcranial Direct Current Stimulation (tDCS) of the visual cortex: A proof-of-concept study based on interictal electrophysiological abnormalities in migraine. *J. Headache Pain* **2013**, *14*, 23. [CrossRef]
7. Pinchuk, D.; Pinchuk, O.; Sirbiladze, K.; Shuhgar, O. Clinical effectiveness of primary and secondary headache treatment by transcranial direct current stimulation. *Front. Neurol.* **2013**, *4*, 25. [CrossRef]
8. DaSilva, A.F.; Pt, M.E.M.; Zaghi, S.; Lopes, M.; DosSantos, M.F.; Spierings, E.L.; Bajwa, Z.; Datta, A.; Bikson, M.; Fregni, F. tDCS-induced analgesia and electrical fields in pain-related neural networks in chronic migraine. *Headache* **2012**, *52*, 1283–1295. [CrossRef]
9. Martin, T.V.; Lipton, R.B. Epidemiology and biology of menstrual migraine. *Headache* **2008**, *48* (Suppl. 3), S124–S130. [CrossRef]
10. Wickmann, F.; Stephani, C.; Czesnik, D.; Klinker, F.; Timäus, C.; Chaieb, L.; Paulus, W.; Antal, A. Prophylactic treatment in menstrual migraine: A proof-of-concept study. *J. Neurol. Sci.* **2015**, *354*, 103–109. [CrossRef]
11. Andrade, S.M.; Aranha, R.E.L.D.B.; De Oliveira, E.A.; De Mendonça, C.T.P.L.; Martins, W.K.N.; Alves, N.T.; Fernández-Calvo, B. Transcranial direct current stimulation over the primary motor vs prefrontal cortex in refractory chronic migraine: A pilot randomized controlled trial. *J. Neurol. Sci.* **2017**, *378*, 225–232. [CrossRef]
12. Rahimi, M.D.; Fadardi, J.S.; Saeidi, M.; Bigdeli, I.; Kashiri, R. Effectiveness of cathodal tDCS of the primary motor or sensory cortex in migraine: A randomized controlled trial. *Brain Stimul.* **2020**, *13*, 675–682. [CrossRef] [PubMed]
13. Feng, Y.; Zhang, B.; Zhang, J.; Yin, Y. Effects of Non-invasive Brain Stimulation on Headache Intensity and Frequency of Headache Attacks in Patients With Migraine: A Systematic Review and Meta-Analysis. *Headache* **2019**, *59*, 1436–1447. [CrossRef]
14. Leao, A.A. Spreading depression. *Funct. Neurol.* **1986**, *1*, 363–366. [PubMed]
15. Moliadze, V.; Atalay, D.; Antal, A.; Paulus, W. Close to threshold transcranial electrical stimulation preferentially activates inhibitory networks before switching to excitation with higher intensities. *Brain Stimul.* **2012**, *5*, 505–511. [CrossRef] [PubMed]
16. Haigh, S.M.; Karanovic, O.; Wilkinson, F.; Wilkins, A. Cortical hyperexcitability in migraine and aversion to patterns. *Cephalalgia* **2012**, *32*, 236–240. [CrossRef] [PubMed]
17. Mickleborough, M.J.; Hayward, J.; Chapman, C.; Chung, J.; Handy, T.C. Reflexive attentional orienting in migraineurs: The behavioral implications of hyperexcitable visual cortex. *Cephalalgia* **2011**, *31*, 1642–1651. [CrossRef] [PubMed]
18. Chen, W.-T.; Lin, Y.-Y.; Fuh, J.-L.; Hämäläinen, M.S.; Ko, Y.-C.; Wang, S.-J. Sustained visual cortex hyperexcitability in migraine with persistent visual aura. *Brain* **2011**, *134 Pt 8*, 2387–2395. [CrossRef]
19. Höffken, O.; Stude, P.; Lenz, M.; Bach, M.; Dinse, H.R.; Tegenthoff, M. Visual paired-pulse stimulation reveals enhanced visual cortex excitability in migraineurs. *Eur. J. Neurosci.* **2009**, *30*, 714–720. [CrossRef]
20. Chadaide, Z.; Arlt, S.; Antal, A.; Nitsche, M.A.; Lang, N.; Paulus, W. Transcranial direct current stimulation reveals inhibitory deficiency in migraine. *Cephalalgia* **2007**, *27*, 833–839. [CrossRef]
21. Angelini, L.; De Tommaso, M.; Guido, M.; Hu, K.; Ivanov, P.C.; Marinazzo, D.; Nardulli, G.; Nitti, L.; Pellicoro, M.; Pierro, C.; et al. Steady-state visual evoked potentials and phase synchronization in migraine patients. *Phys. Rev. Lett.* **2004**, *93*, 038103. [CrossRef] [PubMed]
22. Martín, H.; Del Río, M.S.; De Silanes, C.L.; Álvarez-Linera, J.; Hernández, J.A.; Pareja, J.A. Photoreactivity of the occipital cortex measured by functional magnetic resonance imaging-blood oxygenation level dependent in migraine patients and healthy volunteers: Pathophysiological implications. *Headache* **2011**, *51*, 1520–1528. [CrossRef] [PubMed]
23. Coppola, G.; Di Lorenzo, C.; Parisi, V.; Lisicki, M.; Serrao, M.; Pierelli, F. Clinical neurophysiology of migraine with aura. *J. Headache Pain* **2019**, *20*, 42. [CrossRef] [PubMed]
24. Valiengo, L.; Benseñor, I.M.; Goulart, A.C.; De Oliveira, J.F.; Zanao, T.A.; Boggio, P.S.; Lotufo, P.A.; Fregni, F.; Brunoni, A.R. The sertraline versus electrical current therapy for treating depression clinical study (select-TDCS): Results of the crossover and follow-up phases. *Depress. Anxiety* **2013**, *30*, 646–653. [CrossRef] [PubMed]
25. Hagenacker, T.; Bude, V.; Naegel, S.; Holle, D.; Katsarava, Z.; Diener, H.; Obermann, M. Patient-conducted anodal transcranial direct current stimulation of the motor cortex alleviates pain in trigeminal neuralgia. *J. Headache Pain* **2014**, *15*, 78. [CrossRef]

26. Alonzo, A.; Fong, J.; Ball, N.; Martin, D.; Chand, N.; Loo, C. Pilot trial of home-administered transcranial direct current stimulation for the treatment of depression. *J. Affect. Disord.* **2019**, *252*, 475–483. [CrossRef]
27. Dobbs, B.; Pawlak, N.; Biagioni, M.; Agarwal, S.; Shaw, M.; Pilloni, G.; Bikson, M.; Datta, A.; Charvet, L. Generalizing remotely supervised transcranial direct current stimulation (tDCS): Feasibility and benefit in Parkinson's disease. *J. Neuroeng. Rehabil.* **2018**, *15*, 114. [CrossRef]
28. Im, J.J.; Jeong, H.; Bikson, M.; Woods, A.J.; Unal, G.; Oh, J.K.; Na, S.; Park, J.S.; Knotkova, H.; Song, I.U.; et al. Effects of 6-month at-home transcranial direct current stimulation on cognition and cerebral glucose metabolism in Alzheimer's disease. *Brain Stimul.* **2019**, *12*, 1222–1228. [CrossRef]
29. Bikson, M.; Hanlon, C.A.; Woods, A.J.; Gillick, B.T.; Charvet, L.; Lamm, C.; Madeo, G.; Holczer, A.; Almeida, J.; Antal, A.; et al. Guidelines for TMS/tES clinical services and research through the COVID-19 pandemic. *Brain Stimul.* **2020**, *13*, 1124–1149. [CrossRef]
30. Olesen, J.; BOusser, M.G.; Diener, H.C.D.; Dodick, D.; First, M.B.; Godsby, P.J.; Gobel, H.; Láinez, M.J.A. Headache Classification Subcommittee of the International Headache Society. The International Classification of Headache Disorders: 2nd edition. *Cephalalgia* **2004**, *24*, 9–160. [CrossRef]
31. Herd, C.P.; Tomlinson, C.L.; Rick, C.; Scotton, W.J.; Edwards, J.; Ives, N.J.; Clarke, C.E.; Sinclair, A.J. Cochrane systematic review and meta-analysis of botulinum toxin for the prevention of migraine. *BMJ Open* **2019**, *9*, e027953. [CrossRef] [PubMed]
32. Charvet, L.; Kasschau, M.; Datta, A.; Knotkova, H.; Stevens, M.C.; Alonzo, A.; Loo, C.; Krull, K.R.; Bikson, M. Remotely-supervised transcranial direct current stimulation (tDCS) for clinical trials: Guidelines for technology and protocols. *Front. Syst. Neurosci.* **2015**, *9*, 26. [CrossRef] [PubMed]
33. Neuling, T.; Wagner, S.; Wolters, C.; Zaehle, T.; Herrmann, C.S. Finite-Element Model Predicts Current Density Distribution for Clinical Applications of tDCS and tACS. *Front. Psychiatry* **2012**, *3*, 83. [CrossRef]
34. Poreisz, C.; Boros, K.; Antal, A.; Paulus, W. Safety aspects of transcranial direct current stimulation concerning healthy subjects and patients. *Brain Res. Bull.* **2007**, *72*, 208–214. [CrossRef] [PubMed]
35. Ferrari, M.D.; Goadsby, P.J.; Roon, K.I.; Lipton, R.B. Triptans (serotonin, 5-HT1B/1D agonists) in migraine: Detailed results and methods of a meta-analysis of 53 trials. *Cephalalgia* **2002**, *22*, 633–658. [CrossRef] [PubMed]
36. Puledda, F.; Goadsby, P.J. An Update on Non-Pharmacological Neuromodulation for the Acute and Preventive Treatment of Migraine. *Headache* **2017**, *57*, 685–691. [CrossRef]
37. Cho, S.-J.; Song, T.-J.; Chu, M.K. Treatment Update of Chronic Migraine. *Curr. Pain Headache Rep.* **2017**, *21*, 26. [CrossRef]
38. Facco, E.; Liguori, A.; Petti, F.; Fauci, A.J.; Cavallin, F.; Zanette, G. Acupuncture versus valproic acid in the prophylaxis of migraine without aura: A prospective controlled study. *Minerva Anestesiol* **2013**, *79*, 634–642.
39. Varkey, E.; Cider, Å.; Carlsson, J.; Linde, M. Exercise as migraine prophylaxis: A randomized study using relaxation and topiramate as controls. *Cephalalgia* **2011**, *31*, 1428–1438. [CrossRef]
40. Sutherland, A.; Sweet, B.V. Butterbur: An alternative therapy for migraine prevention. *Am. J. Health Syst. Pharm.* **2010**, *67*, 705–711. [CrossRef]
41. Schoenen, J.; Roberta, B.; Magis, D.; Coppola, G. Noninvasive neurostimulation methods for migraine therapy: The available evidence. *Cephalalgia* **2016**, *36*, 1170–1180. [CrossRef] [PubMed]
42. Bhola, R.; Kinsella, E.; Giffin, N.; Lipscombe, S.; Ahmed, F.; Weatherall, M.; Goadsby, P.J. Single-pulse transcranial magnetic stimulation (sTMS) for the acute treatment of migraine: Evaluation of outcome data for the UK post market pilot program. *J. Headache Pain* **2015**, *16*, 535. [CrossRef] [PubMed]
43. Brighina, F.; Piazza, A.; Vitello, G.; Aloisio, A.; Palermo, A.; Daniele, O.; Fierro, B. rTMS of the prefrontal cortex in the treatment of chronic migraine: A pilot study. *J. Neurol. Sci.* **2004**, *227*, 67–71. [CrossRef] [PubMed]
44. Conforto, A.B.; Amaro, E., Jr.; Goncalves, A.L.; Mercante, J.P.; Guendler, V.Z.; Ferreira, J.R.; Kirschner, C.C.; Peres, M.F. Randomized, proof-of-principle clinical trial of active transcranial magnetic stimulation in chronic migraine. *Cephalalgia* **2014**, *34*, 464–472. [CrossRef] [PubMed]
45. Stilling, J.M.; Monchi, O.; Amoozegar, F.; Debert, C.T. Transcranial Magnetic and Direct Current Stimulation (TMS/tDCS) for the Treatment of Headache: A Systematic Review. *Headache* **2019**, *59*, 339–357. [CrossRef] [PubMed]

46. Rocha, S.; Melo, L.; Boudoux, C.; Foerster, Á.; Araújo, D.; Monte-Silva, K. Transcranial direct current stimulation in the prophylactic treatment of migraine based on interictal visual cortex excitability abnormalities: A pilot randomized controlled trial. *J. Neurol. Sci.* **2015**, *349*, 33–39. [CrossRef]
47. Ahdab, R.; Mansour, A.G.; Khazen, G.; Khoury, C.E.; Sabbouh, T.M.; Salem, M.; Yamak, W.; Ayache, S.; Riachi, N. Cathodal Transcranial Direct Current Stimulation of the Occipital cortex in Episodic Migraine: A Randomized Sham-Controlled Crossover Study. *J. Clin. Med.* **2019**, *9*, 60. [CrossRef]
48. Zaehle, T.; Rach, S.; Herrmann, C.S. Transcranial alternating current stimulation enhances individual alpha activity in human EEG. *PLoS ONE* **2010**, *5*, e13766. [CrossRef]
49. Neuling, T.; Rach, S.; Herrmann, C.S. Orchestrating neuronal networks: Sustained after-effects of transcranial alternating current stimulation depend upon brain states. *Front. Hum. Neurosci.* **2013**, *7*, 161. [CrossRef]
50. Sava, S.L.; De Pasqua, V.; Magis, D.; Schoenen, J. Effects of visual cortex activation on the nociceptive blink reflex in healthy subjects. *PLoS ONE* **2014**, *9*, e100198. [CrossRef]

**Publisher's Note:** MDPI stays neutral with regard to jurisdictional claims in published maps and institutional affiliations.

*Article*

# Noisy Galvanic Vestibular Stimulation (Stochastic Resonance) Changes Electroencephalography Activities and Postural Control in Patients with Bilateral Vestibular Hypofunction

**Li-Wei Ko [1,2,3,4,†], Rupesh Kumar Chikara [1,2,3,†], Po-Yin Chen [5,6], Ying-Chun Jheng [5,6,7], Chien-Chih Wang [8,9], Yi-Chiang Yang [6], Lieber Po-Hung Li [7,10], Kwong-Kum Liao [11], Li-Wei Chou [3,5,*] and Chung-Lan Kao [3,6,7,*]**

1 Institute of Bioinformatics and Systems Biology, National Chiao Tung University, Hsinchu 300, Taiwan; lwko@nctu.edu.tw (L.-W.K.); rupesh.bt01g@g2.nctu.edu.tw (R.K.C.)
2 Department of Biological Science and Technology, National Chiao Tung University, Hsinchu 300, Taiwan
3 Center for Intelligent Drug Systems and Smart Bio-Devices (IDS2B), National Chiao Tung University, Hsinchu 300, Taiwan
4 Drug Development and Value Creation Research Center, Kaohsiung Medical University, Kaohsiung 807, Taiwan
5 Department of Physical Therapy and Assistive Technology, National Yang-Ming University, Taipei 112, Taiwan; azxd32@gmail.com (P.-Y.C.); cycom1220@gmail.com (Y.-C.J.)
6 Department of Physical Medicine and Rehabilitation, Taipei Veterans General Hospital, Taipei 112, Taiwan; yichiang2312@hotmail.com
7 School of Medicine, National Yang-Ming University, Taipei 112, Taiwan; phli@gm.ym.edu.tw
8 Department of Physical Medicine and Rehabilitation, Taipei Veterans General Hospital Yuli Branch, Hualien 98142, Taiwan; candycandywang@gmail.com
9 Institute of Clinical Medicine, National Yang-Ming University, Taipei 112, Taiwan
10 Department of Otolaryngology, Cheng Hsin General Hospital, Taipei 112, Taiwan
11 Department of Neurology, Neurological Institute, Taipei Veterans General Hospital, Taipei 112, Taiwan; kkliao0730@gmail.com
* Correspondence: lwchou@ym.edu.tw (L.-W.C.); clkao@vghtpe.gov.tw (C.-L.K.)
† These first authors have contributed equally to this work.

Received: 23 September 2020; Accepted: 13 October 2020; Published: 15 October 2020

**Abstract:** Patients with bilateral vestibular hypofunction (BVH) often suffer from imbalance, gait problems, and oscillopsia. Noisy galvanic vestibular stimulation (GVS), a technique that non-invasively stimulates the vestibular afferents, has been shown to enhance postural and walking stability. However, no study has investigated how it affects stability and neural activities while standing and walking with a 2 Hz head yaw turning. Herein, we investigated this issue by comparing differences in neural activities during standing and walking with a 2 Hz head turning, before and after noisy GVS. We applied zero-mean gaussian white noise signal stimulations in the mastoid processes of 10 healthy individuals and seven patients with BVH, and simultaneously recorded electroencephalography (EEG) signals with 32 channels. We analyzed the root mean square (RMS) of the center of pressure (COP) sway during 30 s of standing, utilizing AMTI force plates (Advanced Mechanical Technology Inc., Watertown, MA, USA). Head rotation quality when walking with a 2 Hz head yaw, with and without GVS, was analyzed using a VICON system (Vicon Motion Systems Ltd., Oxford, UK) to evaluate GVS effects on static and dynamic postural control. The RMS of COP sway was significantly reduced during GVS while standing, for both patients and healthy subjects. During walking, 2 Hz head yaw movements was significantly improved by noisy GVS in both groups. Accordingly, the EEG power of theta, alpha, beta, and gamma bands significantly increased in the left parietal lobe after noisy GVS during walking and standing in both groups. GVS post-stimulation effect changed EEG activities in the left and right precentral gyrus, and the right parietal lobe. After stimulation, EEG activity changes were greater in healthy subjects than in patients. Our findings reveal noisy GVS as a non-invasive therapeutic alternative to improve postural stability in patients with BVH. This novel approach provides insight to clinicians and researchers on brain activities during noisy GVS in standing and walking conditions in both healthy and BVH patients.

**Keywords:** electroencephalography (EEG); independent component analysis (ICA); galvanic vestibular stimulation (GVS); bilateral vestibular hypofunction (BVH)

---

## 1. Introduction

Vestibular systems sense linear and angular movements of the head, keeping the body in an upright position to maintain gaze and postural control. Through the peripheral vestibular organs, i.e., the semicircular canals and otoliths, vestibular afferents continuously provide precise information to the brain, so that individuals can explore the environment without losing balance [1]. Patients with bilateral vestibular hypofunction (BVH) often experience a variety of symptoms, including dizziness, oscillopsia, spatial disorientation, and unsteadiness during standing and walking [2,3]. Until now, the primary treatment for BVH has been physical therapy, with the optimization of its efficacy becoming an eminent issue, as patients with BVH suffer from a higher risk of falls.

The vestibular system is known to exhibit high neuroplasticity [4]. Although there is no primary cortical area responsible for vestibular functions, the parieto-insular-vestibular-cortex is known to be the most robust area modulating the vestibular system [5,6]. By applying galvanic vestibular stimulation (GVS), the firing activity of the eighth cranial nerve is enhanced on the side with the cathode electrode and decreased on the side with the anode electrode [7,8]. This current input is a non-invasive method that has long been applied in the investigation of vestibular functions. Along with functional imaging tools, such as functional Magnetic Resonance Imaging (fMRI) and electroencephalography (EEG), GVS over both mastoids has helped scientists to reveal the complex network of the vestibular system. Only recently, zero-mean GVS current delivered to the mastoids has been used to improve the postural control of patients with BVH and has gained growing attention. Zero-mean GVS, also called noisy GVS (or stochastic resonance, SR) has been shown to improve the function of detection in sensory neurons by reduce the threshold of sensory input [9–11]. Noisy GVS has been applied in various patient populations to improve neuroplasticity, e.g., in enhancement of spatial memory, development of motor control in patients with Parkinson disease, improvement of cognitive deficiencies in those with Alzheimer's disease, and improvement of recovery of visual deficits in patients after stroke. In addition, it has been previously applied in patients with vestibular disorders and in healthy subjects [1,12,13], and shown to reduce center of pressure (COP) sway in patients with bilateral vestibulopathy [13]. Unlike caloric testing or conventional GVS, noisy GVS provides stimulation without directional specificity, acting as a perfect means for enhancing vestibular sensory inputs in both vestibular afferents in patients with BVH. There is a growing body of literature exploring the clinical applications of noisy GVS for improving static and dynamic postural control in these patients [1,14]. The applications of this novel methodology have been expanded by the discovery of a non-invasive prosthetic device used for rehabilitation of patients with BVH.

EEG signals clean from artifacts may be obtained by independent component analysis (ICA) methods, which separate the various types of artifact. An earlier study using ICA processing investigated the amplitude modulations of EEG signals associated with gait. During active walking, the upper alpha (10–12 Hz) and beta (18–30 Hz) oscillations in the central sensorimotor areas of the brain were suppressed compared to those during the standing condition [15]. Another ICA study reported that EEG beta band activity in the premotor cortex is higher during stabilized than during normal gait [16]. To date, the cortical effects of noisy GVS during walking have not been thoroughly investigated, while changes in neural activities after noisy GVS are still unclear. Therefore, the aim of this study was to observe and compare the differences in cortical stimulation mappings during standing and walking with head turning, before and after the noisy GVS, in order to elucidate the underlying neural mechanisms. To this end, we utilized an EEG neuroimaging method that detects neuromodulations in the human brain in the above-mentioned conditions.

## 2. Materials and Methods

### 2.1. Participants

In this prospective, observational study, we recruited healthy subjects as well as BVH patients from hospital. The BVH patients were recruited from the Department of Physical Medicine and Rehabilitation, Taipei Veterans General Hospital as well as Department of Otolaryngology, Cheng Hsin General Hospital. The patients presented to ENT clinic with complaints of dizziness/vertigo, oscillopsia or unsteady gait. The ENT doctor (Liber PH Li) performed the Caloric test to confirm the diagnosis of bilateral vestibular hypofunction. The patients were then referred to PM&R (CL Kao) to receive the video head impulse test (vHIT) to evaluate the vestibular ocular reflex (VOR) gains. The diagnosis of BVH was based on the results of air irrigation caloric test as well as vHIT results. A total response in the caloric test <20 degrees per second was defined as BVH. The gain of VOR in healthy people ranges from 0.9 to 1.0. VOR gain less than 0.9 on either side is considered hypofunction

The exclusion criteria included any history of neurological or orthopedic disorders, and any visual, auditory or cognitive impairments. At the beginning of the trial, each participant signed an informed consent, approved by the Ethics Committee of Taipei Veterans General Hospital and the Food and Drug Administration, Taiwan, Republic of China. (trial no. NCT0355494).

### 2.2. Experimental Design

We designed an experimental scenario to examine neural activity changes in healthy controls and patients with BVH under both a walking and a standing condition, and in relation to noisy GVS (Figure 1A). In each trial, a green "circle" symbol and an alarm sound (65 dB, 500 Hz) were presented to instruct subjects to perform 5 s of walking with horizontal head movement. As they walked, they were instructed to turn their head horizontally every 500 ms at the speed of 2 Hz according to the auditory cues, since the vestibular system was previously shown to play a role in 2 Hz head yaw movements [17]. Then, a red "circle" symbol was presented to instruct the subjects to perform 5 s of standing. We projected all stimuli onto a screen in front of the participants for them to perceive the instructions more easily during the trial conditions. Each participant was required to complete 70 trials (35 trials pre-stimulation, and 35 trials post-stimulation). The entire experimental process for each participant lasted approximately one hour. The placement of EEG electrodes is shown in Figure 1B.

A VICON computer-assisted video motion analysis system (VICON Motion Systems, Oxford, UK) was utilized for motion capture in this experiment. For studying the motion trials, we used VICON standard mode (Full body modeling with Plug-in Gait), with 8 MX T-020 VICON cameras (Vicon Motion Systems Ltd., Oxford, UK) and one AMTI force plate (Advanced Mechanical Technology Inc., Watertown, MA, USA) in a 6 × 10 m space. Subjects wore standard motion-capture suits with a transcranial direct-current stimulation (tDCS) main box fixed on their back, as appropriate for capturing the light markers (Figure 1C). All signals were simultaneously recorded by a personal computer (PC) with a signal sampling rate of 100 Hz. To determine the optimal GVS intensity for each subject, a range of stimulation strengths was used (peak amplitude 0, 200, 400, 600, 800, or 1000 μA). When subjects received noisy GVS with different random intensities, they were asked to stand on the AMTI force plate as steadily as possible for 30 s. After recording the COP (X, Y) in the standing position, we averaged all COP positions during that period of standing and calculated the root mean square (RMS) from the averaged (X, Y) point. Then, we compared the COP RMS under different stimulations to determine the minimal value and defined that value as the optimal intensity, since higher RMS values indicate worse standing stabilities. Each subject then walked with a 2 Hz head yaw, with and without noisy GVS.

The 2 Hz frequency was provided by a metronome, and the subjects were requested to make their best effort to keep up with the rhythm. The VICON system captured each subject's corresponding gait motion.

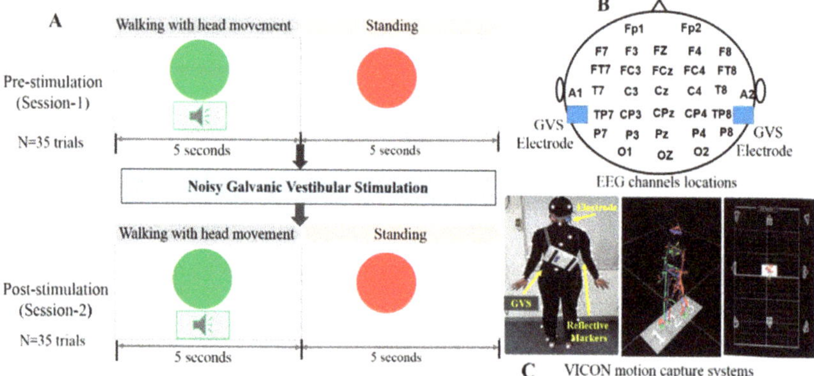


**Figure 1.** Experimental design. (**A**) Presentation of green and red circle stimuli for walking and standing in pre-stimulation and post-stimulation conditions. (**B**) Placement of electroencephalography (EEG) electrodes. EEG cap with 32 channels was placed on the scalp according to the International 10–20 System. Galvanic vestibular stimulation (GVS) electrodes were placed with one electrode on the mastoid process behind each ear denoted by blue square. (**C**) VICON motion capture (Vicon Motion Systems Ltd., Oxford, UK) process: Subjects wore standard motion capture suits with plug-in-gait model setting. (Left and middle figures). The wireless transcranial direct-current stimulation (tDCS) box was fixed on the subjects' backs. The subjects were tested in the capture volume with eight cameras (right figure).

### 2.3. Noisy Galvanic Vestibular Stimulation Process

Noisy GVS was provided using DC-STIMULATOR PLUS (Eldith, NeuroConn GmbH, Ilmenau, Germany). During the stimulation period, noisy GVS was delivered at its optimal intensity for 6 min in the walking and standing conditions, through carbon rubber electrodes, bilaterally and bipolarly. An electrode coated with Tac gel was placed over the mastoid process, behind each ear, to optimize conductivity and adhesiveness. Analog command voltage signals were subsequently passed to a constant current stimulator connected to the stimulation electrodes, as described previously [12,13,18].

### 2.4. EEG Acquisition and Analysis

The EEG signals were acquired from all subjects using a 32-channel EEG cap connected to a neuro Scan NuAmps system (Compumedics USA Inc., Charlotte, NC, USA). They were down-sampled from 1000 to 250 Hz and filtered through a 1 to 50 Hz band-pass finite impulse response (FIR) filter using the EEGLAB toolbox (Version 13.6.5b, UC San Diego, Swartz Center for Computational Neuroscience (SCCN), La Jolla, CA, USA. [16,19]. Then, they were re-referenced to the mean of the A1 and A2 electrodes before further event-related spectral perturbation (ERSP) analysis [20]. EEG signals were analyzed by MATLAB R2014 (The MathWorks Inc., Natick, MA, USA).

In this study, ICA was used to separate brain region dipole sources of brain activity during walking and standing [21]. The extracted EEG signals were analyzed using time-frequency analysis of ERSP [21,22]. Seven clusters with independent components from all subjects were selected for ERSP analysis. After ICA processing, each dipole source was investigated using DIPFIT2 routines in EEGLAB, to find the 3D location of an equivalent dipole source, based on a four-shell spherical head model [16,19]. After manually removing the artifact components, component clustering was performed using k-means (k = 7) criteria and dipole-fitting coordinates to identify the most representative clusters [19]. The value of k was obtained to select the independent components that were found before and after GVS.

After completing EEG-ICA pre-processing, each epoch was extracted from −2 to 8 s (i.e., baseline −2 to 0 sec; walking 0 to 5 sec.; standing 5 to 8 sec) before and after noisy GVS. From the resting condition (i.e., standing position), we used 2 s EEG signals as baseline in each trial. The power spectrum of EEG signals was divided into five frequency bands, delta (1–4 Hz), theta (4–8 Hz), alpha (8–12 Hz), beta (13–30 Hz), and gamma (30–50 Hz), in order to observe the changes in neural activity before and after the noisy GVS.

### 2.5. Statistical Analysis

In the VICON experiment, the COP in the XY-plane and the RMS of the trajectory to the mean point were measured [17]. During the 2 Hz head yaw walking task, we measured the absolute difference of the subjects' head rotation. The differences in COP RMS and head rotation velocity with and without GVS were compared using Wilcoxon signed-rank test, with alpha set at 0.05. Statistically significant ($p < 0.05$) differences before and after noisy GVS in ERSP analysis were estimated using bootstrap statistical processing [23] in the EEGLAB toolbox [19]. To calculate multiple comparisons, the significance values were corrected using the false discovery rate (FDR) method [24] in EEGLAB [19].

## 3. Results

### 3.1. Demographic Data

Ten healthy participants (seven male, three female; age range 23–53 years, mean ± SD age: 29.1 ± 8.4 years) and seven patients with BVH (all female; age range 22–68 years, mean ± SD age: 53.4 ± 15.7 years) completed the trials. The subjects' demographic data is shown in Table 1.

**Table 1.** Demographic data for patients with bilateral vestibular hypofunction (BVH).

| | Sex | Diagnosis | Onset Time | Clinical Presentations | Training |
|---|---|---|---|---|---|
| P1 | F | idiopathic BVH | 2.5 m | Unsteady gait, dizziness | NA |
| P2 | F | idiopathic BVH | 1.5 m | Unsteady gait, dizziness | NA |
| P3 | F | idiopathic BVH | 2 m | oscillopsia | 1 m |
| P4 | F | idiopathic BVH | 2 m | Hearing loss, vertigo, unsteady gait | 1 m |
| P5 | F | idiopathic BVH | 1 m | Unsteady gait, dizziness | 1 m |
| P6 | F | idiopathic BVH | 2.5 m | Dizziness, hearing loss, unsteady gait | 0.5 m |
| P7 | F | idiopathic BVH | 1.5 m | Vertigo, unsteady gait | NA |

### 3.2. Behavioral Results

In the healthy group, the mean RMS displacement was 5.27 ± 1.79 mm RMS of averaged COP point without noisy GVS and 3.36 ± 0.80 mm RMS of mean COP with noisy GVS for the standing condition ($p = 0.005$, Wilcoxon signed rank test). In the patient group, the mean RMS was 8.86 ± 3.31 mm and 6.19 ± 2.29 mm without and with noisy GVS, respectively ($p = 0.018$, Wilcoxon signed rank test, Figure 2A). A plot can be seen of typical COP displacements over the course of a trial for two individual subjects (one healthy subject and one patient, Figure 2B).

When walking with a 2 Hz head yaw, both groups showed a closer approximation to the 2 Hz head yaw with noisy GVS than without (healthy subjects with GVS: 0.48 ± 0.32 Hz, without GVS: 0.31 ± 0.26 Hz, $p = 0.005$; patients with GVS: 0.46 ± 0.28 Hz, without GVS: 0.35 ± 0.24 Hz, $p = 0.018$; Figure 2C). The head turning frequency for each subject, with and without noisy GVS, is shown in the form of scatter plots in Figure 2D. These behavioral results indicated that the RMS of COP sway was significantly reduced during noisy GVS in patients and healthy subjects while standing. In addition, both groups showed significantly better control in performing the 2 Hz head yaw movements with noisy GVS while walking.

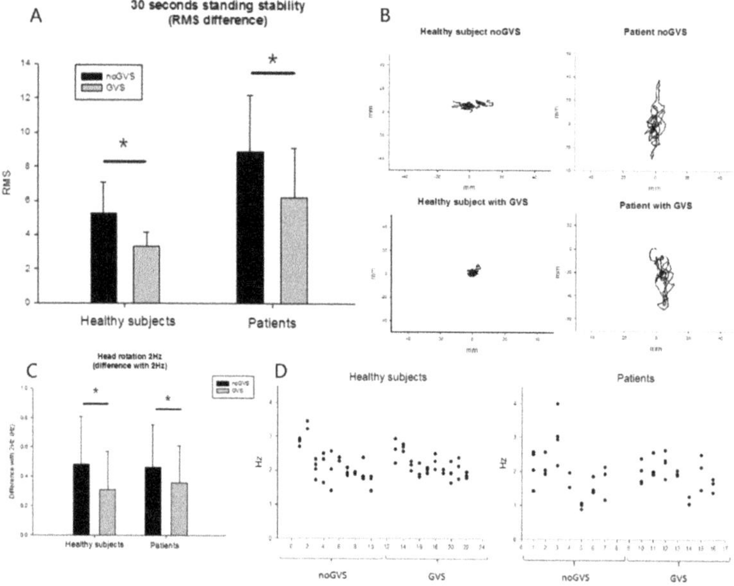


**Figure 2.** Behavioral results of subjects with and without noisy galvanic vestibular stimulation (GVS) during walking and standing conditions. (**A**) Shows the center of pressure (COP) displacements mean root mean square (RMS) in healthy and patient groups. In healthy group, the mean RMS was 5.27 ± 1.79 without noisy GVS and 3.36 ± 0.80 with noisy GVS in standing condition ($p = 0.005$, Wilcoxon signed rank test). In patient group, the mean RMS was 8.86 ± 3.31 without noisy GVS and 6.19 ± 2.29 with noisy GVS ($p = 0.018$, Wilcoxon signed rank test). (**B**) Displays the COP displacements in one single healthy & one patient subjects. When walking with 2 Hz head yaw, both healthy and patient groups showed a tendency towards a closer proximity to 2 Hz head yaw with noisy GVS. (Healthy subjects with GVS = 0.48 ± 0.32 Hz; without GVS = 0.31, ± 0.26 Hz, $p = 0.005$. In patients, with GVS = 0.46, ± 0.28 Hz; without GVS = 0.35, ± 0.24 Hz, $p = 0.018$, (**C**). The head turning frequency for each subject with and without noisy GVS is shown in scattered plot (**D**). * no GVS: the condition without noisy GVS stimulation. * GVS: the condition with noisy GVS.

### 3.3. EEG Results: EEG Scalp Map and Dipole Source Locations

The independent components (IC) obtained from all subjects with similar scalp maps and dipole source locations, clustered into the same groups, is shown in Figure 3. The value of k was obtained by considering the potential number of dipole sources activated after GVS in the walking and standing conditions. The seven clusters of IC in Figure 3 were common among all subjects [25,26]. The seven activated brain regions were identified as left and right frontal gyrus, left and right precentral gyrus (LPG and RPG, respectively), left and right parietal lobe (LPL and RPL, respectively), and occipital lobe (OL). The activation of these seven brain regions verifies that the resulting mean of the independent components within each cluster was highly similar for each scalp map and dipole source location (Figure 3).

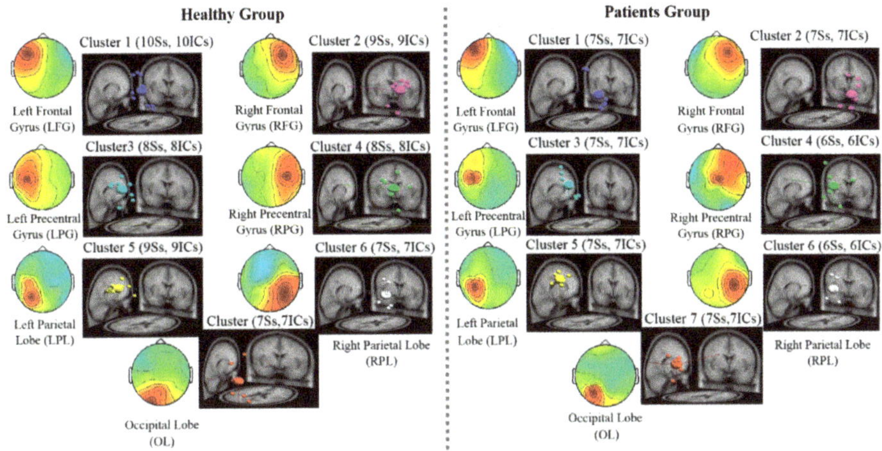


**Figure 3.** Independent component clusters of dipole locations for the analysis of brain dynamics. The activated seven clusters in healthy subjects and patients with bilateral vestibular hypofunction (BVH) were identified including left frontal gyrus (LFG), right frontal gyrus (RFG), left precentral gyrus (LPG), right precentral gyrus (RPG), left parietal lobe (LPL), right parietal lobe (RPL) and occipital lobe (OL).

### 3.4. Noisy GVS Increases EEG Activities in Patients with BVH and Healthy Subjects

In the EEG study, we investigated the average ERSP during the standing condition in patients with BVH and healthy subjects after noisy GVS. EEG activities were observed in the LPG, RPG, LPL, and RPL during pre- and post-stimulation conditions in healthy subjects and patients. For the standing condition in healthy subjects and patients, beta and gamma band activities at the LPG, RPG, LPL, and RPL increased more after than before the stimulation (Figure 4A,B).

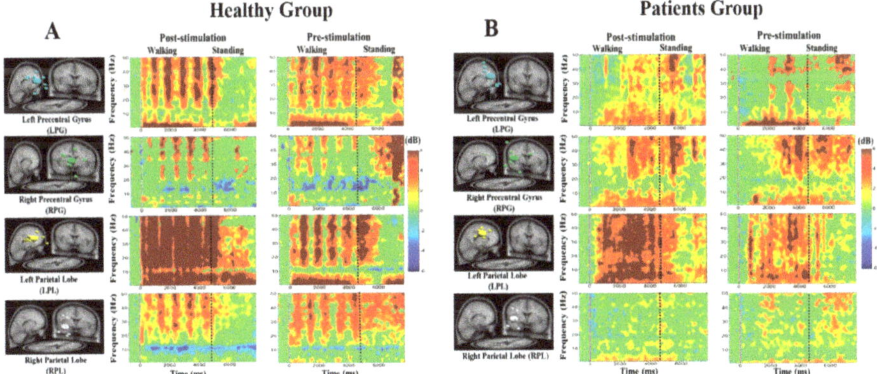


**Figure 4.** (**A**) The event-related spectral perturbation (ERSP) images of healthy subjects in left and right precentral gyrus (LPG, RPG), and left and right parietal lobe (LPL and RPL) of the brain. (**B**) The ERSP of patients with BVH in LPG, RPG, LPL and RPL of the brain after GVS stimulation. The vertical x-axis reveals the frequency (Hz), horizontal y-axis shows the time (ms), color bars indicate the power (dB) of the ERSPs, in which red indicates significant power increase and green shows power decrease relative to baseline (false discovery rate (FDR)-adjusted, $p < 0.05$); statistical threshold at $p < 0.05$.

The mu rhythm and alpha power in the LPG, RPG, LPL, and RPL were suppressed during post-stimulation and pre-stimulation during walking both in healthy subjects and patients (Figure 4A,B). This suppression is related to movement of the feet or legs. However, we observed less mu rhythm suppression in patients than in healthy controls.

During head turning in healthy controls and patients, the power of theta, beta, and gamma bands increased more after than before the stimulation at the LPG, RPG, LPL, and RPL (Figure 4A,B). The EEG activities increased during head turning possibly because of the neuroplasticity induced by noisy GVS in these vestibular regions.

The comparison of each component power spectrum in terms of the "difference between post-stimulation and pre-stimulation" effects is shown in Figure 5. EEG power increased more in the LPL than in the LPG, RPG, and RPL regions of the brain (Figure 5A–H). The power spectra of theta, beta, and gamma bands increased significantly after noisy GVS (i.e., post-stimulation–pre-stimulation) in the LPL both during walking and standing in patients with BVH (Figure 5E,F).

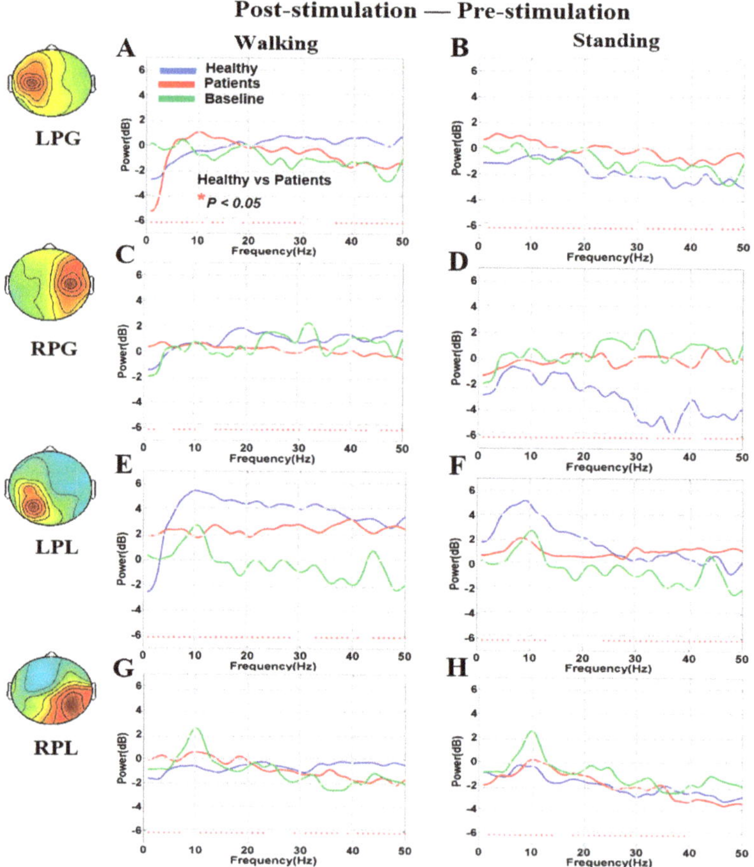


**Figure 5.** The power spectral density results of the brain with healthy and BVH patients. These result in LPG, RPG, LPL, and RPL shows the effect of noisy GVS (post-stimulation–pre-stimulation) in healthy vs. patients during walking and standing conditions. The power spectra of patients are indicated by red traces and those of healthy subjects are indicated by blue traces. The baseline power spectra are presented by green traces (Figure 5**A–H**). An asterisk shows the significant difference between healthy subject's vs patients (Wilcoxon signed-rank test, $p < 0.05$).

## 4. Discussion

In our study, we utilized motion capture analysis and EEG recordings to delineate the post-stimulation effects of noisy GVS during standing and walking with horizontal head turning in healthy subjects and patients with BVH. By this combined approach, we demonstrated that GVS enhances the standing postural stability and improves the head rotation rhythm during walking in both subject groups. These performance improvements are possibly attributed to neuroplasticity changes induced by noisy GVS in the vestibular cortex, as indicated by signal changes in the LPG, RPG, LPL, and RPL. These areas were specifically explored in our study as they have been identified as the main vestibular cortex regions in humans together with the central sulcus and insular lobe [18,27].

The vestibular system is the sensory system that primarily contributes to the detection of angular motion and sense of spatial orientation, necessary to manage movement while maintaining balance and to stabilize postural control of the body [18,27]. Previous research has demonstrated that noisy GVS improves static balance in patients with BVH [2]. Only recently, this specific method of vestibular stimulation has shown promising results, enhancing postural and walking stability, both in older people and in patients with BVH [1,3,12,13]. Wuehr et al. proposed that the mechanism of noisy GVS effects lies in its SR nature [3]. SR is the phenomenon by which a noisy input with a mean value other than 0 (below the intensity of human perception), through chaotic numbers, may optimize the sensory nervous system and facilitate the incorporation of incoming information from the outside world [28]. The basis of this theory was established in rat and cat models [29,30], and SR has been practically applied to various sensory receptors throughout the human body, allowing more acute hearing [31] and improving the control of lower limb posture [32,33]. When SR occurs in conjunction with vestibular nerve stimulation, a very low-intensity electrical current is sufficient to assist balance function. A study by Pal et al. showed that vestibular nerve stimulation with an intensity of 0.1 mA improves the balance of patients with Parkinson's disease [34]. Similarly, Iwasaki et al. found that 200–400 μA of noisy electrical stimulation of the vestibular nerve promotes the maintenance of the center of gravity position in healthy people and patients with bilateral vestibulopathy [2].

Conventionally, bilateral bipolar GVS enhances the firing rate of vestibular afferents by depolarization on the cathodal side and reduces their firing rate by hyperpolarization on the anodal side. Usually, the anodal stimulation site is depolarized, but the frequency and intensity of different stimulations can change the polarization to hyperpolarization [9–11]. Zero-mean noisy GVS has a very important advantage over conventional GVS, in that it does not induce unilateral oculomotor and postural responses. Moreover, in comparison to other suggested treatment modalities for patients with BVH, such as vestibular implants that excite the peripheral vestibular nerves through inserted probes or electrodes [34,35], noisy GVS is non-invasive and easy to apply, with fewer side effects, such as the loss of hearing related to cranial surgical operations.

Analysis of spectral changes in EEG power in the motor cortex showed that the alpha power was suppressed by GVS and that this suppression was significantly higher in healthy subjects than in patients with BVH. This may be explained by the fact that bilateral damage to the vestibular system impacts the ability to maintain body balance, particularly when patients are walking [13]. The LPG and RPG, also referred to as the primary motor area of the brain, control foot movements related to walking. Our findings of alpha band power suppression in the LPG, RPG, LPL and RPL are consistent with those of previous studies in healthy subjects during walking [15,36–39]. Spectral power analyses of theta, alpha, beta, and gamma bands showed a gradual increase in the EEG power in the RPG, LPG, LPL, and RPL, which correspond to the motor and vestibular cortices of the brain, both in healthy subjects and in patients. We speculate that these increased EEG activities may indicate neuroplasticity induced by noisy GVS in the vestibular cortex (LPL, RPL) via the peripheral vestibular system. Synaptic plasticity in the vestibular cortex involves both vestibular nuclei and cerebellar circuits. Voluntary movements such as posture, balance, and coordination are represented in cerebellar fibers parallel to Purkinje cell synapses, which trigger the vestibular nuclei [40]. In turn, the vestibular nuclei send excitatory inputs to the motor cortex (i.e., extensor motor neurons of the legs or feet) [41].

The stimulation of neurons in the vestibular cortex probably induces neuroplasticity changes that enhance postural stability [42]. By measuring COP sways, Fujimoto et al. suggested that noisy GVS enhances postural stability for 4 h after discontinuing stimulation, indicating that post-stimulation effects and vestibular neuroplasticity may exist [12,14]. Through vestibular spinal reflex, standing stability improved in our subjects. Our findings are in contrast to those of Helmchen et al. [43], who observed no brain responsiveness during imperceptible noisy GVS by functional magnetic resonance imaging. This difference may be attributed to the fact that our experiments were performed during movement, with EEG signals being recorded in real time. Moreover, noisy GVS has been shown to improve sensory neuron sensitivity and enhance vestibular sensory afferent inputs during balancing tasks [1,12,13], therefore, any related cortical changes are more likely to be observed during functional tasks rather than when lying down for prolonged image acquisition.

Notably, the behavioral and EEG results of our study showed that, with noisy GVS, the synchronization and precision of the horizontal head rotation rhythm was more pronounced in healthy subjects than in patients. Although both patients and healthy subjects reduced the head yaw error, healthy subjects showed greater improvements. The intact peripheral vestibular afferents may explain the more pronounced enhancement in healthy subjects. Our findings are opposite to those of previous studies that suggested that individuals with elevated vestibular motion perception thresholds, i.e., patients with BVH, would benefit more from noisy GVS [44,45]. However, these previous study results were mostly based on eyes-open, quiet stance experiments. Instead, our experiment combined multiple motor tasks, including walking with a 2 Hz head yaw, and optimal stimulation location frequency may vary due to different sensory-motor demands. Furthermore, we observed that, after noisy GVS, the power of beta and gamma bands increases in the somatosensory cortex (LPG and RPG) of healthy subjects during walking. These findings are consistent with those of several previous studies on GVS [15,16,18]. These earlier studies have demonstrated the post-stimulation effects of GVS in healthy subjects who use facial stimuli, albeit in sitting positions [18,46]. In contrast, our study utilized a more realistic experimental scenario, including both walking and standing conditions. In addition, we showed that the EEG power of theta, alpha, beta, and gamma bands increases significantly after noisy GVS in the vestibular cortex (LPL) both in healthy subjects and in patients, in line with the results of previous studies on GVS and gait-training [5,15,16,47,48]. The decrease, damage or decreased functionality of the vestibular system bilaterally affects the brain, making it difficult in maintaining body balance, particularly in patients during walking. This explains the observation of less mu rhythm suppression in patients than in healthy controls. Several studies have shown that the post-stimulation effects of GVS and tDCS increase the EEG power of theta, alpha, and beta bands in the frontal, temporal, posterior, parietal, and occipital lobes in normal subjects [18,46,49–52], in agreement with our findings in healthy subjects. By using ICA with dipole source localization for EEG analysis, our method goes beyond those used in previous studies, resulting in cleaner EEG signals; thus, a more precise spatial temporal resolution in the vestibular cortex could be determined.

This study had some limitations. First, we recruited a limited number of subjects, and we did not perform vestibular myogenic evoked potentials (VEMP) to assess the otolithic system. Because the aim of this study was to investigate the EEG signal changes after noisy galvanic stimulation during 2 Hz horizontal head movements, we only performed a caloric test for the evaluation of bilateral horizontal semicircular canals. Second, we did not use a control group with sham stimulation, and pre-stimulation trials might carry over some habituation effects to the post-stimulation trials. Because the stimulation intensities used in the pre-stimulation trials were sensory-imperceptible, we believe the habituation effects were minimal. In the future, we will conduct experiments with sham controls to further investigate the habituation effects of noisy GVS. Third, the effects of noisy GVS on posture stability were investigated under a laboratory setting, with controlled conditions, and a regular 2 Hz stimulus, as opposed to a real setting with more complex stimuli. Future real-world studies are required to confirm our findings. Designing new paradigms for imaging the human brain during walking, such as mobile brain-body imaging, will provide further insight regarding GVS effects.

## 5. Conclusions

EEG recordings with dipole source localization provide the unique ability to evaluate neural activities in patients with BVH. Noisy GVS changed brain activities both in healthy individuals and in patients with BVH, both when walking with 2 Hz horizontal head rotation and while standing. Our behavioral and EEG results reveal that noisy GVS improves postural stability in healthy subjects, as well as in patients, and this improvement could be due to the induction of neuroplasticity in the vestibular cortex. Noisy GVS may be used as a non-invasive adjuvant therapy for patients with BVH.

**Author Contributions:** Conceptualization, L.-W.K.; R.K.C. and C.-L.K.; methodology, L.-W.K.; R.K.C. and C.-L.K.; software, R.K.C.; P.-Y.C.; Y.-C.J.; C.-C.W.; Y.-C.Y.; L.P.-H.L.; K.-K.L. and L.-W.C., validation, L.-W.K.; R.K.C. and C.-L.K.; formal analysis, R.K.C.; P.-Y.C.; Y.-C.J.; C.-C.W.; investigation, R.K.C.; P.-Y.C.; Y.-C.J. and C.-C.W.; resources, L-W.K.; and C.-L.K. data curation, R.K.C.; P.-Y.C. and Y.-C.J.; writing—original draft preparation, L.-W.K.; R.K.C.; C.-L.K.; L.-W.C.; P.-Y.C.; Y.-C.J.; C.-C.W. and Y.-C.Y.; writing—review and editing, L.-W.K.; R.K.C. and C.-L.K.; visualization, L.-W.K.; R.K.C. and C.-L.K., supervision, L.-W.K. and C.-L.K. project administration, L.-W.K. and C.-L.K.; funding acquisition, L.-W.K. and C.-L.K. All authors have read and agreed to the published version of the manuscript.

**Funding:** This work is sponsored in part by the Ministry of Science and Technology of Taiwan for funding this research under MOST 108-2314-B-010 -042 -MY3107-2314-B-010-010-, MOST 105-2314-B-010-006-. Taipei Veterans General Hospital-National Yang-Ming University Excellent Physician Scientists Cultivation Program, No. 108-V-B-008. Taipei Veterans General Hospital No. VN109-118C-152, and is particularly supported by the "Center For Intelligent Drug Systems and Smart Bio-devices (IDS2B)" from The Featured Areas Research Center Program within the framework of the Higher Education Sprout Project by the Ministry of Education (MOE) in Taiwan.

**Acknowledgments:** All authors would like to thank you all the patients for joining in this study.

**Conflicts of Interest:** The authors declare no conflict of interest.

## Abbreviations

| | |
|---|---|
| GVS | galvanic vestibular stimulation |
| EEG | electroencephalography |
| SR | stochastic resonance |
| RMS | root mean square |
| COP | center of pressure |
| ICA | independent component analysis |
| BBS | blind source separation |
| BVH | bilateral vestibular hypofunction |
| ERSP | event-related spectral perturbation |
| LFG | left frontal gyrus |
| RFG | right frontal gyrus |
| LPG | left precentral gyrus |
| RPG | right precentral gyrus |
| LPL | left parietal lobe |
| RP | right parietal lobe |
| OL | occipital lobe |

## References

1. Wuehr, M.; Decker, J.; Schniepp, R. Noisy galvanic vestibular stimulation: An emerging treatment option for bilateral vestibulopathy. *J. Neurol.* **2017**, *264*, 81–86. [CrossRef] [PubMed]
2. Iwasaki, S.; Yamamoto, Y.; Togo, F.; Kinoshita, M.; Yoshifuji, Y.; Fujimoto, C.; Yamasoba, T. Noisy vestibular stimulation improves body balance in bilateral vestibulopathy. *Neurology* **2014**, *82*, 969–975. [CrossRef] [PubMed]
3. Wuehr, M.; Nusser, E.; Decker, J.; Krafczyk, S.; Straube, A.; Brandt, T.; Jahn, K.; Schniepp, R. Noisy vestibular stimulation improves dynamic walking stability in bilateral vestibulopathy. *Neurology* **2016**, *86*, 2196–2202. [CrossRef] [PubMed]

4. Gittis, A.H.; du Lac, S. Intrinsic and synaptic plasticity in the vestibular system. *Curr. Opin. Neurobiol.* **2006**, *16*, 385–390. [CrossRef]
5. Lopez, C.; Blanke, O.; Mast, F. The human vestibular cortex revealed by coordinate-based activation likelihood estimation meta-analysis. *Neuroscience* **2012**, *212*, 159–179. [CrossRef]
6. Zu Eulenburg, P.; Caspers, S.; Roski, C.; Eickhoff, S.B. Meta-analytical definition and functional connectivity of the human vestibular cortex. *Neuroimage* **2012**, *60*, 162–169. [CrossRef]
7. Goldberg, J.; Smith, C.E.; Fernandez, C. Relation between discharge regularity and responses to externally applied galvanic currents in vestibular nerve afferents of the squirrel monkey. *J. Neurophysiol.* **1984**, *51*, 1236–1256. [CrossRef]
8. Minor, L.B.; Goldberg, J.M. Vestibular-nerve inputs to the vestibulo-ocular reflex: A functional-ablation study in the squirrel monkey. *J. Neurosci.* **1991**, *11*, 1636–1648. [CrossRef]
9. Yamamoto, Y.; Struzik, Z.R.; Soma, R.; Ohashi, K.; Kwak, S. Noisy vestibular stimulation improves autonomic and motor responsiveness in central neurodegenerative disorders. *Ann. Neurol.* **2005**, *58*, 175–181. [CrossRef] [PubMed]
10. Pan, W.; Soma, R.; Kwak, S.; Yamamoto, Y. Improvement of motor functions by noisy vestibular stimulation in central neurodegenerative disorders. *J. Neurol.* **2008**, *255*, 1657–1661. [CrossRef] [PubMed]
11. Soma, R.; Kwak, S.; Yamamoto, Y. Functional stochastic resonance in human baroreflex induced by 1/f-type noisy galvanic vestibular stimulation. In: Fluctuations and Noise in Biological, Biophysical, and Biomedical Systems. *Int. Soc. Opt. Photonics* **2003**, 69–76.
12. Fujimoto, C.; Yamamoto, Y.; Kamogashira, T.; Kinoshita, M.; Egami, N.; Uemura, Y.; Togo, F.; Yamasoba, T.; Iwasaki, S. Noisy galvanic vestibular stimulation induces a sustained improvement in body balance in elderly adults. *Sci. Rep.* **2016**, *6*, 1–8. [CrossRef] [PubMed]
13. Fujimoto, C.; Egami, N.; Kawahara, T.; Uemura, Y.; Yamamoto, Y.; Yamasoba, T.; Iwasaki, S. Noisy galvanic vestibular stimulation sustainably improves posture in bilateral vestibulopathy. *Front. Neurol.* **2018**, *9*, 900. [CrossRef] [PubMed]
14. Iwasaki, S.; Karino, S.; Kamogashira, T.; Togo, F.; Fujimoto, C.; Yamamoto, Y.; Yamasoba, T. Effect of noisy galvanic vestibular stimulation on ocular vestibular-evoked myogenic potentials to bone-conducted vibration. *Front. Neurol.* **2017**, *8*, 26. [CrossRef] [PubMed]
15. Seeber, M.; Scherer, R.; Wagner, J.; Solis-Escalante, T.; Müller-Putz, G.R. EEG beta suppression and low gamma modulation are different elements of human upright walking. *Front. Hum. Neurosci.* **2014**, *8*, 485. [CrossRef]
16. Bruijn, S.M.; Van Dieën, J.H.; Daffertshofer, A. Beta activity in the premotor cortex is increased during stabilized as compared to normal walking. *Front. Hum. Neurosci.* **2015**, *9*, 593. [CrossRef]
17. Lee, M.H.; Durnford, S.J.; Crowley, J.S.; Rupert, A.H. Visual vestibular interaction in the dynamic visual acuity test during voluntary head rotation. *Aviat. Space Environ. Med.* **1997**, *68*, 111–117.
18. Kim, D.J.; Yogendrakumar, V.; Joyce Chiang, E.T.; Wang, Z.J.; McKeown, M.J. Noisy galvanic vestibular stimulation modulates the amplitude of EEG synchrony patterns. *PLoS ONE* **2013**, *8*, e69055. [CrossRef]
19. Delorme, A.; Makeig, S. EEGLAB: An open source toolbox for analysis of single-trial EEG dynamics including independent component analysis. *J. Neurosci. Methods* **2004**, *134*, 9–21. [CrossRef]
20. Makeig, S.; Debener, S.; Onton, J.; Delorme, A. Mining event-related brain dynamics. *Trends Cogn. Sci.* **2004**, *8*, 204–210. [CrossRef]
21. Jung, T.P.; Makeig, S.; Westerfield, M.; Townsend, J.; Courchesne, E.; Sejnowski, T.J. Removal of eye activity artifacts from visual event-related potentials in normal and clinical subjects. *Clin. Neurophysiol.* **2000**, *111*, 1745–1758. [CrossRef]
22. Makeig, S.; Inlow, M. Lapses in alertness: Coherence of fluctuations in performance and EEG spectrum. *Electroencephalogr. Clin. Neurophysiol.* **1993**, *86*, 23–35. [CrossRef]
23. Efron, B.; Tibshirani, R.J. *An Introduction to the Bootstrap*; CRC Press: Boca Raton, FL, USA, 1994.
24. Benjamini, Y.; Hochberg, Y. Controlling the false discovery rate: A practical and powerful approach to multiple testing. *J. R. Stat. Soc. Ser. B* **1995**, *57*, 289–300. [CrossRef]
25. Ko, L.W.; Shih, Y.C.; Chikara, R.K.; Chuang, Y.T.; Chang, E.C. Neural mechanisms of inhibitory response in a battlefield scenario: A simultaneous fMRI-EEG study. *Front. Hum. Neurosci.* **2016**, *10*, 185. [CrossRef]

26. Chikara, R.K.; Chang, E.C.; Lu, Y.C.; Lin, D.S.; Lin, C.T.; Ko, L.W. Monetary reward and punishment to response inhibition modulate activation and synchronization within the inhibitory brain network. *Front. Hum. Neurosci.* **2018**, *12*, 27. [CrossRef]
27. Guldin, W.; Grüsser, O. Is there a vestibular cortex? *Trends Neurosci.* **1998**, *21*, 254–259. [CrossRef]
28. Collins, J.; Chow, C.C.; Imhoff, T.T. Stochastic resonance without tuning. *Nature* **1995**, *376*, 236–238. [CrossRef]
29. Collins, J.J.; Imhoff, T.T.; Grigg, P. Noise-enhanced information transmission in rat SA1 cutaneous mechanoreceptors via aperiodic stochastic resonance. *J. Neurophysiol.* **1996**, *76*, 642–645. [CrossRef]
30. Fallon, J.B.; Carr, R.W.; Morgan, D.L. Stochastic resonance in muscle receptors. *J. Neurophysiol.* **2004**, *91*, 2429–2436. [CrossRef]
31. Zeng, F.G.; Fu, Q.J.; Morse, R. Human hearing enhanced by noise. *Brain Res.* **2000**, *869*, 251–255. [CrossRef]
32. Priplata, A.A.; Niemi, J.B.; Harry, J.D.; Lipsitz, L.A.; Collins, J.J. Vibrating insoles and balance control in elderly people. *Lancet* **2003**, *362*, 1123–1124. [CrossRef]
33. Reeves, N.P.; Cholewicki, J.; Lee, A.S.; Mysliwiec, L.W. The effects of stochastic resonance stimulation on spine proprioception and postural control in chronic low back pain patients. *Spine* **2009**, *34*, 316–321. [CrossRef] [PubMed]
34. Pal, S.; Rosengren, S.M.; Colebatch, J.G. Stochastic galvanic vestibular stimulation produces a small reduction in sway in Parkinson's disease. *J. Vestib. Res.* **2009**, *19*, 137–142. [CrossRef] [PubMed]
35. Guyot, J.P.; Gay, A.; Izabel, K.M.; Pelizzone, M. Ethical, anatomical and physiological issues in developing vestibular implants for human use. *J. Vestib. Res.* **2012**, *22*, 3–9. [CrossRef]
36. Jasper, H.; Penfield, W. Electrocorticograms in man: Effect of voluntary movement upon the electrical activity of the precentral gyrus. *Arch. Psychiatr. Nervenkrankh.* **1949**, *183*, 163–174. [CrossRef]
37. Crone, N.E.; Miglioretti, D.L.; Gordon, B.; Sieracki, J.M.; Wilson, M.T.; Uematsu, S.; Lesser, R.P. Functional mapping of human sensorimotor cortex with electrocorticographic spectral analysis. I. Alpha and beta event-related desynchronization. *Brain* **1998**, *121*, 2271–2299. [CrossRef]
38. Miller, K.J.; Leuthardt, E.C.; Schalk, G.; Rao, R.P.; Anderson, N.R.; Moran, D.W.; Miller, J.W.; Ojemann, J.G. Spectral changes in cortical surface potentials during motor movement. *J. Neurosci.* **2007**, *27*, 2424–2432. [CrossRef]
39. Severens, M.; Nienhuis, B.; Desain, P.; Duysens, J. Feasibility of measuring event related desynchronization with electroencephalography during walking. In Proceedings of the 2012 Annual International Conference of the IEEE Engineering in Medicine and Biology Society, San Diego, CA, USA, 28 August–1 September 2012; pp. 2764–2767.
40. De Zeeuw, C.I.; Hansel, C.; Bian, F.; Koekkoek, S.K.; Van Alphen, A.M.; Linden, D.J.; Oberdick, J. Expression of a protein kinase C inhibitor in Purkinje cells blocks cerebellar LTD and adaptation of the vestibulo-ocular reflex. *Neuron* **1998**, *20*, 495–508. [CrossRef]
41. Grillner, S.; Hongo, T. Vestibulospinal effects on motoneurones and interneurones in the lumbosacral cord. In *Progress in Brain Research*; Elsevier: Amsterdam, The Netherlands, 1972; Volume 37, pp. 243–262.
42. Grassi, S.; Pettorossi, V.E. Synaptic plasticity in the medial vestibular nuclei: Role of glutamate receptors and retrograde messengers in rat brainstem slices. *Prog. Neurobiol.* **2001**, *64*, 527–553. [CrossRef]
43. Helmchen, C.; Rother, M.; Spliethoff, P.; Sprenger, A. Increased brain responsivity to galvanic vestibular stimulation in bilateral vestibular failure. *Neuroimage Clin.* **2019**, *24*, 101942. [CrossRef]
44. Galvan, G.R.; Clark, T.; Mulavara, A.; Oman, C. Exhibition of stochastic resonance in vestibular tilt motion perception. *Brain Stimul.* **2018**, *11*, 716–722. [CrossRef] [PubMed]
45. Priesol, A.J.; Valko, Y.; Merfeld, D.M.; Lewis, R.F. Motion perception in patients with idiopathic bilateral vestibular hypofunction. *Otolaryngol Head Neck Surg.* **2014**, *150*, 1040–1042. [CrossRef] [PubMed]
46. Wilkinson, D.; Ferguson, H.J.; Worley, A. Galvanic vestibular stimulation modulates the electrophysiological response during face processing. *Vis. Neurosci.* **2012**, *29*, 255–262. [CrossRef] [PubMed]
47. Ertl, M.; Moser, M.; Boegle, R.; Conrad, J.; zu Eulenburg, P.; Dieterich, M. The cortical spatiotemporal correlate of otolith stimulation: Vestibular evoked potentials by body translations. *NeuroImage* **2017**, *155*, 50–59. [CrossRef]
48. Lobel, E.; Kleine, J.F.; LEROY-WILLIG, A.; VAN DE MOORTELE, P.F.; Bihan, D.L.; GRÜSSER, O.J.; Berthoz, A. Cortical areas activated by bilateral galvanic vestibular stimulation. *Ann. N. Y. Acad. Sci.* **1999**, *871*, 313–323. [CrossRef] [PubMed]

49. Song, M.; Shin, Y.; Yun, K. Beta-frequency EEG activity increased during transcranial direct current stimulation. *Neuroreport* **2014**, *25*, 1433–1436. [CrossRef]
50. Ray, W.J.; Cole, H.W. EEG alpha activity reflects attentional demands, and beta activity reflects emotional and cognitive processes. *Science* **1985**, *228*, 750–752. [CrossRef]
51. Keeser, D.; Padberg, F.; Reisinger, E.; Pogarell, O.; Kirsch, V.; Palm, U.; Karch, S.; Möller, H.J.; Nitsche, M.; Mulert, C. Prefrontal direct current stimulation modulates resting EEG and event-related potentials in healthy subjects: A standardized low resolution tomography (sLORETA) study. *Neuroimage* **2011**, *55*, 644–657. [CrossRef]
52. Stam, C.; Montez, T.; Jones, B.; Rombouts, S.; Van Der Made, Y.; Pijnenburg, Y.; Scheltens, P. Disturbed fluctuations of resting state EEG synchronization in Alzheimer's disease. *Clin. Neurophysiol.* **2005**, *116*, 708–715. [CrossRef]


**Publisher's Note:** MDPI stays neutral with regard to jurisdictional claims in published maps and institutional affiliations.

MDPI
St. Alban-Anlage 66
4052 Basel
Switzerland
Tel. +41 61 683 77 34
Fax +41 61 302 89 18
www.mdpi.com

*Brain Sciences* Editorial Office
E-mail: brainsci@mdpi.com
www.mdpi.com/journal/brainsci